Dinosaur Footprints &
Trackways of La Rioja

Life of the Past   James O. Farlow, editor

Indiana University Press   Bloomington & Indianapolis

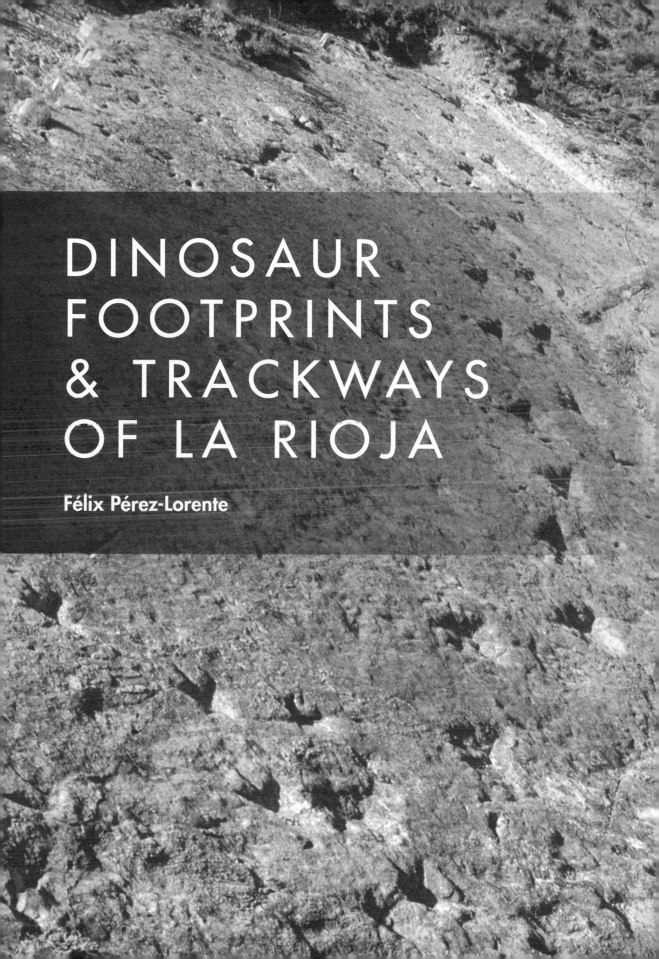

DINOSAUR FOOTPRINTS & TRACKWAYS OF LA RIOJA

Félix Pérez-Lorente

This book is a publication of

Indiana University Press
Office of Scholarly Publishing
Herman B Wells Library 350
1320 East 10th Street
Bloomington, Indiana 47405 USA

iupress.indiana.edu

© 2015 by Félix Pérez-Lorente

*Manufactured in the
United States of America*

*Library of Congress
Cataloging-in-Publication Data*

Pérez-Lorente, Félix.
  Dinosaur footprints and trackways of La Rioja / Félix Pérez-Lorente.
    pages cm. – (Life of the past)
    Includes bibliographical references and index.
    ISBN 978-0-253-01515-0 (cl : alk. paper) – ISBN 978-0-253-01541-9 (eb)
  1. Footprints, Fossil – Spain – La Rioja.
  2. Dinosaur tracks – Spain – La Rioja.
  3. Dinosaurs – Spain – La Rioja.
  4. Paleontology – Spain – La Rioja.
  5. La Rioja (Spain) I. Title.
    QE845.P465 2015
    567.90946'354 – dc23

                                    2014030006

1 2 3 4 5   20 19 18 17 16 15

Contents

Dinosaur Footprints &
Trackways of La Rioja

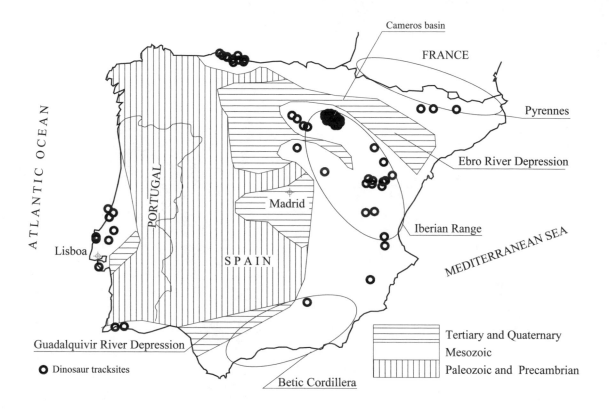

1.1. Schematic geological and geographical representation of Iberian Peninsula showing location of dinosaur tracksites (modified from Pérez-Lorente, Romero, and Torcida, 2002). Locations of outcrops of Mesozoic rocks in Iberian Peninsula that potentially contain dinosaur fossils are shown in white; areas marked with lines show those that cannot contain fossils because they are the wrong age. Regions marked with vertical lines have rocks from before the time of dinosaurs (Precambrian and Paleozoic); regions marked with horizontal lines are post-Cretaceous (Cenozoic: after the dinosaurs).

La Rioja Footprints

THE SPANISH PROVINCE OF LA RIOJA IS AN AREA OF THE WORLD where a huge number of dinosaur footprints have been found, with many more likely yet to be discovered. This hilly region has many rock slopes with layers so full of tracks that, if the vegetation, loose rock, and debris could be removed, would yield from 8000 to as many as 25,000 footprints. Using the best estimates from some slopes – that is, the maximum estimate from that partial data – there may be as many as 70,000 footprints.

Many of the footprints are so easy to see that the first people to discover them were likely shepherds or hunters who passed through the area. However, the identity of the first person to correctly interpret them is another question. The footprints are so evocative that the inhabitants of the region have long associated them with animals. In the villages of Enciso, El Villar, and Poyales, there were people who thought the footprints now understood to be those of theropod dinosaurs had been made by giant chickens. In the village of Navalsaz, it was said that the ornithopod footprints of the Cuesta de Andorra had been made by huge lions. It is difficult to know exactly how long such claims have been made, whether the local population even knew about wild animals such as lions, or whether this interpretation was offered by visitors to the region.

The footprints have also been attributed to animals from medieval mythology, in some cases inspired by religion. For example, in Igea, it was said that the footprints had been left by the horse of the apostle James on his travels. Popular tradition has it that James helped the Christians in their wars against the Muslims. As with the legendary "mule tracks" of Setubal in Portugal, there is no end of imaginative interpretations. In some cliffs to the south of Lisbon (Portugal) there are some dinosaur footprints which the ancient Portuguese interpreted as being miraculous. Miguel Telles Antunes (1976) says that according to legend, the Virgin Mary "Santa Maria da pedra da mua" (or an image of her from the 18th century) had come out of the sea and ascended to the top of the cliff while sitting on a mule. During the ascent, the mule left the footprints on the wall. According to Antunes, this tradition may date back to the 13th century. In Igea, the "horse's" footprints are visible near the Santa Ana chapel, at the place where the apostle's horse was said to have jumped 3 km to land near the shrine of the Virgen del Villar, where it also left footprints. Interpretations such as these are to be expected, given the knowledge of the population. For example, the presence of marine fossils in many places had to be explained as a whim of nature. Even if no one could explain why, the sea must have been there. Nature is capable of wonderful things! Even today, if you ask an old farmer or shepherd how the sea could ever have been so far inland, although he may not be able to provide an explanation, no one will be able to convince him it was not so. For the people, the rocks, rivers, sea, and mountains have always been where they are at present. In their village, there was never a sea.

Nor would any of the local inhabitants have been able to predict that the sloping rock strata on which the tracks appear had once been horizontal mudflats. It is understandable that people did not consider that normal animals could leave their footprints in such hard rock, whereas the horse of Saint James might have had magical properties that allowed it to make an impression in solid rock. It may have been the wish of the saint – or even the horse – that the footprint left behind had a shape very different from that left by a typical horse.

It is likely that almost all settlements with dinosaur footprints had traditions and legends surrounding them that have since been lost or that the older generation do not want to tell to strangers.

History of Discoveries

The first publication about dinosaur footprints in the Iberian Peninsula that I have in my possession is from Jacinto Pedro Gomes (1915–1916) for Cabo Mondego (Portugal), and the first in Spain is from Albert F. de Lapparent (1965) for the east (province of Valencia). Although geological research began in La Rioja a long time ago, the footprints were not recognized until 1969, when the first publication about dinosaur footprints in La Rioja appeared in the newspaper *El Correo Español – El Pueblo Vasco*. The authors were Moisés Iglesias Ponce de León, a geologist, and Luis Vicente Elias, an ethnologist, who found the footprints while doing fieldwork on the customs of the people from the Cameros region. After this discovery, a number of favorable events occurred.

The first was that the news was not published just in the newspaper. Almost immediately, a learned researcher, Blas Ochoa, who was a schoolteacher from Enciso, began collaborating with a team of vertebrate paleontologists, Maria Lourdes Casanovas and José Vicente Santafé. The team described five sites in two publications in 1971 and 1974. The two publications and comments from townspeople who had known the paleontologists inspired two groups of amateurs to search for new sites in their spare time and publish their findings as a challenge to other researchers. By 1979, nine sites had been identified, all near the village of Enciso. (These groups are still working, partly because they have active members and partly because we have followed in their footsteps in looking for new track sites.) Later, a schoolteacher from Igea, Angel Gracia, taught his pupils the importance of fossils and showed them how to search for, classify, and preserve them. His students found sites with footprints near the village, one of which remains to be studied.

More recently, traces of other vertebrates have been discovered: birds, turtles, and pterosaurs (Moratalla and Hernán, 2009; Moratalla and

Sanz, 1992; Moratalla, Sanz, and Jiménez, 1992), crocodiles (Ezquerra and Pérez-Lorente, 2002, 2003), and fishes (Costeur and Ezquerra, 2008; Ezquerra and Costeur, 2009; Ezquerra and Pérez-Lorente, 2002, 2003). Today, the followers of Blas and Angel have their own specialist centers. In Enciso, there is a paleontological museum, and the first phase of a learning center on "the lost ravine" has been built, as well as another center in Igea that houses some interesting paleontological material collected mostly by those pupils, who in 2010 are about 40 years old. Both centers are strongly committed to scientific activities and educating tourists.

Just as dinosaur track sites are still being found in La Rioja, so are they also being found in other locations of the Iberian Peninsula, although in rocks of different ages. There are examples of dinosaur footprints from the Late Triassic, Middle and Upper Jurassic, and Early and Late Cretaceous periods.

The geological diagram in Fig. 1.1 shows the locations and ages of outcrops of Mesozoic rocks in the Iberian Peninsula that potentially contain dinosaur fossils. The Iberian sites with dinosaur tracks occur in the regions with Mesozoic outcrops. Thus, this figure indicates the geographic limits for the areas of possible dinosaur remains in Iberia.

The Iberian Peninsula consists of four major geological areas, as follows:

1. A central region of ancient volcanic and sedimentary rocks, metamorphic rocks, and plutonic rocks (Paleozoic and Precambrian), considered a stable zone (not folded) during the Alpine Orogeny. The Alpine Orogeny is the mountain-building event that created the Alps and in Spain the Pyrenees, Iberian Range, and Betic Cordillera. The Alpine Orogeny affects (folds) the Mesozoic strata between or adjacent to the stable Paleozoic and Precambrian zones.
2. A border with outcrops of Mesozoic sedimentary rocks, sometimes mixed (at the eastern edge) with older ones.
3. Two Tertiary depressions filled with Tertiary and Quaternary rocks (depressions associated with the Ebro River to the northeast and the Guadalquivir River to the south).
4. Two external Alpine chains with deformed ancient and modern sedimentary rocks (the Pyrenees bordering France and the Betic Cordillera to the south).

The Cameros Basin is located on the northeastern edge of the stable zone, where the Paleozoic and Precambrian Mesozoic sedimentary rocks, folded by the Alpine Orogeny, are found. It is between the Ebro Basin and the stable zone. The Cameros Basin was named by Brenner and Wiedmann (1974), although Götz Tischer (1966) and Gerhard Richter

Cameros Basin

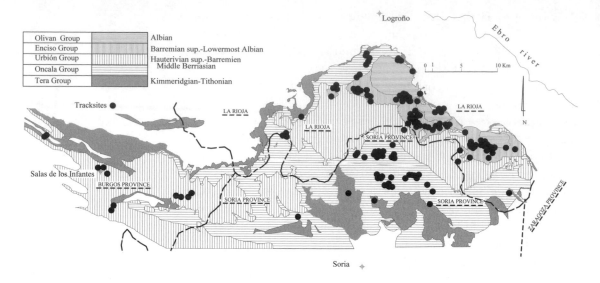

Olivan Group	▤	Albian
Enciso Group	▥	Barremian sup.-Lowermost Albian
Urbión Group	▥	Hauterivian sup.-Barremien
Oncala Group	▦	Middle Berriasian
Tera Group	▓	Kimmeridgian-Tithonian

Tracksites ●

1.2. Upper Jurassic (Kimmeridgian–Tithonian)–Lower Cretaceous (Berriasian–Albian) continental stratigraphic groups of continental Cameros Basin, indicating Cameros Basin groups and locations of dinosaur footprint sites (2003). Area covers parts of La Rioja, Burgos, and Soria provinces.

(1930) have previously used the word "basin" to encompass the Wealdian sediments ranging from the province of Burgos to north of Ricla (Zaragoza Province).

The first stratigraphic summary of this region was published by Palacios and Sánchez Lozano (1885), who divided the "Wealdian formation" into strata of lacustrine sediments divided into different superimposed lithological parts called "sections (B and C), bank (A) and levels (a, b, c)." The terms are literal translations from the Spanish (*sección, banco, nivel*) used in an ancient geological language. The possible equivalents are "unit" or "member," "bed," and "level." Rafael Sánchez Lozano's five superimposed lithological parts (Fig. 1.2) correspond in part to the Tera, Oncala, Urbión, Enciso, and Oliván groups of Tischer (1966). The names of the groups are still used, with virtually the same meaning and positions defined by Tischer.

Long before footprints were discovered, the Early Cretaceous (Wealdian) age of the rocks containing them was already known and had been more precisely classified into five sedimentary groups defined by Tischer and by other followers of Professor Hans Mensink (Bochum University). It was also known that the lowest group (Tera), with abundant conglomerates at its base, was underlain by marine rocks of Late Jurassic age.

The exact age of the base and the top of the sediments in the Cameros Basin is difficult to establish because they lack appropriate fossils. The transition from the Late Jurassic to the Early Cretaceous occurs in continental sediments. The uppermost Jurassic marine sediments are limestone reefs of Kimmeridgian age (Alonso, Meléndez, and Mas, 1986–1987; Mas et al., 2002). The age of the continental sediments in the Cameros Basin ranges from the mid-Kimmeridgian (sensu lato) to the mid–late Albian (see Doublet, 2004) (Fig. 1.3). This period extends from the base of the Tera Group to the top of the Oliván Group. So far, all dinosaur footprints in La Rioja are of Early Cretaceous age (Fig. 1. 2). Two sites, El Encinar and La Vuelta de los Manzanos (Moratalla, Sanz, and

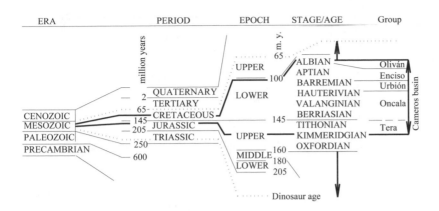

1.3. Position of Cameros Basin in geological time table of dinosaur age.

Jiménez, 1996, 2000b), situated at the boundary of the Tera and Oncala groups, have been assigned to the Late Jurassic (see Cámara and Durantez, 1982) (Fig. 1.4, Tables 1.1, 1.2).

The Cameros Basin is situated at the northeast end of the Iberian Range. Sites with dinosaur footprints or bones occur from this point to the Mediterranean coast. Correlation of the Cretaceous sediments in La Rioja is simple in all directions except toward the north. The Cameros Basin continues into the provinces of Burgos to the west, Soria to the south, and Zaragoza to the east (Fig. 1.2). To the north, the continuity of Mesozoic outcrops is lost because the Tertiary rocks of the Ebro Basin cover them.

Outcrops in the south and southwest of the Cameros Basin reveal rocks of detrital material, the source of which was further south and west. The five stratigraphic groups come together in the northern outcrops in a spectacular pattern of thinning and overlap, and both the limestones and shales and dark sandstones of the Urbión and Enciso groups grade into light (sand, silt, and clay) and red-colored siliciclastic rocks (Doublet, 2004).

The deepest part of the Cameros Basin was initially in the province of Soria (the Oncala and Urbión/Berriasian–Barremian) and later in La Rioja (first the Enciso Group to the southwest/Barremian–lower Albian, and finally in the Oliván Group situated more to the northeast/Albian). It seems, therefore, that the center of subsidence of the basin drifted from Soria to La Rioja during the Early Cretaceous, at least for the Oncala, Urbión, Enciso, and Oliván groups. All fossils found in this sequence are continental freshwater, or brackish at most. Only two stratigraphic levels, located in the Oncala Group, have been shown to contain marine microfossils (Alonso and Mas, 1993; Suárez-González et al., 2011).

In the Cameros Basin area of La Rioja, dinosaur bones have been found that are attributed (Fig. 1.5) to *Baryonyx* (Torres and Viera, 1997; Viera and Torres, 1995a) and *Hypsilophodon* (Torres and Viera, 1994); also found were eroded vertebra and bones from sauropods and ornithopods related to *Iguanodon*. In addition, theropod and ornithopod teeth have been discovered, and some theropod phalanges are being studied. In addition to dinosaur bones, fossils include osteodermal and

1.4. Diagrammatic map of location of footprint sites in La Rioja. (Top) Location of sites 3.1 through 3.24 as numbered in this book. (Bottom) More detailed key to location of tracksites. Each published tracksite is labeled with letter abbreviation, as identified in tracksite description that follows (e.g., VDP indicates La Virgen del Prado) (Table 1.1). Black footprints without labels indicate unpublished tracksites.

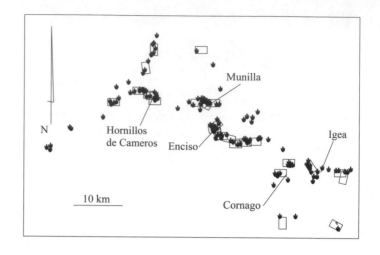

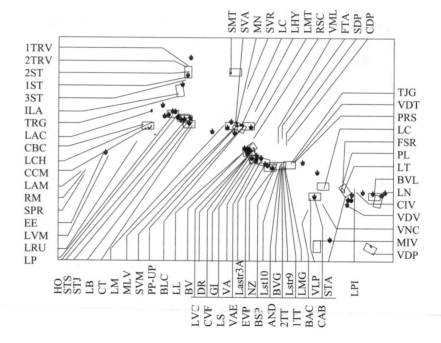

other bony elements of turtles, crocodiles, pterosaurs, and various kinds of fish (*Lepidotes, Hybodus*); charophytes, ostracods, and shells from bivalves and turritellid gastropods. Many sedimentary layers are composed of algal laminations, and in some layers there are higher plants, such as transported conifers and ginkgoales, as well as abundant roots of variable thickness, but less than 10 cm in diameter.

Skeletal remains of dinosaurs are abundant in the Early Cretaceous of Spain; they perhaps constitute the richest fossil association in the sediments of continental Europe in the Hauterivian–Aptian interval. Proof of this diversity is that remains from 15 different dinosaur taxa have been found at a single site from the Early Barremian of Teruel (Canudo et al., 2009, 2010; Ruiz-Omeñaca, 2011). Another significant aspect of

Table 1.1. La Rioja ichnological sites.

No. of outcrops	Abbreviation	Site name	No. of outcrops	Abbreviation	Site name
1	AND	La Cuesta de Andorra	1	RSC	Las Riscas
1	BAC	Barranco de Acrijos	1	LRU	La Rueda
1	BLC	Barranco de la Canal	1	LS	La Senoba
1	BSP	Barranco de la Sierra del Palo	1	LSN	Las Navas
1	BV	Barranco de Valdeño	9	LT	La Torre
1	BVG	Barranco de Valdegutiérrez	4	LVC	La Virgen del Campo
1	BVL	Barranco de Valdebrajés	1	LVM	La Vuelta de los Manzanos
1	CAB	Cabezuelos	1	LPI	Los Piojos
1	CBC	Valdemayor	1	MLV	Malvaciervo
5	CCM	Cabezón de Cameros	1	MIV	Mina Victoria
1	CDP	Cuesta del Peso	1	MN	Munilla
1	CHS	Chorrón del Saltadero	1	NZ	Navalsaz
1	CIV	Camino de Igea a Veldebrajés	7	PL	Era del Peladillo
1	CT	El Contadero	1	PP	Peñaportillo
1	CVF	Corral de Valdefuentes	1	PER	Perosancio
1	DR	Del Rio	1	PEÑ	Peña Untura
1	ENC	El Encinar	1	RMG	Rio Maguillo
1	EV	La Virgen del Villar	1	SJA	Santa Juliana
1	EVP	El Villar Poyales	1	SLP	Sol de la Pita
1	FAM	Fuenteamarga	3	SM	San Martín de Jubera
1	FSR	Fonsarracín	1	SPR	San Prudencio
1	GI	Guilera	1	1ST	Soto 1
5	HR	Hornillos	1	2ST	Soto 2
1	RI	Icnitas 3	1	3ST	Soto 3
1	ILA	La Ilaga	1	STA	Santa Ana
4	LAC	La Cela	1	STS	Santisol
1	Lastra 3	Lastra 3	14	SV	San Vicente de Munilla
1	Lastra 9	Lastra 9	1	SVA	El Sobaquillo
1	Lastra 10	Lastra 10	1	SVR	San Vicente de Robres
1	LBG	La Barguilla	1	TJG	Tajugueras
6	LCA, LCB, . . .	Los Cayos	1	TR	Camino a Treguajantes
11	LC	La Canal	1	TT	Corral del Totico
1	LCH	Los Chopos	2	TRV	Trevijano
1	LCÑ	La Cañada	1	UP	Umbria del Portillo
1	LHY	La Hoya	1	VA	Valdecevillo
1	LL	Las Losas	1	VAE	Valdecevillo Este
1	LM	Las Mortajeras	1	VBB	Barranco de Valdebrajés abajo
1	LMG	La Magdalena	1	VDM	Valdemurillo
1	LMO	La Moga	1	VDP	La Virgen del Prado
1	LMT	La Mata	1	VLD	Valdeté
1	LN	Las Navillas	1	VNC	Valdenocerillo
1	LP	La Pellejera	4	VLP	Valdeperillo

Table 1.2. Stratigraphic groups, lithology, age, and tracksite distribution.

Group	Section	Lithology	Sites	Age
Oliván		Siliciclastic		Albian
Enciso	Upper	Siliciclastic	LP, HR, STS, SJA, LBG, SVA, SV, PP, BLC, LC, LCA-S, RSC	Lower Albian
	Middle	Carbonatic	LN, BVL, CIV, LPI, STA, LT, LSN, PL, VLD, TJG, PER, LMG, BVG, TT, AND, BSP, NZ, EVP-3I, DR, GI, BV, CVF, MLV, LM, CT	Aptian
	Lower	Siliciclastic	LCÑ, EV, VAE, VA, LS, LVC, LL	Upper Barremian–Aptian
Urbión		Siliciclastic	LCH, CCM, CBZ, LRU, MIV, VNC, VLP	Upper Hauterivian–Barremian
Oncala		Carbonatic	LAC, TR, ST, TRV, SM, SVR, VDM, FAM, SLP, CDP, VDP	Berriasian
Tera		Siliciclastic	ENC, LVM	Middle Kimmeridgian–Tithonian

Note: Site abbreviations as in Table 1.1.

the Spanish dinosaurs of the Early Cretaceous is their great paleobiogeographical complexity, including taxa with Asiatic, Gondwanan, and North American affinities as well as European ones (Canudo, Royo-Torres, and Cuenca-Bescós, 2008; Canudo et al., 2009; Ortega, Escaso, and Sanz, 2010; Pereda-Suberbiola et al., 2007; Ruiz-Omeñaca et al., 2004). The best-represented dinosaurs are the sauropods. Representatives of neosauropods have been found, both diplodocimorphs such as *Demandasaurus* (Torcida et al., 2011) and macronarians such as *Tastavinsaurus* and *Aragosaurus* (Canudo, Royo-Torres, and Cuenca-Bescós, 2008; Sanz et al., 1987). Nonavian theropods are represented by basal tetanurans such as allosauroids and spinosaurids and by derived tetanurans such as ornithomimosaurs and dromaeosaurids. To date, two taxa have been described: the ornithomimosaur *Pelecanimimus* (Pérez-Moreno et al., 1994) and the carcharodontosaur *Concavenator* (Ortega, Escaso, and Sanz, 2010). Thyreophorans are scarce, though remains of "polacanthid" ankylosaurs and stegosaurs have been described (Pereda-Suberbiola and Galton, 2001; Pereda-Suberbiola et al., 2007). Ornithopods are represented by basal iguanodontoids, dryosaurids, and "hypsilophodontid"-like basal euornithopods. The iguanodontoid *Delapparentia* has been described (Ruiz-Omeñaca, 2011). The basal iguanodontoids are the most abundant dinosaurs in terms of the number of dinosaur specimens in the Early Cretaceous of Spain.

Sedimentological studies indicate that a large portion of the sedimentary succession is lake or marsh deposits, with both siliceous and limestone strata, as well as other fluvial siliciclastic (silt, sand, and conglomerate) sediments. In some places there are sebka-type deposits containing gypsum and other salts.

The Enciso Group was the subject of a study that detected the presence of a large lake (about 500 km²) (Doublet, 2004), the existence of which is demonstrated by, among other things, the great lateral continuity (in composition and thickness) of certain distinctive sedimentary layers. After the sedimentary filling of the subsiding Cameros Basin in the earliest Cretaceous came the compressive and thermal stage. The Alpine Orogeny began in the Albian period with compression and

1.5. Reproduction of *Baryonyx walkeri*. Paleontological Center of Igea (La Rioja).

metamorphism. Silt and loose sand were compacted and their minerals recrystallized. Where the temperature was greater, veins of quartz and phyllosilicates grew. Chloritoid was formed, which indicates that the maximum temperature reached nearly 400°C, with large pyrite crystals forming at the same locations. The high temperature had a positive impact on the preservation of the tracks because it hardened the rocks in which they are located without destroying structures directly or indirectly related to them.

Dinosaur footprints occur in all the stratigraphic groups of the Cameros Basin (Fig. 1.2) except the Oliván Group. In La Rioja, the distribution of the sites is approximately linear in association with outcrops of the Enciso Group (Fig. 1.4) (Blanco et al., 1999b; Casanovas and Santafé, 1995; Moratalla, 2002; Moratalla and Hernán, 2007; Pérez-Lorente, 2006). The number of outcrops containing footprints is difficult to quantify. In some cases, multiple sites were recognized along the same layer as they were being discovered (such as six in Era del Peladillo). In other cases (e.g., La Torre in the same area), multiple sites were grouped together because they all occur in a particular exposure (e.g., one side of a hill), even though there may be no continuity between them and they may occur in different sedimentary layers. The number is in the range of 110 to 156

Number, Distribution, and Area of Sites

sites or exposures with footprints (Table 1.1) (Caro, Pavía, and Pérez-Lorente, 1997; Pérez-Lorente, 2003a, 2003b).

Outcrops with easily recognizable footprints are more abundant in the Enciso Group than in the other groups. However, this is not necessarily an indication of the number of prints that actually exist because the footprint-bearing stratification surfaces may be obscured as a result of the nature of the rocks. The interaction between the composition and structure of the rock and weathering is the main factor in how well footprints are displayed. Many of the rocks in the Urbión Group, for example, contain many footprints that go unnoticed by observers. Isolated footprints in single rock surface fragments on the slopes are not considered to be sites.

The size of the sites depends on what is shown and the source of the information–area with footprints, area cleaned and prepared, area with a surrounding protection zone, lateral areas where more footprints are expected, or whether the area is considered a heritage site and is protected, and so forth. Most of the outcrops could be extended by additional digging, so the presently exposed outcrop surface provides little indication of the total surface area that may contain footprints. Indeed, the area of the site occupied by tracks has changed in many outcrops as a result of the incorporation of lateral areas with newly discovered footprints.

The number of footprints in the rocks of the Cameros Basin is difficult to calculate. The superimposition of layers with footprints is spectacular. For example, at El Villar-Poyales there are seven surfaces with tracks at the tops of the seven strata in the site. Similarly, in Los Cayos A–C, the tracks are visible not only on the two surfaces of the strata (the top and the bottom/hollows, casts, and undertracks) but also in the transversal sections that show the interior in the deformed sedimentary laminas of the strata. In Los Cayos A, B, and C there are strata formed by superimposed flat laminated structures.

Age of Sites

All authors who have written recently about the Cameros Basin (Doublet, 2004; Mas et al., 2002) agree on the age of the sediments that fill it. Kimmeridgian reef limestone deposits are the most recent marine sediments. Continental sedimentation began during the mid-Kimmeridgian–Tithonian (Tera Group) and did not end until the Cameros Basin was filled. During the mid-Berriasian, a regional discontinuity was created when deposition of the Oncala Group ended; there is, however, a large deposit of siliceous sediments of this group to the south, outside of La Rioja. Missing from much of the basin are rocks of mid-Berriasian through Late Hauterivian age. The Urbión Group is Late Hauterivian to Upper Barremian in age. The age of the Enciso Group ranges from Late Barremian through Early Albian. The Oncala Group is Early Albian in age. The correlation between the groups and the subdivision of the Enciso Group, with the sites, are detailed throughout the text (Table 1.2).

All the groups thin out toward the north; this is clear on the maps (Fig. 1.2). The outcrops of the groups are much larger to the south, which indicates that they are thicker there. In the northeast of the Cameros Basin, the geological map shows that the outcrop of the Tera and Oncala groups is a very narrow strip, and that of the Urbión and Enciso groups is almost unnoticeable or has disappeared. The Urbión and Enciso groups end toward the north in detrital sediments, which are sometimes quite thick and show evidence of total emergence above lake level, erosion, and processes of pedogenesis and colonization by roots. Exposures of these five groups south of the basin are more than 6000 m thick, while to the north they are reduced to 500 to 600 m. The bases of the younger strata moving northward rest on the lower layers, forming an onlap.

Erosion of layers beneath the Oliván Group has, in the most north-easterly part of the basin, established direct contact between the sandstones of this group and the underlying Oncala Group limestones (Fig. 1.2). The last of the outcrops in the Oncala, Urbión, and Enciso groups, or the fact that these are so thin, is the reason why there are not many dinosaur footprint sites in the northeastern part of the basin. The only rocks appropriate for preserving fossil footprints in this region are the limestone and some of the intervening sandstone levels in the Oncala Group.

Most of the tracksites in the area are accompanied with placards and information panels. Tourist offices, centers, and museums in La Rioja (Igea Paleontological Center, Museum of Enciso, "Barranco Perdido" Park) provide brochures and situation maps. Several sites are prepared for visitors and include fences, walkways, informational panels, and even dinosaur reproductions: *Allosaurus*, *Tarbosaurus*, *Brachiosaurus*, *Iguanodon*, and *Stegosaurus* (Barranco de Valdecevillo [Fig. 3.112], La Virgen del Campo [Fig. 3.126], Peñaportillo [Fig. 3.165], and La Pellejera [Fig. 3.193]).

Many popular-science articles have been published in addition to scholarly ones, so information on the track sites may be accessed via many levels. Field guides suggest excursions; books suggest educational or teaching activities; and texts provide general informative descriptions (Blanco et al., 1999b; Brancas, Martínez, and Blaschke, 1979; Casanovas et al., 1996; Casas et al., 1996; Jiménez, 1978; Moratalla and Hernán, 2005; Moratalla, Sanz, and Jiménez, 1990, 1997a, 1997b; Moratalla et al., 1988; Pérez-Lorente, 1992b, 2003a, 2007; Pérez-Lorente, Fernández, and Uruñuela, 1986; Torcida, 1996, 2003). Some works have been published with support from public institutions or private companies.

Field Guides

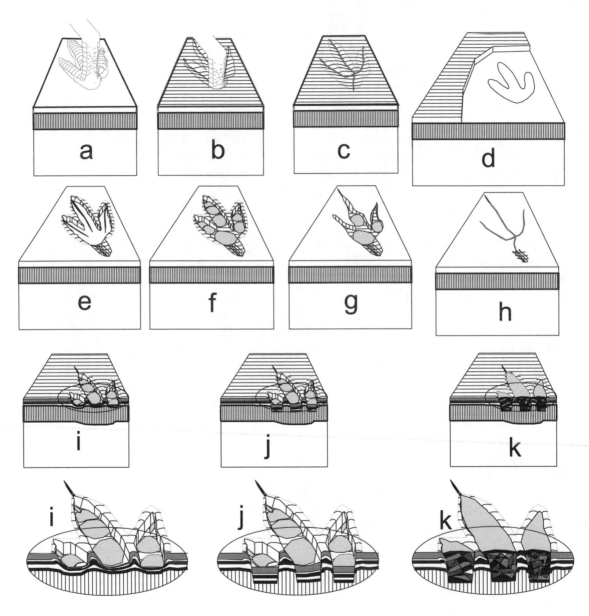

2.1. Different footprint features related to phases of foot interaction with substrate identified by Thulborn and Wade (1989). T phase features are made during initial touchdown of foot with substrate (a). W phase features are made as the animal's weight is supported by the foot, causing the foot to exert maximum pressure against the ground (b). K phase features are made during final kickoff as the foot leaves the ground (not illustrated). Footprint features made during T phase include slash-like incisions (c), slide marks (e–h), and smooth axial downfolds (i). W phase features include extrusion rims (e, f, i–k), lateral folds (i–k), undertracks (d), downfolds (i–k), stamps (f), and dead zones (j). K phase includes collapse track walls (g, h) and claw marks (e, g–k). Subtracks (k) can be created during all three phases.

Ichnology

2

ACCORDING TO THE MOST RECENT PUBLISHED COUNT (PÉREZ-Lorente, 2003b), the number of footprints in La Rioja is 7967. Subsequent studies of additional sites have provisionally increased that number to 9150. This number only includes footprints that have been studied and that have data available regarding their form and dimensions. The 9150 footprints are distributed among 866 theropod, 146 ornithopod, 22 sauropod, and 34 unidentified trackways, as well as a number of isolated prints. There are 5236 theropod footprints, 1059 ornithopod prints, 1198 sauropod prints, and 950 as yet unidentified prints. Many footprints at other sites remain to be studied. For example, the site of La Pellejera contains more than 700 footprints in about 70 trackways.

The distribution of the tracksites in the Cameros Basin is heterogeneous. They occur in the rocks of the Tera, Oncala, Urbión, and Enciso groups. The rocks with footprints are limestone, sandstone, and shale. The neighboring outcrops may be adjacent to each other or separated by hundreds of meters.

Lockley (1991) defines megatracksites as ichnologic strata covering a large geographic area of hundreds or thousands of square kilometers. Such megatracksites often consist of a series of small outcrops, which are part of extensive strata with footprints. There may be only sporadic outcrops of strata containing footprints. Lockley does not consider a megatracksite as having to be exclusively on the same bed. Footprints may therefore be located in strata of the same geological interval (formation, unit, or stratigraphic sequence) in terms of their stratigraphic situation, sedimentary environment, or lithological composition.

In La Rioja, for both sandstone and limestone (detrital and chemically precipitated rocks, respectively), it can be difficult to ascertain whether there are many sites on the same stratum for several reasons. First, the continuity has not been studied. For example, within each mappable member in the Enciso Group, the layers are often part of large sheets of cross-bedding or linsen (or lenticular) bedding. Second, in each mappable member, there tend to be multiple layers containing tracks, even at the same site. It cannot be determined which outcrops with footprints from the same member reveal the same stratum. Finally, the more widely separated the sites are, the more difficult it is to correlate the beds and even the stratigraphic units.

If it is assumed that the megatracksites are large accumulations of traces that occur in the same sedimentary event level (sedimentary

Number of Footprints and Their Distribution

sequence, stratigraphic unit) and in rocks of the same composition, one can speak of several types of megatracksites in the Cameros Basin area.

Enciso Group Carbonate Megatracksite

The Era del Peladillo site (PL) consists of several layers of dark limestone, which overlie a colmatation sequence in the Enciso Group. In 1PL–6PL, the top two of the strata containing tracks can be seen. The two strata are in the upper part of a sedimentation sequence (Meléndez and Pérez-Lorente, 1996). The 1PL–6PL sequence is the third of five such super-imposed sequences that would comprise a stratigraphic unit. The five sedimentation sequences, which begin with a period of increased energy (evidenced by small peaks of sandstone dispersed in a shaly matrix) end in large lakes (limestone) that cover more than 500 km^2 (Doublet, 2004). I have followed this limestone member and the two lower sedimentation sequences (shaly sandstone and limestone members) uninterruptedly for 15 km from this point eastward. The strata continue in the two directions, and it is not known where they stop. The two upper sequences are prob-ably less extensive because they are eroded by the Oliván Group. The entire unit (five sedimentation sequences) is between 80 and 90 m thick. Underneath the unit are at least three additional levels with footprints.

In a stratigraphic log, the vertical separation between the upper and lower track-bearing limestone horizons is in tens of meters, but the dis-tribution of their ichnological outcrops may be separated by hundreds of meters. The separation between tracksites in the same unit verifies the same pattern. Great geographical distance does not imply that the track sites are stratrigraphically, paleogeographically, and temporally widely separated.

Enciso Group Sandstone Megatracksites

In the total Enciso Group there are different limestone- and sandstone-rich members. In addition to the sites in limestone there are many track-sites in sandstone-rich parts of the same Enciso Group (above all in the western part of the basin) in two different stratigraphical units.

The dinosaur tracksites in La Rioja could be considered as forming part of a megatracksite at several alternative scales. If the megatracksite is considered a facies, La Rioja and two neighboring provinces (Soria and Burgos) contain it. This Weald Facies ranges in age from the Late Jurassic (part of the Kimmeridgian) to the end of the Early Cretaceous (Albian), covering an area of 1100 km^2, with more than 500 outcrops. If it is considered a formation or group, the entire Enciso Group would be a megatracksite exposed over an area in excess of 100 km^2, with more than 200 outcrops. Finally, if it is a sedimentary unit or depositional sequence, then there are at least three potential candidates for megatracksites (>100 km^2/35 tracksites): the Enciso Group carbonate sediments (Aptian), and the lower (Barremian–Aptian) and upper (Aptian–lower Albian) siliceous sediments (Table 1.2).

Most of the described and published footprints and tracksites are in the Enciso Group, which is second to the Oncala Group in the number of sites. There are some tracksites in the Urbión Group, and one site is referred to in the Oliván Group. The Tera Group has hardly any outcrops in La Rioja, and only one locality, a place with casts, is known from the wall of a ravine.

Pérez-Lorente (2002a) attributed the difference in the number of sites between the Enciso Group and the other groups to the difference in the outcrop exposure area and the resistance to weathering and erosion of the rocks (which is less in the remaining four groups). The Enciso Group contains strata surfaces that are wider, have fewer joints, and are less weathered. However, some rocks in all the groups are resistant to erosion. These include sandstones and conglomerates in the Tera Group, limestone in the Oncala Group, sandstone in the Urbión Group, limestone and sandstone in the Enciso Group, and sandstone in the Oliván Group.

Paleoenvironment

The dinosaur footprints in La Rioja occur in sandy mudstones or shales, fine-grained sandstones, and carbonates (Pérez-Lorente, 2002a). Doublet (2004) and Doublet and Garcia (2004) provided data on sedimentology, environment, and climate that enabled the paleoenvironment and footprints to be linked. Most of the sediments with footprints accumulated in lakes and marshy areas with a continuous sheet of shallow water and some more deeply flooded zones. The lakes may have been freshwater or brackish as a result of drying.

The oldest footprints in La Rioja occur in an as yet unstudied Tera Group site in the Leza River Canyon (probably fluvial sediments). The oldest published footprints in La Rioja, however, are in carbonate rocks of the Oncala Group, a dolomitic unit known as the Inestrillas Formation at the site of Valdeprado. This formation consists of fine-grained deposits of carbonate and shale, with gypsum and salts. These sediments were deposited in a lake where the rate of evaporation exceeded the rate of water replacement. Other lake deposits with little salt in the same group (Leza Formation) contain many sites with footprints (Camino a Treguajantes; Soto 1, 2, and 3; Trevijano; San Martín 1, 2, and 3). Additional localities in this group with footprints that are not described in this book are Fuenteamarga, Sol de la Pita, Cuesta del Peso, Valdemurillo, San Vicente de Robres, and Barranco de Antoñanzas (Casanovas et al., 1990b; Moratalla, 1993; Moratalla, Garcia-Mondejar et al., 1994a; Moratalla, Sanz, and Jiménez, 1999, 2000a; Pérez-Lorente, 1993b, 1993c).

In the Urbión Group, the rocks include uncommon carbonates and more abundant sandstone, mudstone, and shale. The sedimentary environment was fluvial (deltas, channels, and floodplains) as well as open, slightly haline lakes. Floodplain deposits show soil horizons with root traces. Sites in this group discussed in the book are Mina Victoria, Valdeperillo, Cabezón de Cameros, and Los Chopos. Additional sites not described in the text are Cabezuelos, Barranco de Acrijos, La Moga, San Prudencio, Rio Maguillo, and Barranco de la Muga (Caro et al., 1995; Caro, Pérez-Lorente, and Requeta, 2002; Casanovas et al., 1995e; Moratalla, 1993; Pérez-Lorente, 2001c).

In the Enciso Group, the sedimentary rocks are sandstone, mudstone, shale, and abundant limestone. These were deposited primarily in freshwater rivers and lakes. The river environments are represented by channels, floodplains, and marshy areas with abundant vegetation, deltas, and emergent coastal areas (beaches). The lakes were shallow and extensive in area (several hundred square kilometers), with large sublittoral areas. Fossil footprints are found in almost all environments except in the open lake carbonate zones (which were deeper) and the swampy areas.

The Oliván Group is entirely fluvial. There are very few sites with footprints; they are located near the base of the unit. There is oral reference to a site with four footprints near the village of Oliván, but these footprints were stolen.

The climate was warm, with periods of both semiarid and humid tropical conditions during the depositions of the Oncala and Enciso groups, and a humid tropical climate during the deposition of the Urbión Group (Doublet, 2004).

Description of the Tracks

Nomenclature

For identification and location of tracksites, trackways, and footprints, a combination of letters and numbers is used in the following order: first, identification of the tracksite (or outcrop) through a combination of letters that may be preceded by a number; second, identification of the trackway with a number; and third, identification of the footprint with an ordinal number. For example, 7PL1.4 means the following: 7PL is outcrop 7 of the Era del Peladillo tracksite (PL); 7PL1 indicates that it is trackway 1; and 7PL1.4 is the symbol of footprint number 4 of trackway 1 from outcrop 7 of the Era del Peladillo tracksite.

In quadrupedal trackways, the pes footprint is distinguished from the manus mark with the letters p (pes) and m (manus). Thus, 7PL1.4p refers to the print of the fourth pes in the 7PL1 sauropod trackway.

In studies done by our team, values are taken from footprints and trackways, then analyzed to infer characteristics of gait and trackmaker. Data, ratios, and abbreviations used in the book are provided in Table 2.1.

Types of Features: Tracksite Surfaces

The sites and the tracks they contain are complex sets of structures that record sequences of events in time and need to be examined carefully. We will first consider the outcrops and tracks, followed by the relationship

Table 2.1. Key to abbreviations for description of footprint and trackway measurements.

Trackway or footprint	Identification
l	Footprint length
a	Footprint width
O	Orientation
Ar	Trackway deviation
Lr	Trackway width
P	Pace length
Ap	Pace angle
z	Stride length
II, III, IV	Digit length
II^III^IV	Interdigital angles
H	Acetabular height
z/H	Relative stride
v_1	Speed (Alexander, 1976; Thulborn, 1990)
v_2	Speed (Demathieu, 1986)
(l-a)/a	Relative footprint length
Ar/a	Relative trackway deviation
z/l	Stride/footprint length
Track depth	N hollow print wide; S with sediments (cast, subtraces) in the hollow
Ichnotype	t, theropod; o, ornithopod; s, sauropod; n, unidentified; S, semiplantigrade
Number of tracks	Number of present traces (complete trackway, including eroded and inexistent trackway tracks)
Direction	Trackway direction

Notes: Identification symbols for site, trackway, and/or footprint identification. Footprint length (l) is measured along axis of print; footprint width (a) at right angle to footprint length; orientation (O) angle between axis of footprint and midline of trackway (if footprint points inward to midline of trackway, it is negative rotation or inward divarication [Leonardi, 1979]); trackway deviation (Ar) distance between footprint midpoint and trackway midline; trackway width (Lr) distance between outer edges of footprints (external trackway width of Thulborn, 1990); pace length (P) distance between midpoints of two successive footprints; pace angle (Ap) angle between two successive pace length lines; stride length (z) distance between corresponding midpoints in two successive footprints of single foot; digit length (II, III, IV) length of digits measured from hypex to apex of digit (some tables also include digit I length); interdigital angles (II^III^IV) between II–III, III–IV digits (some tables also include I–II and II–IV angles); hip height (H) calculated after Thulborn (1990); relative stride (z/H) after Thulborn (1990); speed (v1) estimate after Alexander (1976) and Thulborn (1990) method; speed (v2) estimate after Demathieu (1986) method; relative footprint length [(l – a)/a] after Pérez-Lorente (2001a); relative trackway deviation (Ar/a) after Pérez-Lorente (2001a); limb thickness (z/l) or relative cursorial or graviportal hind limb character after Pérez-Lorente (2001a) from Sternberg (1926, 1932) suggestion and Haubold (1971) data. Footprint and trackway tables in this book may provide some or all of the data in these columns.

between these features and the nature of the rock, as well as phases of the foot/sediment interaction. Finally, we will examine the classification of the footprints and identification of the trackmakers.

The study surface is the visible surface on the site (Requeta et al., 2006–2007). It may be the tracking surface (the surface on which the dinosaurs walked; Fornós et al., 2002), or it may be a layer above or below it. At many sites, the study area comprises the tracking surface as well as the tops or vertical sections of higher or lower levels. Sometimes the study area contains both real tracks and undertracks, i.e., it is the tracking surface for some footprints but not for others (undertracks), even if the study surface is deformed by them.

Each stage of passing of dinosaurs may leave the following types of structures. On the tracking surface may exist true footprints, stamps (Brown, 1999), and natural casts (infilling by later sediment). On the top of strata below the tracking surface may exist direct structures (casts) and subtracks (if the foot goes through the upper bed), and undertracks on the top of strata below the tracking surface. A dinosaur foot can completely pass through a sedimentary bed and leave its print on an underlying layer; such footprints are true traces even if not printed on the tracking surface. Finally, on surfaces above the tracking surface, there may be overtracks. The tracking surface and those below it may contain one sector undeformed by the foot and another folded and/or broken. These are the pretrack and posttrack surfaces of Gatesy (2003).

Modes of Footprint Preservation

The footprint can register many types of structures regardless of the shape of the foot producing it, which makes trackmaker identification complicated (Table 2.2). The simplest footprints are composed of a shaft left by the foot as it penetrated the substrate, along with surrounding structures (Allen, 1997). Gatesy (2003) classified the structures associated with footprints as direct or indirect. Direct structures are those that were made by contact with the skin or nails (the walls and bottom of the footprint shaft, for example), while indirect structures are induced "and do not participate in the skin/sediment interface" (Gatesy, 2003:94). These are features such as extrusion rims, undertracks, and side folds.

An ideal footprint consists of the following parts:

1. Direct structures, or the faithful reproduction of the sole, or stamp (Requeta et al., 2006–2007), as well as sliding structures and other marks made as the foot entered and exited the substrate.
2. Indirect structures, or more diffuse replicas of the footprint shaft as the prints become deeper along the trackway, as in undertracks (Hitchcock, 1858), side and bottom folds, extrusion rims, and radial and circular cracks (Allen, 1997; Avanzini, 1998; Demathieu et al., 2002; Falkingham et al., 2009; Thulborn, 1990).
3. Composite structures, such as subtracks and deformation structures within the footprint due to movement of the foot through the soft sediment.
4. Structures not directly induced by the movement of the foot, such as collapse of the walls and filling in footprints with sediment.

Many workers have described structures that are included in the first three categories, most notably Allen (1997), Avanzini (1998), Boutakiout et al. (2006), Demathieu et al. (2002), García-Ramos, Piñuela, and Lires (2002b), Gatesy (2003), Hitchcock (1858), Leakey (1987), Lockley (1991), Marty (2008), Pérez-Lorente (2003c), Romero-Molina, Pérez-Lorente, and Rivas (2001), and Thulborn (1990). The following discussion is based on their work.

Table 2.2. Names of more common direct (interface between mud–skin/claw), indirect (features induced by pes/manus action), and mixed structures of vertebrate footprints.

Direct Structures	Indirect Structures	Mixed Structures
Base striae or base incision	Axial downfold	Dead zone
Breccia, base breccia	Collapse and mudflow	Deformed cast or shaft breccia (subtrack)
Narrow cleft	Cast	Shaft
Shaft	Convolutions	
Slash-like incision	Extrusion mud, extrusion rim	
Stamp, true footprint, strict sense footprint, underprint	Marginal fold, marginal thrust	
Skid mark, slide mark, skim mark, drag mark	Shaft	
Toe mark retroversion	Subcircular and linear tension cracks	
Skin and nail mark	Undertrack, ghost mark, transmitted print	

The formation of a footprint is not an instantaneous process, like stamping a piece of paper. Thulborn and Wade (1989) showed that particular structures form at different times during the impressing of a footprint. Following these authors, this study identifies three distinct phases of print creation, one static and two dynamic. The static (weight-bearing) W phase occurs when the dinosaur exerts maximum pressure on the substrate. The two dynamic phases are the T phase, the interval between when the foot touches the ground until the W phase, when it exerts maximum pressure, and the K (kickoff) phase, the interval between the W phase and when the foot lifts off the ground. Each of these phases has structures typically associated with it, as seen in the back, middle, or front part of the footprints.

The structures can also differ depending on the depth of the footprint. Footprints in which the foot scarcely sinks into the substrate are different from those where the foot penetrates so deeply that both the toes (Boutakiout et al., 2006; Gatesy et al., 1999) and part of the metatarsus (Pérez-Lorente and Herrero, 2007) sink in deeply. Deformation of the substrate walked upon depends on its structure and physical properties, so that the following features may be found: folds and gaps caused by substrate flow (Avanzini, 1998; Demathieu et al., 2002); rock fragments from higher levels that are impacted deeper (García Ramos et al., 2002b); and the subtracks of Romero-Molina, Pérez-Lorente, and Rivas (2003). The latter can be considered as the dead zone of Allen (1997), separated entirely from their level and intruded into the sediment. Marks of incision, stretching, and folding, as well as others, are also found; they can help in deducing the footprint formation conditions in each case. The structures associated with the track formation phases which have been found in La Rioja include slides, incisions, folds, extrusion rims, collapse, undertracks, and stamps (Table 2.3).

The depth of footprints depends on the physical properties of the ground and the pressure applied. Variations in depth within a print therefore depend on the ground's resistance and the dinosaur's foot exerting more or less pressure. Sometimes dinosaurs were moving across pools

Table 2.3. Most common structures (direct, indirect, and mixed) produced during each phase of formation of a footprint. They are grouped according to values of mud fluidity.

Phase T	Phase W	Phase K	Mud fluidity
Entry slide marks (skid marks)	Circular extrusion rims	Collapse of walls	High
Pushing of mud in front of and between toes	Flow folds	Toemarks retroversion	
Slashlike incision	Collapse	Skid and skim marks	
Fluidal folds		Cleft due to metatarsus	
		Narrow cleft	
Skid or slide marks	Circular extrusion rims	Claw marks	Medium
Slashlike incisions	Subcircular tension tracks.	Skid or slide marks	
Base breccia	Lateral folds	Subtracks	
	Subtracks and undertracks		
	Dead zone		
Skid and skim marks	Stamps	Claw marks	Low
Fluidal folds	Extrusion rims		Complex (changing values)
	Dead zone		
	Subtracks and undertracks		
	Downfolds		

of water of variable depth, which would mean the foot pressure applied was not always the same. It is possible that shallow tracks were caused by dinosaurs with part of their body submerged in water (Bird, 1944; Wilson and Fisher, 2003).

According to Alexander (1989), the pressure of the foot on the ground may have varied greatly from one kind of dinosaur to another, in the same way as in extant mammals. However, in this book, I will assume minimal differences between the pressures exerted by the feet of different kinds of dinosaurs. The first reason for this caution is the difficulty of determining the body mass of extinct animals such as sauropods (Ganse et al., 2011). The second is that in studying the depth of dinosaur footprints there are a number of things we need to know irrespective of the physical conditions of the mud, such as variation of pressure exerted by the sole of the pes; the size and morphology of the pes sole or the displacement distance of the mud underfoot (Fig. 3.85) (the particles cover a greater distance under a sauropod footprint than a theropod footprint); and the duration of time that pressure is applied on each (variability in the length of step and stride is a function of this time). The most sedentary animals, which remain standing for a longer time and do not have to carry out rapid movements, may produce more strain in the substrate. The rate of plastic deformation depends on the magnitude and duration of the stress applied.

It is unknown whether the tissues of the sole of the foot are similar enough that equal areas would support the same weight and the same forces (stopping and running). The ontogenic changes in the size of the foot or the trackmaker's weight are likewise unknown. We cannot ever affirm that different dinosaur species from the same groups (e.g., sauropods) must have exerted the same pressure. Finally, if the same tracksite contains shallow theropod footprints and deep sauropod footprints (or the reverse), can one safely conclude that this is due to differences in sole pressure? Should this instead be interpreted as reflecting the times when the prints were made, so the physical conditions of the ground mud were different? In La Rioja there are tracksites with deep theropod footprints and shallow sauropod and/or ornithopod footprints, and vice versa.

Shape of the Footprint

From the beginning of the study of the ichnology of dinosaurs, there has been considerable debate about the shapes of dinosaur tracks and their implications. Many authors have discussed the problem of how footprints should be illustrated. Thulborn (1990) showed that the same footprint drawn by three experts had significant differences in the way the print was depicted. Sacchi (2004) provided examples of drawings of the same footprint made by an expert and nonexperts, also showing marked differences. Additionally, in the field, many footprints will look different under different lighting conditions, and some tracks can only be seen at certain times of day, such as at sunrise and sunset.

One of the main problems is drawing the outer boundary of the print. There would be no problem if the pretrack surface and the posttrack surface (Galesy, 2003) ended at the edge of a stamp, but this is never the case. Marking the boundary between the surface not affected or deformed by the tread (pretrack surface) and the deforming print (posttrack surface) is rarely feasible. Drawing the footprint boundary at the border of the direct surface (where the extrusion rim begins) is not easy. It is also impossible to use the same drawing approach consistently on all outcrops. Therefore, what is it that is represented by a line drawn as an outline of a footprint?

Not all authors attempt to draw a footprint in the same way. Some look for the best imprint produced by the foot, so they select the stamps and remove all undertracks and distorted footprints. An ideal footprint can be reconstructed by synthesizing partial observations in each set of footprints. This may be an appropriate technique for recognizing potential trackmakers, but it is not suitable for studying either large sites or the likely behavior patterns of dinosaurs obtained from studying isolated trackways or associated trackways made by groups of animals.

The accuracy attached by the researcher to drawing the boundaries of footprints is also variable; it may be based on what is being studied or represented. For example, there is no doubt that if one is simply recording trackways or marks at a site, the outline of the footprints does not have the same importance as when trying to identify the anatomic features of the soles of trackmakers.

As a starting point in drawing the outline, we used the internal portion of the footprint that would have resulted from the intersection of the shaft with the tracking surface (Casanovas et al., 1989; Pérez-Lorente, 2001a). However, this was not a fixed rule because sometimes there are elevations above the tracking surface produced by the mud being dragged in areas where the toes exit, which of course must be represented. Also, undertracks do not follow this rule, as this line would have to be drawn in the air and would be entirely imaginary.

If the shape of the footprints depends on many factors, a statistical treatment can be applied. There are study methods used primarily by French authors that are also based on statistics of the dimensions of the footprints (stamps) to establish the trackmaker type (Demathieu, 1987a; Demathieu et al., 2002). The problems involved in deciding how such measurements should be made, however, are similar to those involved in illustrating prints.

A number of different features can be drawn in a footprint, each with a different meaning, such as outlines of digital pads, marks left by the claws, pressure ridge boundaries, slide and skid marks, and fall lines caused by mud collapsing. One possible approach is for the footprints to always be drawn by the same observer in each team, even if this means the same mistakes are always being made.

Undertracks

Many La Rioja sites have undertracks of varying form and depth. Although in some the edges of impressions made by toes are easily distinguishable, in others the margins are more difficult to recognize, or there may be merely a slight depression in the surface of the bed. This poses a difficult question. The depths and shapes of undertracks reflect both time and anatomic variables. If undertracks are going to be studied along with true footprints, which ones should be investigated? The answer is related to the objectives of this study.

Sedimentary layers may contain undertracks that are deep or shallow, with smooth or abrupt edges. There are three main reasons for the variations: the physical state of the substrate, the vertical distance separating the tracking surface and study surface, and the shape of the dinosaur's foot. The physical properties depend on the mineral composition, the organic content, and the amount of water in the mud. If all the sediment levels dehydrate to the same extent over time, it means that the oldest undertracks are the deepest. In a surface with undertracks of several generations, the deepest impressions should be the oldest (the first to be marked) and the shallowest the most recent (those marked later). In sedimentary sequences with overlapping levels and with several tracking surfaces, the oldest impressions must be those in the oldest sediments. It seems logical to suppose that graviportal dinosaurs will leave undertracks with rounded outlines and that dinosaurs with long, thin toes will leave more elongated tracks, but "with increasing depth over the first sedimentary layer, the footprint becomes shallower and less clearly defined, but also bigger" (Farlow et al., 2006:21).

The depth of undertracks and indistinctness of their outlines are not sufficient to establish the depth at which they are formed—that is, the vertical distance between the undertrack and the sole of the real track of the same footprint, or the undertrack depth with respect to the tracking surface.

Just as trackways consisting of true footprints in many cases may end in a particular ichnite for reasons that are not obvious, undertrack trackways may also be interrupted, or they may be so faint that the researcher has difficulties drawing accurate print outlines. As a result, not all undertracks (or even true footprints) in a site are described or represented.

Shapes of Dinosaur Feet and Footprints

The shape of the sole of dinosaur's feet can be deduced in many cases. However, some preliminary considerations need to be made. First, a distinction must be made between true footprints and undertracks. Many authors recognize that undertracks are often more rounded footprints than those directly impressed (Lockley, 1991; Thulborn, 1990). A theropod footprint may be shaped like an ornithopod footprint in many undertracks. In exceptional cases, however, the opposite is also true, with rounded true tracks transmitted as undertracks with more tapering toes (as at the La Pellejera site). In this case, beneath the true impression sole is a sandwich of sedimentary layers with a fragile level between two plastic beds. The intermediate level breaks, creating a dead zone with sharp edges. Beneath this layer, undertracks are produced that appear more tapered. Changes occurring in the footprints after they were made can also have an influence. Some prints become more open, while others become narrower. Footprints with long, tapering toes can be reduced to ones with shorter and rounder toes by synsedimentary or modern erosion. The opposite (a narrowing of the toe prints) is produced by the collapse of the shaft walls of the track. All these caveats complicate the interpretation of the shape of the trackmaker's foot and consequently the dinosaur that produced the print.

The basic features of the ichnites in La Rioja are simple. Tetrapod footprints made by turtles, pterosaurs, crocodiles, and dinosaurs have been found. So far, no footprints have appeared that can be associated with the ichnogenera *Ceratopsipes*, *Tetrapodosaurus*, *Stegopodus*, or *Deltapodus* (cf. Lockley, 2002), which were made by quadrupedal ornithischian dinosaurs. Large footprints from four-legged animals with rounded or elongated hind feet clearly accompanied by smaller forefeet do occur and were made by sauropods. From such prints, the following ichnogenera were established (Farlow, 1992; Farlow, Pittman, and Hawthorne, 1989; Lockley, Farlow, and Meyer, 1994): narrow trackways (*Breviparopus* or *Parabrontopodus*) and wider ones (*Brontopodus*). Other footprints produced by quadrupeds but with tridactyl hind feet (as at the Valdemayor site) were made by ornithopods (Moratalla, Sanz, and Jiménez, 1992).

Trackways with rounded footprints made by bipeds have been attributed to various types of iguanodont ornithopods (Casanovas et al., 1993b; Moratalla, 1993) and assigned such names as *Hadrosaurichnoides*, *Iguanodontipus*, and *Brachyiguanodontipus*, which take into account the shape of digital pads, the size of the prints, and possible web marks between the toes. There is one trackway of a biped with rounded footprints and no toe or sole marks that has been interpreted as having been produced by a stegosaur walking bipedally (Peñaportillo site) by the Polish ichnologist Gerard Gierlinski (pers. comm.).

Finally, trackways with long, narrow toes prints allow several ichnotypes, some still unclassified, to be distinguished. The largest of these were more than 25 cm long and have been assigned to the ichnogenera *Bueckeburgichnus*, *Megalosauripus*, and *Therangospodus*, as well as indeterminate theropod prints (Casanovas and Santafé, 1974; Lockley, Meyer, and Moratalla, 1998; Moratalla, 1993; Moratalla and Sanz, 1997). Among smaller prints with narrow toes, some can be assigned to indeterminate theropods, and others (from the sites of Barranco de Valdebrajes and Era del Peladillo [PL] 5) have been assigned to the ichnogenus *Dineichnus* Lockley, Santos, Meyer, and Hunt, 1998, which is thought to have been made by small hypsilophodont ornithopods (Aguirrezabala, Torres, and Viera, 1985; Gierlinski, Niedzwiedzki, and Nowacki, 2009; Lockley et al., 1998).

However, my research group is skeptical about how useful the ichnotaxonomic names applied to footprints from La Rioja really are. We assigned footprints to the basic dinosaur groups (theropod, sauropod, ornithopod), pending the results of research in progress by our team or by other specialists to establish more solid foundations (Casanovas et al., 1989; Díaz-Martínez et al., 2008; Haubold, 1971; Pérez-Lorente, 1988, 1996b; Pérez-Lorente and Romero-Molina, 2001b; Romero-Molina, Pérez-Lorente, and Rivas, 2003; Romero-Molina et al., 2003). The measurements proposed by Lockley (2009) of the "anterior triangle," or degree of mexasony, are not applicable because of the great variability found in footprints from some trackways.

We base our skepticism about the usefulness of ichnotaxonomic names on two considerations. First, ichnotaxa should not be discriminated or separated on the basis of size alone. Second, footprints should not be assigned to a specific ichnotaxon or trackmaker only because they were found in a particular time period. To elaborate on the first point, it must be recognized that small dinosaurs could either be the young of large-bodied species or adults of small-bodied species. Consequently, small footprints should not all be grouped together, and small footprints should not be separated from large footprints only on the basis of size. As for the second point, allocating all large theropods of Late Triassic or Early Jurassic age to *Eubrontes*, or all Jurassic and Early Cretaceous large theropod prints to *Megalosauripus*, or all Late Cretaceous large theropod prints to *Tyrannosauripus* does not seem scientifically justifiable.

There are many examples of the same ichnotaxa attributed to quite different kinds of dinosaurs. These were differences not merely of species or genus but rather were at high taxonomic level such as suborder. There are examples of such divergent assignment in La Rioja. There are also ichnotaxonomic identifications that must be abandoned. Before 1990, our group associated some of the footprints to well-known dinosaurs or footprint such as *Megalosaurus*, *Megalosauripus*, Tyrannosauroidea, and *Eubrontes*, and some of these footprints are not stamps or have deformations. Since then, we have stopped identifying our footprints with defined names because of the uncertainty over the features used and because we assume that many of the footprints described in the world were not named on the basis of appropriate specimens (stamps). For example, in La Rioja, the small footprints of the Barranco de Valdebrajés tracksite were attributed by Aguirrezabala, Torres, and Viera (1985) to *Hypsilophodon*; by Casanovas et al. (1991a) to *Coelurosaurus*; by Lockley and Meyer (1999) to an unknown hypsilophodontid or *Dryosaurus*; and by Romero-Molina, Pérez-Lorente, and Rivas (2003) to theropod footprints of an unknown dinosaur.

Influence of Mud

A footprint sinks in sediment to the depth at which the resistance to penetration of the foot is equal to the pressure applied. These conditions can be reached at various depths depending on the type of substrate and according to the pressure applied by the sole of the foot. Assuming a constant pressure, the depth will depend only on the physical properties of the substrate. These properties can be distributed homogeneously over the ground (as with a seamless layer of pastry) or heterogeneously (as in surfaces of different composition or degrees of water saturation). The properties that regulate the ground's response to treading are viscosity, consistency, and adherence.

The first property, viscosity, is a variant of plasticity, in that the original form does not recover after deformation. The plasticity limit can be said to have been exceeded if a mark is left on the ground. The term *viscosity* is used as an indicator of a scale whose extremes are fluidity and rigidity. It measures the substrate's ability to provide resistance (infinite rigidity against foot pressure) or not (infinite flow providing no resistance). The value of viscosity is applied only to the penetration resistance and not to the property of flow, as it is with a liquid or gas.

Consistency allows further clarification of the preceding concept. A foot can penetrate deeply into the substrate, but the walls can still remain vertical in the hollow after it is withdrawn. If the walls do not collapse, the consistency is high, but if they slump back completely, the consistency is very low. At tracksites in La Rioja, the maximum depth of footprints with vertical walls is 30 cm, which was measured using a large ornithopod natural cast from the Urbión Group on display in the Enciso Center (Fig. 2.2). The cast was formed when the footprint was filled with sand, so the compression of the sediment over time should not have modified the depth of the footprint hollow by more than 10%. However, the bottom surface of the layer of sandstone (filling and covering the cast) is erosive. The process by which this cast was created consisted of the formation of the hollow in clayey mud, after which it was filled with sand. During the filling process, the tracking surface and part of the substrate may have been eroded, so the original depth reached by the sole of the ornithopod is likely greater than the present depth and consequently is not known. After forming the cast, the current carrying the sand eroded part of the top of the underlying mud bed in which the print was originally impressed. Similarly, La Rioja has examples of the opposite situation: prints impressed in what was mud of very low consistency so that after the foot was removed, impressions of toe prints and metatarsi totally collapsed, to be preserved only as the closure line of the footprint walls.

Adherence is a deformation usually found in the center of the footprint or the toes. Sometimes mud sticks to the sole of the foot and remains stuck during withdrawal. There may be deformation of the base of the footprint shaft as a result of suction by the foot as it rises.

The possible variations in heterogeneous ground response are several because there are differences in mineralogical composition, fabric (i.e., number, type, and composition of laminites), and water content between layers. In heterogeneous substrates, the response to the footprint always differs depending on the angle of attack of the foot.

Influence of Dinosaur Behavior

CROUCHING DINOSAURS

Dinosaurs did not usually move with the metatarsus flat on the ground. Models of dinosaurs with their feet in a normal position include those in which the metatarsus was vertical or inclined with respect to the base of the toes; those whose metatarsus is inclined but with large fleshy pads underneath its distal end (ornithopods: Langston, 1960; Thulborn, 1990); and those with both the metatarsus and the toes inclined, resulting in a subdigitigrade footprint (Wilson, 2005). A dinosaur resting the metatarsus horizontally on the ground is considered to occur only while it is in a special position (Kuban, 1989; Romero-Molina et al., 2003). Because the dinosaur tarsus is associated with the zeugopodium, Sarjeant et al. (2002) ("plantigrade . . . should be used only when the basipodium is also impressed, which is not the case in these footprints"; Romero-Molina et al., 2003:248) used the term *semiplantigrade* rather than *plantigrade* to characterize such tracks, although William Antony S. Sarjeant (pers. comm.) notes, "Personally I think semiplantigrade is preferable." There are several sites around the world where prints of this kind occur. In La Rioja, all such semiplantigrade footprints are associated with trackways, some of which include both normal digitigrade and unusual semiplantigrade prints.

Not all footprints with metatarsal marks are semiplantigrade ichnites. Dinosaurs with a normally inclined metatarsus could leave the mark of both the metatarsus and the hallux if the foot sank deeply enough into the mud (Aguirrezabala and Viera, 1980). The rear end of such footprints is angular in shape. In dinosaurs with a fully supported horizontal meta-tarsus, the rear end of the footprint is rounded.

SWIMMING DINOSAURS

There has been much discussion about whether dinosaurs could swim. Footprints have been attributed to swimming theropod dinosaurs (Coombs, 1980; Ezquerra, Costeur, and Pérez-Lorente, 2010; Ezquerra et al., 2007; Galton et al., 2003), and trackways thought to have been made by dinosaurs entering or leaving the water have been described (Casanovas et al., 1993d; Ellenberger, 1974; Gierlinski and Potemska, 1987). Currently there is an unresolved debate about whether sauropod trackways with prints of only the forefeet were made by swimming or floating dinosaurs (Bird, 1944; Ishigaki, 1989; Nouri, 2007; Wilson and Fisher, 2003), or whether this kind of preservation occurs because the footprints are undertracks (Lockley and Rice, 1990).

Ellenberger (1974) illustrated tracks that he interpreted as having been made by dinosaurs that were swimming, exiting the water, and walking across solid ground. Prints that he thought were made as the animal exited the water are semiplantigrade, and sometimes the trackway indicates quadrupedal locomotion with a tail mark registering. Classic examples of tail marks (cf. Pérez-Lorente and Herrero, 2007) are in tracks that leave a hallux mark, and it has been suggested that dinosaurs adopted this position in special situations. Tail marks cited in the literature are always in trackways with some irregularities (Pérez-Lorente and Herrero, 2007; Torcida et al., 2003). This means that dinosaurs did not normally leave such tail marks and therefore did not normally drag their tails along the ground, although there may be an exception in the example reported by Hunt and Lucas (2004), with various parallel tracks showing tail marks.

Trackways possibly made by swimming dinosaurs occur at various sites in La Rioja. Footprints attributed to swimming theropods occur at La Virgen del Campo 1 (1LVC) (Pérez-Lorente, 2001a; Pérez-Lorente et al., 2001); 4LVC, where there is a 40 m trackway of a dinosaur that left slide marks through the mud at the base of the claws of the two hind legs (Ezquerra et al., 2004, 2007); and El Villar-Poyales, where there is a trackway of a dinosaur whose footprints became irregular as it progressed. The last two prints are of the left and right autopod, with slits made by the tips of the toes and claws. After that, the trail disappears because the water depth became greater than the reach of the theropod claws (Casanovas et al., 1993d; Pérez-Lorente et al., 2001; Romero-Molina, Pérez-Lorente, and Rivas, 2001).

PIGEON-TOED WALK

Wade (1989) addressed the problem of seeming incompatibility between the narrow, pigeon-toed trackways of dinosaurs and their makers' erect posture. Thulborn (1990) suggested that walking dinosaurs probably rotated around a vertical axis passing through the hips. For tracks in La Rioja (Pérez-Lorente, 1993a), it was postulated that the pigeon-toed orientation of the feet is due to nonparasagittal rotation between the metatarsal sector and the digital part of the limb (Fig. 3.190D).

In their study of *Theroplantigrada encisensis* Casanovas et al., 1993d, Pérez-Lorente (1993a) noted that the axes of the metatarsus and the digitigrade foot are not aligned (Fig. 3.108). It was found that this arrangement was normal in other semiplantigrade footprints in the rest of the world, and the joints between the bones of the first metatarsus and phalanges were reviewed. It was suggested that the digitigrade foot (acropodium) moved in a different plane than the longitudinal plane of symmetry of the metatarsus (Fig. 3.190). This fact is also confirmed because digitigrade footprints show inward rotation (Pérez-Lorente, 2003c).

Bipedal dinosaurs thus did not necessarily move their bodies around a vertical axis as Thulborn (1990) states, even though they made tracks with an inward orientation of footprints congruent with the gait of pigeons. Instead, they moved with the longitudinal axis of the body parallel to the direction of the march. The movement of the foot and its segments in bipedal dinosaurs is reflected in three types of footprints: digitigrade (Fig. 3.79A), or supported only by the toes (inward rotation of the footprint increasing with size); semiplantigrade, or supported by the toes and metatarsus (Fig. 3.79C); and footprints where the toes and part of the metatarsus sink into the mud (Fig. 3.79B).

CAUTIOUS DINOSAURS

One of the assumptions we make in La Rioja is that the pressure applied by the feet of dinosaurs was similar for almost all of them. The depth of the prints therefore does not depend on the weight of the dinosaur but rather on the resistance of the sediment.

Almost all the deep footprints described in La Rioja are from herbivorous dinosaurs (sauropods and ornithopods), contrary to Alexander's (1989) data. Given this fact, the Rioja sites suggest a "cautious dinosaur" hypothesis (García Ramos, Piñuela, and Lires 2002a). Herbivorous dinosaurs were safer in the mud where their movements were as slow as they were on land, while carnivorous dinosaurs–agile and fast on land–were increasingly awkward the more their feet sank into the mud. Consequently, herbivorous dinosaurs are presumed to have preferred muddy situations, where their customarily slow movements would have been less impeded than those of their carnivorous enemies (Farlow, pers. comm.). However, this is contrary to assumptions about the speed of theropod dinosaurs on soft and hard ground. The theropod semiplantigrade and irregular (soft ground) trackways provide faster estimates of trackmaker speed than the digitigrade (hard ground) trackways. This is a paradox.

The interpretation of cautious herbivorous dinosaurs based on the depth difference of the footprints has other contradictions: there are abundant shallow footprints made by herbivorous dinosaurs; dinosaurs

with broad toes displaced a lot of mud and left wide hollows, so the walls of the prints were therefore unlike to collapse totally (and so would not change the original depth of prints); and there are many theropod footprints that sink into the mud but left shallow hollows; collapse and filling of the hollow was normal.

Although the depth of graviportal dinosaur footprints is apparent in both the track layer surface and in vertical sections, the depth of the footprints of theropod dinosaurs goes unnoticed. The deep structures of theropod footprints have been misinterpreted to such an extent that new dinosaur ichnospecies have been defined and probably sometimes confused with bird footprints (Boutakiout et al., 2006).

2.2. Natural cast of an ornithopod print in collection of Enciso Museum. Height of footprint walls is 30 cm. Dinosaur pes penetrated deeply into sediment, but there is no collapse of the walls of the print.

Tracks, Undertracks, and Subtracks

Most of the footprints found in La Rioja are hollows (true footprints, stamps, and undertracks) (Fig. 2.1). They are followed in number by subtracks (with the filling deformed), then casts, and finally tracks with a mixed filling (one part deformed by the footstep and another with undeformed sediment that has settled in the hollows).

Molds copy the shape of the foot. The mold is the shaft of the footprint after the foot has been lifted from the mud. Molds are direct structures that can be either stamps or true tracks deformed by the behavior of the mud. According to our interpretation, molds are structures that must always be direct. Hollows formed by undertracks are not considered to be molds.

Table 2.4. Summary of data for the selected tracksites.

Tracksite	No. of footprints				No. of trackways				Area m²	Age	Group	References
	Theropod	Ornithopod	Sauropod	Unidentified	Theropod	Ornithopod	Sauropod	Unidentified				
LVP	109			35	12			3	260	Berriasian	Oncala	Moratalla (1993, 2005), Moratalla et al. (1993, 2000a)
LN	102	4	137	45	22		4	1	486	Aptian	Enciso	Casanovas et al. (1995h); Pérez-Lorente (1994,1995,1996b); Pérez-Lorente et al. (2001)
BVG	40	8			9				20	Aptian–Lower Albian	Enciso	Aguirrezabala et al. (1985); Casanovas et al. (1991a,1992c); Lockley et al. (1998); Lockley and Meyer (1999); Pérez-Lorente et al. (2001a, d); Martín Escorza (1992); Viera and Torres (1992)
PL	470	412	913	244	69	30	6	18	2170	Aptian	Enciso	Blanco et al. (1999); Casanovas et al. (1985, 1993a,b,c, 1995j, 1997b, 1998); Meléndez and Pérez-Lorente (1996); Pérez-Lorente (1990, 1991, 1992a, 1993b, 1994, 1995, 1998, 1999a, 2001a); Pérez-Lorente et al. (2001)
MIV, VLP	17	9	2	12								Ansorena et al. (2007–2008); Moratalla (1993); Moratalla et al. (1997b)
LC	1000	70	13	20	52		1		1500	Lower Albian	Enciso	Moratalla (1993, 2005); Moratalla and Hernán (2008, 2009); Moratalla and Sanz (1992a); Moratalla et al. (1988, 1992b, 1994c, 1999, 2001, 2002, 2003, 2004); Sanz et al. (1985)
VT		11				1			20	Aptian	Enciso	Lockley et al. (1994); Moratalla (1993); Moratalla et al. (1988b)
TT, VG	43	112	8	176	7	16		3	31600	Aptian	Enciso	Brancas et al. (1979); Hernández Medrano et al. (2006); Jiménez Vela and Pérez-Lorente (2005, 2006–2007); Pérez-Lorente and Jiménez Vela (2006–2007)
AND	2?	21			1?	1			65t	Aptian	Enciso	Brancas et al. (1979); Casanovas and Santafé (1971); Viera and Torres (1979)
NZ	25	87		33	2	12			530	Aptian	Enciso	Brancas et al. (1979); Casanovas et al. (1993); Pérez-Lorente (1991); Pérez-Lorente et al. (2001)
EVP, 3I	184	3			24	1			380	Aptian	Enciso	Brancas et al. (1979); Casanovas et al. (1992a, 1993d, 1998); Pérez-Lorente, F. (1993a, 1995, 2001a); Pérez-Lorente, F. et al (2001); Viera and Torres (1979)
VL	62	44	58	9	15	6	2	4	1900	Aptian	Enciso	Brancas et al. (1979); Casanovas et al. (1989a, 1996); Pérez-Lorente et al. (2001); Valle (1993); Viera and Torres (1979)

Code	A	B	C	D	E	F	G	Total	Age	Formation	References
LS	143				19			190	Upper Barremian–Aptian	Enciso	Casanovas et al. (1989a); Pérez-Lorente et al. (1986, 2001)
LVC	350	30		150	77	7	7	4900	Upper Barremian–Aptian	Enciso	Blanco et al. (1999b); Brancas et al. (1979); Caro and Pavia (1998); Casanovas et al. (1985, 1989a, 1991c).; Ezquerra and Pérez-Lorente (2002, 2003); Ezquerra et al. (2004, 2007, 2011); Pérez-Lorente (1990, 1991, 1993b, 1994, 1996c, 1997, 1998, 1999b, 2000b, 2001a); Pérez-Lorente et al. (1986, 2001); Requeta (1999)
LL	375				77			406	Aptian	Enciso	Blanco et al. (1999a, 2000); Pérez–Lorente (1999a, 2000b); Romero-Molina et al. (2003b); Sarjeant et al. (2002)
PP, UP	66	49			6	10		600	Lower Albian	Enciso	Blanco et al. (1999b); Casanovas et al. (1993c, 1998); Pérez-Lorente (1991, 1992a, 1996c, 2000b, 2001a); Pérez-Lorente et al. (2001); Viera et al. (1984)
BLC	4	60		1	7			190	Lower Albian	Enciso	Casanovas et al. (1995a); Pérez-Lorente et al. (2001); Viera et al. (1984)
SB	4		19		1	1	1	350	Lower Albian	Enciso	Casanovas er al. (1997a); Pérez-Lorente (1995)
SM	31	8	16	6	3	2	2	70	Berriasian	Oncala	Casarovas et al. (1995b); Pérez-Lorente (1993a)
EC	17	24			1	7		50	Aptian	Enciso	Pérez-Lorente (2000c); Pérez-Lorente et al. (2000)
LB	17	2			1	2		300	Aptian	Erc so	Pérez-Lofrente (200oc, 2003c)
STS	86	4	14		24	1		120	Lower Albian	Erc so	Pérez-Lorente, 200c, 2003c)
LP	528	236		61	51	19		840	Lower Albian	Enciso	Hernández-Medrano et al. (2006); Pérez-Lorente et al. (2009); Requeta et al. (2006–2007)
1ST–2St	41	11	154	3	12	4	1	70–65	Berriasian	Oncala	Casanovas et al. (1990a, 1992b)
TRG	8				1			25	Berriasian	Oncala	Barco et al. (2004); Casancvas et al. (1995f)
CBC		25			2	2		60	Upper Barremian–Aptian	Enciso	Moratalla (1993); Moratalla et al. (1992b, 1992c)

Casts are the undistorted fillings of molds. Most casts found in La Rioja are in the sandstone that fills the shafts (Figs. 2.2, 3.140). Casts can be deep and filled with a lot of sand (up to 30 cm in the ornithopod footprint deposited in the Enciso Paleontological Center), or they can be shallow (less than 5 cm), like certain footprints at La Virgen del Campo (3LVC) (Fig. 3.140).

Sometimes casts preserve complex structures. There are dinosaur footprints where the footprint consists of not just the sole of the foot, but also the side and even the tops of the toes (Huerta et al., 2012). The shaft left by the foot in the mud is not only something that rises and falls. The foot first moves forward, pushing into the mud, and is pulled out backward on exit. Toes leave horizontal openings within the muddy layer.

Casts fill the gaps left by toes in the mud. There are examples where the part above the nails is preserved in the cast (1LVC outcrop). The top of the claw is also sometimes fossilized in the tips of the toes and its connection with the corresponding toe. Casts at La Virgen del Campo record the entry and exit of the hallux and one of the toes. This behavior implies low-viscosity mud (where the feet and toes penetrated easily) and high consistency (neither the walls nor the top of the tracks collapsed).

Footprints can deform the substrate into which they press. As several authors have noted, under the sole of the foot there are, according to the composition of the rock, several types of extreme deformation: ductile, brittle, or mixed (Avanzini, 1998; Boutakiout et al., 2006; Demathieu et al., 2002; García Ramos et al., 2002b; Gatesy et al., 1999; Loope, 1986; Manning, 2004; Milan, Clemensen, and Bonde, 2004). The foot descends into the mud, dragging some of the material it passes through. If the material is fluid, the foot penetrates and forms folds in the substrate. The material the foot passes through can be rigid and break into a sheet that descends in one piece (neutral zone) or in fragments (breccia). If the substrate is mixed, the footprint may contain a chaotic mixture of broken and folded parts formed when the feet entered the ground. In this book, all of these structures are referred to as subtracks.

Selected Tracksites

Selected tracksites of La Rioja are described in Chapter 3. The basis for the development of each section is the special ichnologic content of each of the selected sites. Table 2.4 provides a compendium of the tracksites, which includes the number and type of footprints and trackways, the area of each site, and related publications. The cited articles provide more information and sometimes detailed maps (1:12.5 scale) of the sites or outcrops.

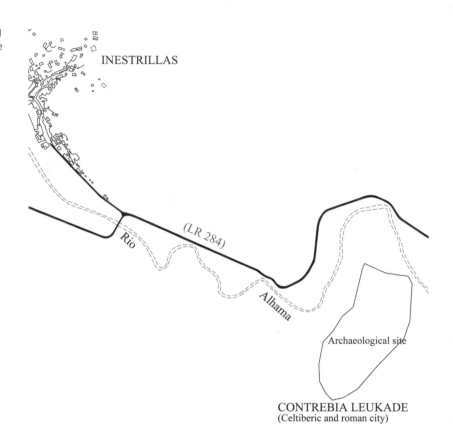

3.1. Location of La Virgen del Prado (VDP) tracksite (also see Fig. 1.4). Local road is labeled LR 284.

INESTRILLAS

Rio

(LR 284)

Alhama

Archaeological site

CONTREBIA LEUKADE
(Celtiberic and roman city)

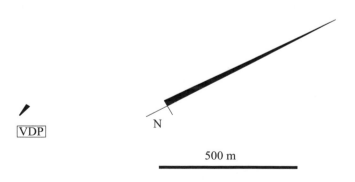

VDP

N

500 m

History

After the discovery of the Virgen del Prado site (Fig. 3.1) in 1991 by an anonymous researcher, a team from the Universidad Autónoma de Madrid and an electrical company (Iberdrola) cleaned and mapped it during subsequent campaigns, finally describing the results in an unpublished thesis by Moratalla in 1993. The thesis stated there were only 36 footprints, of which four were part of a trackway. Years later, Moratalla and Hernán (2005) produced a new map of the site showing 200 footprints and six trackways (see Fig. 3.2A). This new study recognized the existence of a theropod with slender toes and feet, which was apparently abundant in the Oncala Group.

In 2004, Pérez-Lorente carried out a separate unpublished study of the site to better understand the contents of La Virgen del Prado (Fig. 3.2B). This was because the traces in La Virgen del Prado show noteworthy collapse structures, there were types of footprints previously not described, and there were a large number of unrecognized trackways. The map in Fig. 3.2B was not created by surveying the area but rather

La Virgen del Prado

3.2. Map of La Virgen del Prado tracksite. (A) Modified from Moratalla and Hernán (2005). (B) Version made by Pérez-Lorente in 2004. (C) Lines identifying proposed trackway sequences. Footprints enclosed by trapezoid are enlarged in Fig. 3.7. Footprints enclosed in circle can be used to correlate between the two versions of tracksite map (A, B).

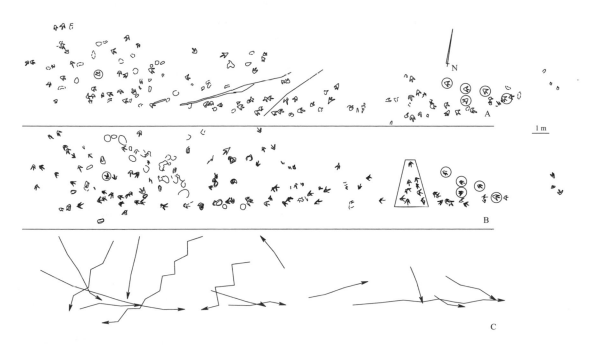

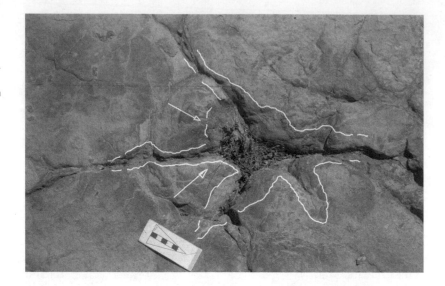

3.3. Typical footprint from La Virgen del Prado, showing mud collapse around impressions of metatarsus (right), three main toes (left), and hallux (bottom). Arrows indicate direction in which mud in area between toes flowed after foot was lifted from substrate. Scale bar in centimeters.

from partial surveys that were superimposed later. The procedure was as follows. First, each of the trackways and adjacent footprints were mapped. Footprints were drawn on the plan to obtain a picture of each trackway and the adjacent footprints. Second, all trackways were placed in their relative position on a single drawing (Fig. 3.2B). Relative position was established by superimposing the common footprints of the trackways. The footprints on the right-hand side of the site were situated according to the 2005 data of Moratalla and Hernán. Both the 2005 and 2004 maps are presented in Fig. 3.2. The Moratalla and Hernán (2005) map (Fig. 3.2A) is the more recently published version, but my version (Fig. 3.2B) provides additional detail. The count of footprints and trackways is taken from the latter (Table 3.1).

Description of Location

The footprints at La Virgen del Prado occur on a layer of limestone outcrops near the Virgen del Prado chapel. The layer dips at an angle of about 35°, which provides an extensive surface. The site is in a desolate area that is green only in the spring or after heavy rains. There are no trees, and the vegetation is arid scrubland. Access is by a road that leads to the chapel above. The road passes near the top of the ruins of Contrebia Leucade, a Celtiberian city razed and reconstructed by the Romans. There are no other places nearby with footprints. The nearest possible site, in the village of Valdemadera (9 km), contains a solitary print. Farther away are the studied sites of La Virgen del Villar (11 km), Las Navillas (11 km), and Navaján (Mina Victoria, 12 km). In the same layer as the tracks are abundant scales and other holostean fish remains (*Lepidotes*); fish fossils also occur at other sites in the same Oncala Group nearby. In addition to the footprints and fish fossils, there is an abundance of ostracods, at least in one of the layers above the site.

3.4. Left footprint from La Vırgen del Prado with extremely long and thin digit impressions. Note strong asymmetry in lengths of side toe marks. Footprint length 30 cm.

Lithology

The tracks occur in a layer of limestone about 40 cm thick, which is dark when freshly cut but brownish-yellow on the surface, with no appreciable sedimentary structures. In adjacent layers there are laminites and cracks due to drying. According to Doublet (2004), the rocks at the site (limestone, dolomite, marl, gypsum) were deposited in a saline lake. The stratigraphic location has been defined as the Inestrillas Formation, one of the formations in the Oncala Group.

Track Features in the Mud

Most of the footprints at this site have structures indicative of soft mud. The mud has collapsed in the toes, the areas between the toes, the metatarsal region, and the hallux (Fig. 3.3).

As the mud slipped back into the hollows of the toes, it enlarged them. Most of the tridactyl footprints have very thin and relatively long toe marks (Fig. 3.4). At the bottom of the prints there are generally longitudinal incisions that are even longer than the sole (autopodium) print (Boutakiout et al., 2006) running through them (Figs. 3.3, 3.4, 3.5). Sometimes the mud collapsed so much that only the incision is visible on the surface of the bed; the opening created by the toes has completely closed. The mud from the interdigital areas also flowed into the hollow of the tracks, sometimes leaving rounded lobes in the hypex (Figs. 3.4, 3.5). There are elongated, subrectangular marks (Figs. 3.6, 3.7, 3.8) in which only impressions made by the metatarsus were retained. The toe holes have sealed over in the same way as similar footprints in the Paluxy River, as cited by others, including Farlow (1987).

3.5. Group of tridactyl and tetradactyl footprints from La Virgen del Prado showing collapsed marks of digits, along with interdigital lobes also due to mud collapse. Note grooves in deepest part of digital impressions. Scale bars in drawing at 10 cm intervals.

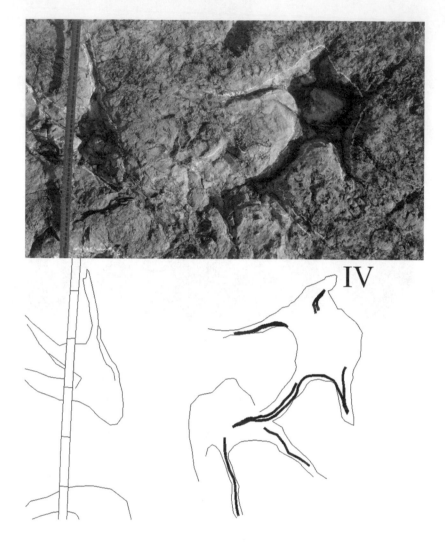

Because the dinosaurs passed through this area when the mud was very soft, many of the prints register both the metatarsus and hallux (digit I). The impressions of the metatarsus, hallux, and the toes (digits II to IV) at the front of the print were narrowed by the collapsing sediment. Almost all of the La Virgen del Prado theropod footprints have this collapsed mud configuration. There are two trackways (Figs. 3.2, 3.8) of prints with a particularly wide metatarsal impression and collapsed front toes.

The deformation produced in the La Virgen del Prado footsteps is striking. Because the mud was fluid, the autopodia penetrated deeply. Moreover, the walls of the prints fell inward during both the T and K phases. The mud flowed into at least some of the interdigital areas, then slipped back after the toes were withdrawn from the sediment. Pérez-Lorente and Herrero (2007) summarized the studies published on footprints showing this deep penetration of the entire foot in mud.

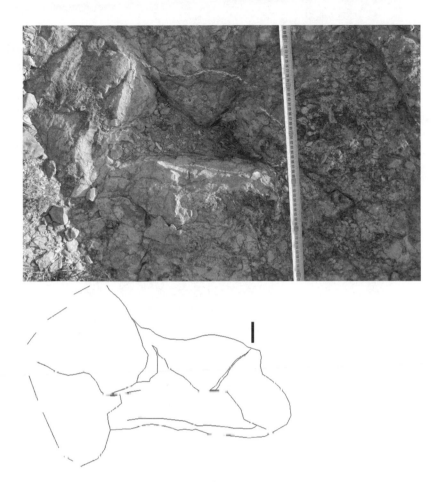

Footprint Features

TRIDACTYL AND TETRADACTYL DIGITIGRADE FOOTPRINTS

In publications relating to this site (Moratalla, 1993; Moratalla and Hernán, 2005; Moratalla, Sanz, and Jiménez, 1993, 1997b, 2000a), the footprints have been interpreted as stamps—that is, footprints whose deformation did not substantially alter the shape of the foot of the trackmaker. The narrowness of the toes was assumed to be due to their anatomic form. A new ichnotaxon, *Filichnites oncalensis*, was defined by Moratalla in a 1993 Ph.D. thesis in the Cameros Basin; it was used to define two associations.

Filichnites Moratalla, 1993, is a tridactyl footprint between 12 and 30 cm in length, with quite thin and long digits, with the marks of digits II and III of comparable length and occasionally curved. The hypex is symmetrical and open. The plantar surface proximal to digit III is small with a slight notch on the medial side. The ratio of digit length III/(footprint length − digit length III) is high, between 2.7 and 3.4, as is the ratio of footprint length/footprint width, between 1 and 1.2. The tracks are narrow and were made by a bipedal animal.

Although Moratalla (1993) assigned the footprints at La Virgen del Prado to the ichnotaxon described, the author acknowledged that some

3.7. Footprints from La Virgen del Prado. (A) Detail from Fig. 3.2B showing tridactyl footprints delimited from trapezoidal outline. (B) Semiplantigrade subrectangular print (circled in Fig. 3.8B). (C) Unidentified footprint (circled in Fig. 3.8C).

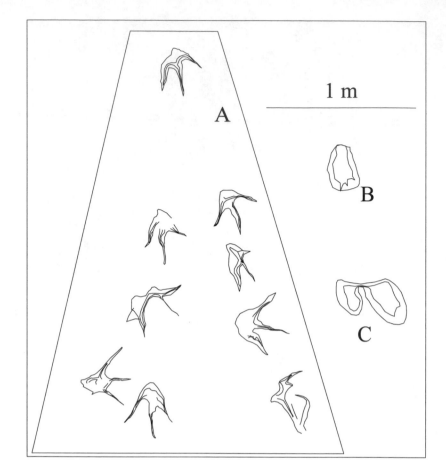

3.8. Trackways from La Virgen del Prado whose makers are difficult to interpret. (A, B) Semiplantigrade trackways. (C) Trackways whose maker is unknown. Possible sauropod trackway cuts across two semiplantigrade trackways.

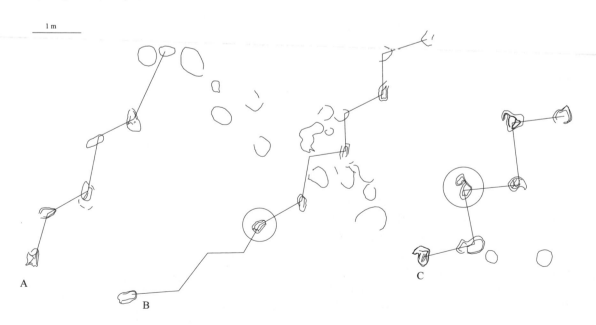

Table 3.1. Measurements of two semiplantigrade trackways and the unidentified trackway of La Virgen del Prado site.

Trackway	l	a	P	z	Ar	Lr	Ap	O
A	38	21	109	192	19	70	134	31
B	34	19	90	108	22	73	112	66
C	21/32	40	115	164	40	126	90	

Note: Abbreviations as in Table 2.1.

features do not match the classification. He wrote that at La Virgen del Prado, some of the footprints are greater than 30 cm in length, with a hallux and without the indentation on the medial side. Moratalla mentioned this ichnotaxon in several publications describing ichnoassociations: a *Therangospodus–Filichnites* association, which is attributed to the Tithonian–Barremian interval (Moratalla, 2008), and a *Therangospodus–Filichnites–Kalohipus* association of Berriasian age (Moratalla, 2009b).

There is no doubt that the thin marks, or "filiforms," of the toes were caused by the collapsing walls of the footprint (Figs. 3.3, 3.4). This process is also responsible for the large values of the two footprint shape ratios. The marks made by the toes and feet are narrowed, but their lengths are unchanged. The collapse also reduces the junction between the toes, causing digit III to increase at the portion of the print proximal to it. This evidence led to the conclusion that the La Virgen del Prado tridactyl and tetradactyl footprints with narrowed digits are not stamps. Footprints of this type could only be made by dinosaurs with long, well-separated toes on their feet. They are therefore identified as theropod dinosaur tracks. The collapsed hallux and metatarsus are not considered to be ichnotaxonomic features but rather are interpreted as having been produced because the foot sank so far into the mud.

SEMIPLANTIGRADE FOOTPRINTS

The semiplantigrade footprints form two parallel trackways (Figs. 3.2, 3.8A, 3.8B). The front toes can be seen in only some of them because those parts of the tracks have significantly collapsed. The footprints are only clear enough to be ascribed to semiplantigrade theropods. For the toe marks to collapse (Casanovas et al., 1993d; Kuban, 1989) but not the metatarsus, the digits must have been long, thin, and highly separated. As with the filiform prints, they are classified as theropod footprints. These are narrow footprints ($[l - a]/a = 0.8$), with a wide trackway ($Ar/a = 0.9–1.1$) and strongly positive orientation (Fig. 3.8) (Table 3.1). The low pace angle ($134–112°$) correlates well with the wide trackway. (See Table 2.1 for definition of these ratios.)

UNIDENTIFIED FOOTPRINTS

At the center-left side of the tracksite (Fig. 3.2) is a trackway of enigmatic footprints consisting of two elongated, oblong lobes (Fig. 3.9) slightly offset from the direction of travel (Fig. 3.8C), with the smaller lobe on

3.9. One bilobed track from trackway of unknown origin (see Figs. 3.7 and 3.8).

the outside of the trackway. After examination, it was concluded that there were no other components recurring throughout the trackway. The average length of the outer lobe is 21 cm, and that of the inner lobe is 32 cm (Table 3.1). The width is greater than the possible length of the longest digit. The ratio between the stride and the length of the foot (z/l) indicates thick limbs (Haubold, 1971; Pérez-Lorente, 2001a; Sternberg, 1932). Ar/a indicates a wide trackway. It might be assumed that the inner lobe of the foot was the hind foot mark of a semiplantigrade (or quadrupedal?) dinosaur, and the outer lobe the forefoot. There is no support for this interpretation, however, because the inner lobes arc rounded at their distal end and straight at their proximal end; the lengthening of the outer lobe is parallel to the direction of the trail; the proximal parts of the two lobes are flat and aligned; the two lobes are interconnected in this trackway; and there is no deformation or flow of the mud to the hollow of the inner or outer lobe. This suggests that the two hollows were formed at the same time.

SAUROPOD FOOTPRINTS?

Other footprints are grouped together in a strip cutting diagonally across the site (Figs. 3.2B [left half], 3.8). The marks are rounded and shallow, and they have no apparent morphological details. They may consist of large and small marks, and the trackway may be associated with a sauropod dinosaur. The direction of the trackway is perpendicular to the three described above. The footprints show no signs of collapse, so the trackmaker probably passed by sometime after the rest of the dinosaurs. There are no clear overlaps between prints in this possibly trackway, and no tridactyl footprints were seen on the inside of this set of tracks.

Distribution of the Tracks According to Ichnotypes

Theropod footprints left two types of tracks: collapsed footprints with impressions of the digits (including the hallux) and a thin metatarsus impression, and semiplantigrade theropod prints with significant collapse of the toes. The theropod footprints formed when the mud was so soft that the feet of the dinosaurs penetrated to the bottom of the layer. At the same time, an unidentified animal (Fig. 3.8C) left its trace. Footprints of the first type of theropod print occur all over the site, with no apparent preferred orientation. The semiplantigrade theropod prints, with a relatively wide depression left by the metatarsus, form two parallel trackways relatively close to one side of the site. The third trackway of unknown origin is near and parallel to the other trackways (Fig. 3.8C). If these rounded, shallow footprints were left by herbivorous dinosaurs, the site would be an exception to the ground hardness affinity theory for certain dinosaurs. The dinosaur prints seen would be contrary to the cautious dinosaur hypothesis, indicating the presence of carnivorous dinosaurs on soft ground and herbivorous dinosaurs on hard ground.

Herd Behavior

The Virgen del Prado tridactyl and tetradactyl footprints have many common features (long, slender digits, reduced footprint size proximal to the toes, low variability in size). The hypothesis that the tridactyl and tetradactyl dinosaurs shared the same space and time cannot be dismissed. This would be due to either their social behavior or habitat. However, there are no grounds to suppose one or the other. It is possible that the tridactyl, tetradactyl, and "semiplantigrade" prints are variants made by the same kind of animal. However, it must be borne in mind that the deformation of the footprints, as a result of the low mud viscosity and consistency, could be responsible for comparable features the tridactyl and tetradactyl forms (general narrowing of the digits for any theropod). It is therefore just possible that dinosaurs that left these thin *Filichnites* traces were quite different.

The orientation of the trackways provides few grounds for supposing group behavior. If this location was visited by related dinosaurs (because of the similarity of the footprints and habitat at that time), then the time of the visits might have been different for each. The two semiplantigrade trackways and the one with the bilobate footprints show the same direction of travel, although this may be coincidental. The semiplantigrade trackways have a lower Ar value than the bilobate trackway. This is of uncertain significance because the trackway width is not fixed—it is variable, even in the middle trackway.

General Comments

The track density is low—about a footprint every 2 m^2, which is normal for many of the La Rioja sites. In the surrounding areas there are marls and crumbling limestones. It is difficult to find footprints in these rocks, even if there even are any. The high-salinity lacustrine sedimentary environment means that the footprint conservation conditions were less favorable than for other areas; the environment seems also to have been an undesirable habitat for terrestrial vertebrates. In La Rioja, the sites with footprints are more numerous in the Enciso Group than in the other groups because the conservation conditions of large stratification surfaces, without weathering or severe erosion, are much better for that group. Therefore, the scarcity of sites in the vicinity of Virgen del Prado may be related to low resistance to weathering of the exposed rocks.

The deformation structures of the footprints due to the collapse of the mud are of interest. The semiplantigrade footprints are similar to those from the Glen Rose Formation of Texas (Kuban, 1989), and like the Texas tracks, some of the Virgen del Prado semiplantigrade prints show the typical theropod marks of three digits facing forward. Like the Glen Rose tracks, semiplantigrade prints from La Rioja have been attributed by some to ancient humans. Jorge Blaschke (a coauthor of Brancas, Martínez, and Blaschke, 1979) showed me some elongated shafts at Barranco de Valdecevillo (VA) that were said to be human footprints, according to a newspaper article published sometime before 1980. More recently, the trackway of a swimming dinosaur at site 4LVC has been claimed by some to have been made during the Great Flood described in the Bible—the same flood that drowned the antediluvians. Unfortunately for such claims, the tracks at Barranco de Valdecevillo were barely recognizable as footprints, let alone human footprints. They were simply shapeless holes produced by fracturing and weathering of the rock and thus possibly not footprints at all. They are not currently visible because they were covered during construction of the Enciso–Los Cayos roadway. The semiplantigrade prints at La Virgen del Prado were clearly made by dinosaurs, not giant humans, indicating that dinosaurs were the most likely makers of any such footprints. Finally, dinosaurs would not have needed a Great Flood in order to go swimming—any suitable body of water would do.

Las Navillas

History

The Las Navillas (LN) site was discovered by a shepherd, Jesús Alvarez, a resident of the town of Igea, who reported it to us in 1993. The first dig was done in the summer of 1994, when the initial data were published, and the second in 1995. It is an easy site to dig because the rocks above the finds are soft as a result of weathering.

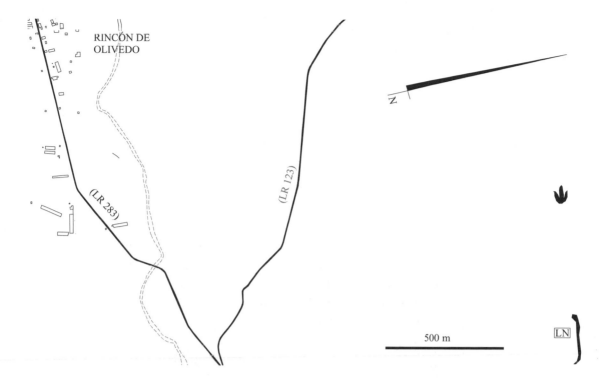

RINCÓN DE
OLIVEDO

(LR 283)

(LR 123)

N

500 m

LN

Description of Location

3.10 Location of site of Las Navillas (LN) and neighboring point with footprints. Local roads are labeled LR 283 and LR 123.

Access is possible by car to a point about 50 m away if there has been no heavy rain. The site can then reached on foot. The layer with tracks stretches from almost the top of the hill to the bottom. Las Navillas is an elongated site located on the side of a small hill (Figs. 3.10, 3.11). The nearest tracksite is Barranco de Valdebrajés, which is 2800 m away. Closer to Las Navillas are small exposures containing tracks (shafts) at the top of some levels and footprint casts in clumps of rock removed by farmers when plowing the fields. The trackways of Las Navillas contain few footprints because they are oblique to the length of the outcrop; trackways cross the site at many locations with varying directions of travel. Because the dip of the layer is low and the overlying rock friable, this site could potentially yield double the number of tracks if a team of about 20 people excavated it manually over the course of about a month. The limestone rocks contain ostracods accumulated in small, almost black layers. Examination of the Las Navillas surface revealed crocodile osteoderms and fish scales. Nearby, Calzada (1977) cited gastropods and bivalves attributed to the Barremian age.

Lithology

The site is on a layer of gray limestone about 30 cm thick interbedded with shale and marl. On the limestone surface is a millimeter-thick pavement of accumulated ostracod shells, giving it a black color. This level is

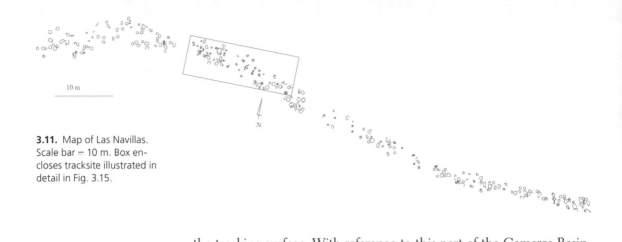

3.11. Map of Las Navillas. Scale bar = 10 m. Box encloses tracksite illustrated in detail in Fig. 3.15.

the tracking surface. With reference to this part of the Cameros Basin, Las Navillas would be stratigraphically beneath the sequences that form the site of Peladillo – that is, below the sites of La Torre and La Era de Peladillo. However, there is no published detailed sedimentary analysis of this part of the basin.

Track Features in Mud

The footprints are of various types and penetrate to various depths in the rock. No preferred trackway orientation was detected by ichnotype or footprint depth. There are deep and shallow tracks of all types, indicating that the viscosity, or ability to penetrate the mud, was highly variable. Some trackways have changes in depth along their path. LN20 begins with shallow tracks, but by the end of the trail, the footprints completely punch through the limestone layer. In other trackways (e.g., LN19), all the footprints are shallow. LN19.1 (the first print of LN19 trackway) is a shallow footprint in the same area where trackway LN20 has footprints so deep that they go through the layer of limestone to reach the marl of the underlying layer. Extrusion rims are often less pronounced. There are no high extrusion rims, even around the footprints that completely pass through the stratum. Although the viscosity of the mud was low (indicting ease of foot penetration), the consistency of the sediment was high. The walls of the deep tracks did not collapse but remained vertical.

Distribution of Tracks by Their Structures

There are no criteria for separating the types of footprints on the basis of the deformation of the mud structures that accompany them. In the studied area at the northern part of the exposure (Fig. 3.11) and its lateral continuation (dug but not studied), deep tracks predominate. Almost all the footprints there are rounded shafts with no features in the contour lines to enable clear identification. The rest of the site has deep and shallow tracks with well-defined contour lines. The mud conditions in LN20 vary according to the location: the first part of the trackway contains

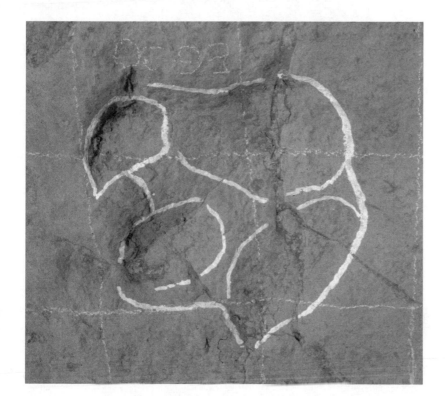

shallow tracks and the tracks are deep at the end. The variation in viscosity of mud of the same composition is probably due to differences in the amount of water it contained. At other tracksites, such variability in print depth has raised speculation this variation indicates the limit of flooded areas. However, at Las Navillas there are no criteria available to evaluate this hypothesis.

All the footprints in LN19 are shallow. LN19.1 is located next to footprints LN20.4 and LN20.5, which are very deep footprints. The mud was softer for LN20 than for LN19. LN19 was initially interpreted as have been made later than LN20 because it is shallower and thus likely made on already-hardened mud. However, this interpretation cannot be correct because recent erosion revealed a shallow footprint (LN19.1), which was apparently deformed by LN20.4p. If the interpretation that LN20 is more recent than LN19 is correct, then either a softening of the lower levels occurred or there was a decrease in the resistance of the upper hard surface levels reflecting a change in substrate conditions, such as the disappearance of algal or microbial mats. There are other examples in which later-formed footprints are deeper than the old footprints on the same study surface at other sites in La Rioja.

Footprint Features (Ichnotypes)

Las Navillas is one of the La Rioja sites that contains footprints from three basic types of dinosaurs: ornithopods, theropods, and sauropods (Table 2.4).

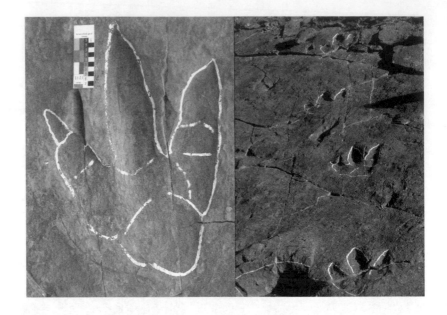

ORNITHOPOD FOOTPRINTS

There are only four footprints with clear ornithopod features. They are large, with a single pad for each toe, and are not very deep compared with the other prints at the site. Footprint LN72 (Fig. 3.12) measures 60 cm in length. Some of the unidentified footprints at the site may be ornithopod prints, so the actual number of such footprints could be greater. These ornithopod footprints must have been left by some type of iguanodontian and were allocated to the ichnogenus *Brachyiguanodontipus* Moratalla, 1993 (cf. Díaz-Martínez et al., 2009).

THEROPOD FOOTPRINTS

Theropod footprints vary in size, with the smallest measuring about 30 cm in length and the largest reaching 55 cm. Generally, they are shallow, with a depth of 5 cm or less. They have clear theropod features: relatively long toes that taper to their tips; a slight offset of digit III with respect to the heel; and interdigital angles II^III less than III^IV. Many have distinct creases delimiting the digital pads, entirely crossing the toes, so the prints are certain to be stamps. All the theropod trackways except LN25 are narrow.

Several theropod trackways (e.g., LN19, LN15) have footprints with features that differ from those of other theropod footprints from La Rioja (Figs. 3.13, 3.15). The footprints are stamps, and so these characters, some considered to be autopomorphies (digital pad soles, mark of claw II), are direct structures:

- Footprint length (51.88 cm) greater than width (37.81 cm)
- Narrow feet ($[l - a]/a = 0.41$) (Table 3.2)
- Toe marks long, clearly defined, and well separated from each other

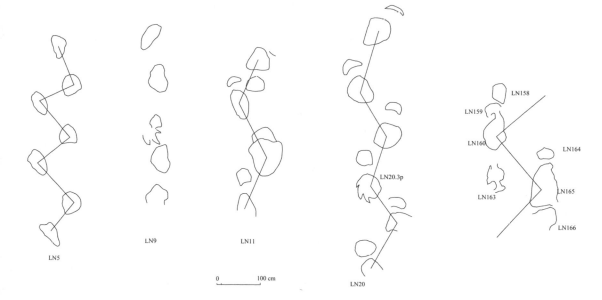

- Digit II not identical with digit IV, but rather wider and separated from the heel by an indentation along the medial margin of the print

- Digital pads soles polygonal rather than oval in shape

- Number of digital pads 2-3-4 – two on digit II, three on digit III, and four on digit IV

- "Heel" pad larger than the others and aligned with those of digit IV (and so is included in the count of digital pads on IV), offset laterally from the long axis of digit III

- Mark of claw II large, wide, and tapered, with length about a third of the total length of digit II (digital pads plus claw)

- Very narrow trackways (Ar/a = 0.3)

- Footprints show positive orientation (cf. Leonardi, 1987), but close to 0°

3.14. Five sauropod trackways from Las Navillas. LN5 is a "manus only" sauropod trackway. Modified from Casanovas et al. (1995h).

This new morphotype has not been defined as a new ichnogenus pending the completion of ongoing studies to establish stable, accepted, and

Table 3.2. Measurements of LN19 trackway.

Track	l	a	O	Ar	Lr	P	Ap	z	II^III^IV	H	z/H	v1	v2	[l-a]/a	Ar/a	z/l
LN19.5		41.25				115			16–23							
LN19.4	52.50		2	20.00	81.25	141.25	146	248.75	17–28	233.60	1.08	4.84	4.58			4.79
LN19.3	51.25	37.50	6	9.75	53.75	130	168	249	18–10	228.86	1.08	4.85	4.59	0.37	0.26	4.80
LN19.2	52.50	37.50	3	6.88	50.00	147	167	270	12–19	233.60	1.17	5.55	4.98	0.40	0.18	5.20
LN19.1	51.25	35.00							21–20	228.86				0.46		
Average	51.88	37.81	4	12.21	61.67	133.44	160	255.92	19–20	231.23	1.11	5.08	4.72	0.41	0.22	4.93

Note: Abbreviations as in Table 2.1

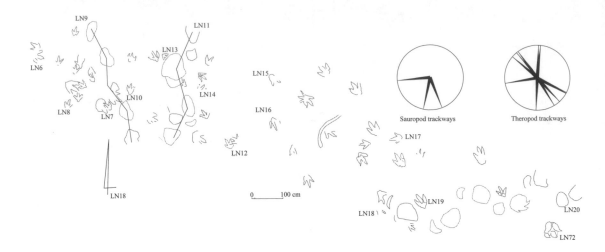

3.15. Map of portion of Las Navillas tracksite. Circles indicate direction of travel of sauropod and theropod trackways at site. LN9 is a "pes only" sauropod trackway.

useful bases of classification, at least for the La Rioja footprints. This ichnotype occurs at several sites in La Rioja and has been given the name of its track, LN19.

SAUROPOD FOOTPRINTS

Several Las Navillas footprints show the typical forefoot–hind foot association of sauropod trackways (Figs. 3.14, 3.15). The width of the parts of the trackways made by the hind feet is less than that made by the forefeet (as in *Breviparopus*; Farlow, 1992; cf. *Parabrontopodus* Lockley, Farlow, and Meyer, 1994). One trackway (LN9) is attributable to sauropod hind feet (Casanovas et al., 1995h), one to a "manus only" sauropod trackway (LN5), and one (footprint association of LN158–LN166) to a "wide-gauge" sauropod. Sauropod footprints at Las Navillas thus show the four variants seen at tracksites around the world: pes only, manus only, wide gauge, and narrow gauge (Fig. 3.14).

There is no difference in depth between tracks of the forefeet and the hind feet for sauropods in La Rioja tracksites. This implies that the pressure on the front and rear autopodia was similar, and so the manus-only sauropod tracks are not undertracks. Consequently, the occurrence of footprint stamps that are shallower for hind feet than forefeet indicates some type of flotation of the sauropods. At Las Navillas, there is variability in the depth of the sauropod footprints along the same trackways, suggesting a difference in sediment viscosity or water depth. Trackway LN20 has shallow footprints (LN20.2p, LN20.3p) as well as footprints that pass through the entire stratum (LN20.5p). The lesser depth of forefeet marks associated with those of deeper hind feet cannot be trusted because the manus prints were distorted by the mud as the animal moved forward.

Most of the sauropod dinosaur tracks in La Rioja are wide gauge (Farlow, 1992). There are only three sites with narrow-gauge sauropod trackways (Las Navillas, Valdemurillo, La Era del Peladillo [PL] 7). Apparently narrow-gauge sauropod trackways (Diplodociforms, following

Wilson and Carrano, 1999) were rare in this environment. LN20.3p (Fig. 3.14) shows three elongated lateral marks corresponding to the three claws attributed to digits I, II, and III of the hind foot, with the longest being the central one. LN20 is one of two trackways in La Rioja that show such long, well-separated claw marks, and it is the only one with claw marks of the *Breviparopus* Dutuit and Ouazou, 1980, narrow gauge (Farlow, 1992) or the *Parabrontopodus* Lockley, Farlow, and Meyer, 1994, type. The other site with such long, separated claw marks is Barranco de Valdecevillo in one trackway (VC13) of the wide-gauge *Brontopodus* type (Farlow, 1992; Farlow, Pittman, and Hawthorne, 1989).

INDETERMINATE FOOTPRINTS

Distributed across the site are isolated, subrounded footprints. None could be grouped into trackways. They cannot be attributed to any ichnotype because they show no clear tridactyl features or forefoot–hind foot groupings. The rounded shape of the majority of these prints is more typical of ornithopod and sauropod footprints, and the prints are more likely to have been made by one of these groups than to be theropod footprints. LN9 (Fig. 3.15) was interpreted as a sauropod trackway by Casanovas et al. (1995h), with the hind foot marks overlapping those of the forefeet. This interpretation is not certain because the hind foot and forefoot marks are on almost the same midline, and there are no claw marks.

General Information

The number of tracks is intermediate, about $0.8/m^2$. Their density, nearly uniform distribution, and lack of a preferred orientation of trackway around the site mean that the water-depth conditions must have been constant over a huge area. There are no emersion marks, as of vegetation, desiccation cracks, or raindrop impact. Above the footprint layer there are no hard levels displaying stratification surfaces. At about 2 m below the track surface, the first hard layer contains large ornithopod and theropod footprints that have not been studied. Trackways LN6 to LN20 (Fig. 3.15) are located near the middle of the site (Fig. 3.11). Of these, three are assigned to sauropods (LN9, LN11, LN20) and the rest to theropods. They do not show the same direction of travel, so nothing can be deduced about the behavior of the trackmakers or the environmental conditions, despite the concentration of trackways.

The Las Navillas site is being restored and maintained. Rock samples from the site were subjected to accelerated aging tests. The main causes of damage to the site are water, plants, and changes in temperature. Water is saturating the lower level, increasing its volume, which results in the expansion and cracking of the rock above. This volume change is most apparent on the lateral edge closest to the ichnological bed. Although it is likely that winter frosts crack the rock, this effect cannot be distinguished

3.16. Conservation of Las Navillas tracksite showing rock surface after cleaning but before application of restoration products.

from the effects caused by expansion of the marl in the lower level. The most abundant vegetation is woody plants, with juniper, gorse, and thyme being the most noticeable. Plants grow between the cracks, then break and separate the blocks formed by the enlargement of the clay at the lower level. Unfortunately, the plants have strong, deep roots and so are hard to remove. Changes in temperature also cause changes in volume sufficient for the surface of the rock to peel and crack. This is due more to daily than seasonal changes, with the ambient temperature on many summer days exceeding 35°C. Work site conservation includes removing plants and roots as far as possible; sealing the cracks with mortar, silicone, and resins where water can penetrate (the sides and open cracks in the stratum); and reattaching loose fragments with mortar and resins. Sometimes the stone blocks are lifted to remove the layer of clay and roots below, which is then replaced with mortar (Fig. 3.16).

General Comments

A previously undescribed large theropod ichnotype has been described from Las Navillas and is now known to occur elsewhere in La Rioja. Because this is a large site, it contains known footprint morphotypes (theropod, sauropod, and ornithopod) as well as others not identified. There are normal sauropod trackways that are broad gauge (*Brontopodus*), narrow gauge (*Breviparopus*, cf. *Parabrontopodus*), and manus only. The only narrow-gauge trackway with claw marks in La Rioja is found here. Differences in preservation suggest that dinosaurs crossed the site in at least two stages: the first with shallow theropod footprints, and the second with deep sauropod footprints. Although it happens frequently, it must not be assumed that all dinosaurs (and sauropods) in a site are always from the same ichnogroup.

The first publication about the Barranco de Valdebrajés (BVL) site was that of Aguirrezabala, Torres, and Viera (1985), who identified its location as the village of Grávalos. In 1990, the mayor of Cervera del Rio Alhama told us about the discovery of dinosaur footprints in his village, which he said had never been described. Our working group did a study, which was published some years later (Casanovas et al., 1991b). The site turned out to be the same as the one identified by Aguirrezabala, Torres, and Viera (1985).

We did not consult or cite the previous publication (we had previously read it but had forgotten about it) because the earlier authors had incorrectly noted the location of Barranco de Valdebrajés as Gravalos, not Cervera del Rio Alhama, and because they attributed the footprints to *Hypsilophodon*. Casanovas et al. (1991b) identified them as theropod prints. Publication of two papers describing the same site led to notes (Martín Escorza, 1992; Viera and Torres, 1992) and a counterreply or justification (Casanovas, Fernández, et al., 1992). Reference to the site was subsequently made by Lockley et al. (1998) and by Lockley and Meyer (1999). These authors suggested that the small footprints in the Barranco de Valdebrajés had been made by ornithopods.

Description of Location

There are two small outcrops: Barranco de Valdebrajés and Lower Barranco de Valdebrajés. Both are located at the bottom of the ravine (*barranco* in Spanish) of the same name and are affected by water passing over the tracks on the surface. They are to one side of a road (Fig. 3.17) and marked by a sign for tourists with the site emblem. In both cases, there is no ready access or parking area, so you have to leave the car a few yards away and jump over the roadside barriers. The sites are about 3 km from Las Navillas and 2.5 km from a site called Camino de Igea a Valdebrajés, which is not discussed in this text (Casanovas et al., 1991a). Las Navillas is east of Barranco de Valdebrajés–Lower Barranco de Valdebrajés, and Camino de Igea a Valdebrajés is to the west. There are outcrops with undescribed tracks at various points nearby. Barranco de Valdebrajés is about 3 km from the village of Igea. The two sites are small. The Barranco de Valdebrajés site is elongated and larger than that of Lower Barranco de Valdebrajés (Fig. 3.18). The footprints are of various types and vary in depth. Nearby is where Calzada (1977) attributed fossils to the Barremian age, as mentioned in the description of the Las Navillas site. The geological section sampled was below Lower Barranco de Valdebrajés.

Lithology

The Barranco de Valdebrajés site is in sandstone. The layer with the footprints is a shaley sandstone with parallel laminations. It is dark in

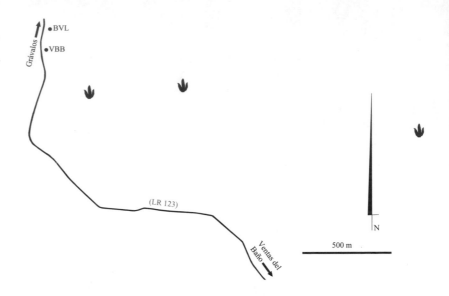

3.17. Location of two sites in Barranco de Valdebrajés (BVL, VBB) with neighboring outcrops containing tracks identified by footprint symbols. Local road is labeled LR 123.

color and greenish gray when freshly exposed. Above the site, the siltstone layers are more abundant than the sandstone, and the weathering of the rock is more intense. The extent to which strata extend laterally beyond the site is unknown as a result of covering by debris rock from higher on the slope. The Lower Barranco de Valdebrajés site is in a layer of limestone belonging to the same stratigraphic unit as the El Peladillo site. This stratigraphic unit occurs in an unbroken stretch from the Era del Peladillo site to at least several kilometers west of Lower Barranco de Valdebrajés. The Barranco de Valdebrajés site is stratigraphically higher than Lower Barranco de Valdebrajés and is close to the contact with the upper Oliván Group. Lower Barranco de Valdebrajés is assumed to be Aptian (it belongs to the limestone interval of Enciso Group), while the Barranco de Valdebrajés site must be categorized as of Aptian–Lower Albian age, possibly closer to the younger age, given its stratigraphic position at the top of the Enciso Group (Table 1.2).

Track Features

BARRANCO DE VALDEBRAJÉS OUTCROP

There are 38 footprints in the Barranco de Valdebrajés site, distributed as follows: eight trackways, BVL1 to BVL8, containing 33 small footprints (Figs. 3.18, 3.19); seven large theropod footprints, three of which are part of a trackway (BVL9) and four of which are isolated prints (from BVL10 to BVL13); and eight large ornithopod footprints (from BVL14 to BVL21). The study surface is divided into two surfaces with footprints. The lower surface contains nine theropod trackways (eight with small footprints and one with larger footprints) and two isolated theropod footprints (BVL10, BVL12). There are also several ornithopod undertracks (BVL14 to BVL18). The upper surface shows a theropod stamp (BVL11) and probably an undertrack (BVL13) not assigned to any ichnotype. The ornithopod footprints

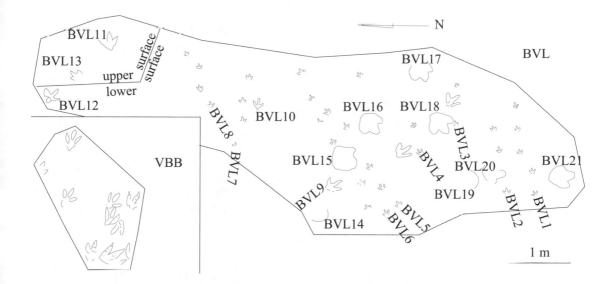

3.18. Maps of Barranco de Valdebrajés (BVL) and Lower Barranco de Valdebrajés (VBB).

were made after the theropod trackways, but their relationship with BVL11 and BVL13 is not known. The eight trackways with smaller footprints (BVL1 to BVL8) are almost parallel, with the same direction of travel. They are the shallowest footprints at the site. The number of footprints in the trackways varies between two and seven.

The depth of the prints varies according to the following pattern: The deepest are the ornithopod undertracks, whose shafts contain mud deformed by the feet; those of medium depth are from large theropod footprints (print length over 25 cm in length); and the shallowest are the footprints from the eight small theropod trackways. The lower surface was deposited before the passage of ornithopod dinosaurs. The greater depth of the ornithopod prints is due to the greater hydration of the sediment at the time when those dinosaurs passed by, which allowed the buried sediment of the lower surface to be deformed as undertracks. This is not the only site where the deepest footprints were made later than more superficial ones. Assuming that the pressure exerted by the soles of the feet is similar for all dinosaurs, the physical condition of the mud was different, and therefore the dinosaurs did not all pass at the same time. If each of the two types of dinosaur (large and small) exerted a different pressure there are no criteria to say whether this happened at the same time. The ornithopod footprints are large and rounded in the toes and the heel. Internal structures are not seen, so additional details used for attributing them to ornithopods cannot be provided.

LOWER BARRANCO DE VALDEBRAJÉS OUTCROP

A search of the area surrounding the Barranco de Valdebrajés site turned up 10 isolated theropod footprints in the limestone at a point called Lower Barranco de Valdebrajés. The rock correlates with one of the limestone strata, Era del Peladillo. The footprints are true tracks that would be considered to be stamps if they were complete. The digital pads and claws are

3.19. Right footprint BVL5.3 from Barranco de Valdebrajés. Squares are 5 cm per side.

well defined, with no slide marks or deformed structures due to the walking phases (e.g., grooves, mudfalls). They are large and narrow theropod footprints, with lengths ranging from about 30 to 45 cm.

SMALL BARRANCO DE VALDEBRAJÉS FOOTPRINTS

There are three footprint morphotypes at the Barranco de Valdebrajés site. According to Casanovas et al. (1991a) and Romero-Molina, Pérez-Lorente, and Rivas (2003), these are two theropod footprint types and one ornithopod type. In contrast, other authors, such as Aguirrezabala, Torres, and Viera (1985), Viera and Torres (1992), Lockley et al. (1998), and Lockley and Meyer (1999), interpret one of the putative theropod ichnotypes as actually being ornithopod tracks. The reason for this contradiction is not clear because the Casanovas team defined a parataxon – theropod footprints – without establishing the connection with the trackmaker (cf. Romero-Molina, Pérez-Lorente, and Rivas, 2003), whereas the other authors used a taxonomic classification because it identified the dinosaurs using geographic and biomorphic correlation criteria (Wilson, 2005).

The features of these footprints are as follows:

· The footprint outline is closed only by the heel in four of the footprints (Fig. 3.20B).

· The length of the footprints ranges between 9 and 11 cm.

· The type of foot is very wide; $(l - a)/a$ has a negative value of around −0.037 (Casanovas et al., 1991a).

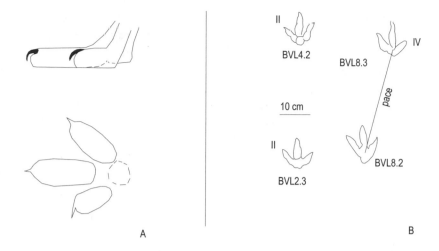

3.20. (A) A 1990 interpretation of foot shape responsible for small prints at Barranco de Valdebrajés. This sketch probably underestimates vertical depth of claws (Farlow, pers. comm.). (B) Examples of small tridactyl footprints of uncertain affinity from Barranco de Valdebrajés. Prints are about 10 cm in length.

- They are tridactyls with tapered toes, longer than they are wide.
- The toes are separate with deep hypexes.
- The mark of the heel pad is shallower or not impressed. Some footprints are direct structures, and the heel pad can be drawn. In others, there is a slight depression corresponding to the front of the pad.
- We cannot ascertain whether the toes have more than one pad.
- The ratio of limb thickness is similar to dinosaurs with thin limbs (z/l is almost 7).
- The footprints tread the midline (Ar/a is between 0.1 and 0.2).

The moving speed for these animals was about 3 km/h (BVL3) and 5 km/h (BVL6). A previously unpublished drawing of the foot, made in 1990 in the course of the fieldwork that resulted in the study of Casanovas et al. (1991b), is reproduced as Fig. 3.20A. The pes shows the elevation of the heel pad. Four full footprints are also drawn (Fig. 3.20B), two of which occur in the same trackway. Lockley et al. (1998) and Lockley and Meyer (1999) supported the attribution to small ornithopods made by Aguirrezabala, Torres, and Viera (1985), and they assigned the footprints to the ichnogenus *Dineichnus* Lockley et al., 1998.

The small footprints were assumed to be made by *Hypsilophodon* (Aguirrezabala, Torres, and Viera, 1985; Viera and Torres, 1992) or by a small, undetermined ornithopod (*Hypsilophodon*, an unknown hypsilophodontid, or dryosaurid; Lockley and Meyer, 1999; Lockley et al. 1998). Following parataxonomic classification, Casanovas and coauthors (Casanovas et al., 1991b; Casanovas, Fernández, et al., 1992) insisted that they were theropod footprints because they are small, so they must inevitably be classified as a small theropod (ancient coelurosaurian parataxon; cf. Casanovas et al., 1989). The assignment of *Hypsilophodon* was based on a comparison between the footprint and Galton's (1974) sketch of the dinosaur's left autopod, and the discovery at a point near the tracksite of the bones of the same genus (Viera and Torres, 1992).

The age and the morphometric and biomorphic characteristics of these footprints pose a problem for being identified as *Dineichnus*; the features show no ornithopod affinity, such as long and slender toes, several pads per toe, acuminate digital termination, marked asymmetry of the lateral toes, and low divarication. In addition, the age of the site is Lower Albian or perhaps Aptian, and the age of *Dineichnus* is Upper Jurassic (Lockley et al. 1998). *Dineichnus* has a symmetrical footprint, with a well-defined circular heel pad. However, the great time differences are not an obstacle for determining the same footprint ichnogenus because of the homoplasty between feet of different dinosaurs (see Farlow et al., 2012). For now, the identity of the trackmaker responsible for these footprints is still unknown.

GREGARIOUSNESS

None of the authors who have written about this site has doubted that the tracks were left by a group of dinosaurs who were walking together. The similarity of the prints, equal depth, parallel orientation, and common direction and separation of individuals, following the reasoning of Bird (1941, 1944) and Ostrom (1972), lead to the conclusion that the tracks are representative of a social group of dinosaurs. However, another question remains unanswered. There is no criterion to determine the degree of maturity of individuals; it is not known whether the dinosaurs were young or adults. There are no indications of dinosaur age groups.

The size of the footprints, about 10 cm, suggests that the acetabulum would be about 40 cm above the ground, which implies that they were not hatchlings. However, and referring to dinosaurs in general, nothing is known of the dependence of the young among themselves and with their parents. In this site there are no criteria to assume that these types of young bipedal, tridactyl dinosaurs would group together, for how long they would walk together, or whether they came from one or more than one family. However, and referring to dinosaurs in general, many remains have been found including bones of adults and juveniles; the dependence of the young among themselves and with their parents is suspected but not proven (Isles, 2009). Among adult dinosaurs, in La Rioja there are examples of gregariousness both among herbivores (sauropods and ornithopods) and among carnivores (theropods). If they were adult dinosaurs, there would have been a concentration of dinosaurs (theropods, ornithopods) whose motivation cannot be deduced by studying the site.

The Barranco de Valdebrajés site has the limitation of being small. It is unknown whether the trackways represent part of a larger group of footprints of the same size or if they are also accompanied by one or two trackways made by larger individuals. In a family group, the offspring would be accompanied by one or two parents, and normally in a crèche group the number of individuals should not exceed a certain limit.

General Information

The footprint density at Barranco de Valdebrajés is almost two footprints per square meter, one of the highest seen in La Rioja. The Barranco de Valdebrajés site was probably at the center of a passing group of small animals whose spacing is as expected for such a group. It is a fragile site at the bottom of a ravine. The rock is impure sandstone susceptible to erosion. Under the lower surface there are probably layers parallel to the top, which are separated as plates as a result of the decomposition of the rock. It is likely that many of the footprints will quickly disappear. The lifting of the layers, the formation of hollows parallel to the surface at a depth of millimeters, the flow of water in storms, and the passage of cattle, as well as the impact of frost and visitors, are factors in accelerating their destruction.

Despite various attempts at site preservation, no work has yet been carried out on these rocks. There are currently campaigns for the rehabilitation and conservation of sites where several teams of university students trained for these tasks work for a month. It is not possible to move these students to the Barranco de Valdebrajés site because of the problems involved in finding technical staff to study the changes in the rocks and the suitability of products for restoration. River specialists and engineers to construct protection for the area are also needed.

Main Features of the Site

This is one of the most controversial tracksites in the region. First there was the inadvertent appropriation of the site by researchers (i.e., my group) who were unaware of its identity. Then there was the issue of the identification of the small footprints—were the trackmakers small theropods or ornithopods? The problem centers on the ornithopod dinosaur to which we attribute them, which is still unresolved. The attribution to *Hypsilophodon*, as suggested in the original study, does not seem defensible, and *Dineichnus* has different features from those in these footprints. They have been allocated to *Hypsilophodon* because it is a small dinosaur and because bones of *H. foxii* were found nearby. Fossil bones are not found in the Enciso Group (in which the tracksite is located), however, but in the younger Oliván Group, and the drawing of the shape of the foot (Viera and Torres, 1992) is not justified. It is hard to distinguish tridactyl (II-III-IV) feet of small theropods from those of small ornithopods (Farlow et al., 2013) on the basis of the pedal skeleton. The "Hypsilophodontids" have a large digit I, and it is hard to see how they could have made a footprint that lacked some impression of the distal end of digit I (Farlow, pers. comm.). The attribution to coelurid dinosaurs made by Casanovas and coauthors (Casanovas et al., 1991a; Casanovas, Fernández, et al., 1992) is opposed by all authors who have investigated this trackmaker.

3.21. Location of Era del Peladillo tracksite (PL), and also nearby sites of La Torre (1–7LTR), La Nava (LNA), and Fonsarracín (FSR), in relation to the village of Igea.

Era del Peladillo

History

The Era del Peladillo (PL) site (Fig. 3.21) was found by the author of this book while searching for black limestone for use in the ceramic industry in 1988. The initial digging campaign on the first exposure (1PL) was done in the summer of 1989, and since then, the site has been worked every year without interruption. The latest (7PL) discovery was made by Isidoro Herce in 1996. La Era del Peladillo is divided into seven outcrops (Fig. 3.22), which were studied and published from 1983 to 1998. It is one of the bigger sites in La Rioja (Fig 3.23), containing several ichnotypes (ornithopod, sauropod, and theropod) (Table 2.4). It is part of a megasite occupying various levels of black limestone, all from the same unit (Meléndez and Pérez-Lorente, 1996). In Igea, the black limestone is called "bell stone" for the bell sound its thin slabs make when struck.

La Era del Peladillo is one part a much larger site called El Peladillo. El Peladillo includes the seven outcrops of La Era del Peladillo (1–7PL) as well as those of La Torre. It is also close to the La Nava and Santa Ana sites, all located on the same hill. La Nava and Santa Ana are the oldest sites, while La Era del Peladillo and La Torre are in limestone at the top part of five sedimentary sequences. The Oliván Group sits unconformably over these. Therefore, they occupy the highest part of the Enciso Group in this geographic zone. Aguirrezabala, Torres, and Viera (1985), Aguirre et al. (2001), and Casanovas et al. (1985, 1991b) studied the sites of Camino de Igea a Valdebrajés, La Torre, Santa Ana, and Las Navas. Several outcrops of La Era del Peladillo (1PL, 2PL, 3PL, 4PL, 5PL, 6PL) are in the same layer. 7PL is below these.

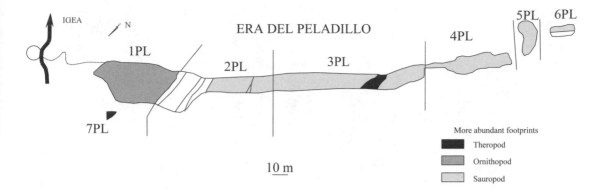

IGEA

N

ERA DEL PELADILLO

1PL

2PL

3PL

4PL

5PL   6PL

7PL

10 m

More abundant footprints

Theropod

Ornithopod

Sauropod

Description of the Location

The Era del Peladillo site is elongated (Figs. 3.21, 3.22, 3.23) and extends from the bottom to the top of a hill. The footprints are of various types and depths. To reach the site, you take a minor road running west from Igea. Three kilometers from the village is an abandoned agricultural work area (the *era* [threshing floor] of Peladillo) where you leave your car. Access from there to the site is on foot. The whole Peladillo area covers a portion of the northern slope of the hill. On the slopes are the sites of La Torre, with samples labeled with numbers (1 to 7) and letters. There are many locations with footprints on the slope between those of La Torre and Era del Peladillo that are not described. On the southern slopes are the sites of Santa Ana and La Nava. The whole Peladillo area (Era del Peladillo, La Torre, Santa Ana, and La Nava) has some 3000

3.22. Distribution of outcrops (subsites) of Era del Peladillo tracksite, with an indication of most abundant footprint type in each outcrop.

3.23. Era del Peladillo tracksite as seen from a distance.

3.24. Bending and breaking of tracking? Surface produced by footstep over flexible ground. Uncatalogued footprint in 1PL.

prints. Fonsarracín is 525 m from Era del Peladillo, and the site of La Nava is 1800 m away.

Abundant fossils have been found in the area, both vertebrates (fish, crocodiles, dinosaurs) and invertebrates (bivalves, gastropods, ostracods), as well as plants and characean algae (Aguirrezabala, Torres, and Viera, 1985; Barale and Viera, 1989, 1991; Calzada, 1975; Torres and Viera, 1994; Viera, 1991; Viera and Torres, 1995a). The dinosaur remains (*Baryonyx*, Viera and Torres, 1995b) were not found close to the footprint site but rather occur in the lower part of the middle section of the Enciso Group; *Hypsilophodon* occurs in the Oliván Group (Torres and Viera, 1994). Also found are charophyte oogonia, both isolated and grouped, as well as indications of flexible ground (algal mudflats) attributed (Fig. 3.24) to the growth of cyanophyta. The trackways cross the site in all directions, recording passage of a group of a few individuals and two herds. There are preferred orientations for some of the various types of tracks.

Lithology

The surface containing outcrops 1PL to 6PL is a layer of dark gray limestone about 30 cm thick, which is on top of a lacustrine sequence. In the same area, immediately below the previous bed, there is at least one other layer of limestone with tracks. 7PL is on limestone at a lower level in the same sequence. The sequence begins with layers of shale, between which is a sandstone layer. Among these are thin shell layers (bivalve

Table 3.3. Types of common structures in footprints of the Era del Peladillo tracksite.

Type of track	Description
Deep	A. Open with vertical walls
	B. Open with high extrusion rims
	C. Open with partially collapsed walls
	D. Collapsed
Superficial	E. With pad marks
	F. Oval with flexible surface
Variable depth	G. Different ichnotypes
	H. Within the same trackway
Erased	I. Due to sliding sections of the track surface

and gastropod), usually vertically separated from each other. There are also smaller numbers of fish scales. Intercalated between the shale levels there are black limestone layers. Limestone layers become thicker near the top of the sequence. There are laminites and desiccation cracks. The dinosaur footprints are in these limestone layers, occurring in several layers. It is assumed that the energy in the environment diminished toward the top of the sequence. In this eastern part of the Cameros Basin, five repeated and overlapping stratigraphic sequences can be counted. The two highest are probably discontinuous as a result of erosion of the base of the overlying rocks (Oncala Group). Era del Peladillo is in the third sequence, and the other sites (La Torre) occupy this and the two lower sequences.

Track Features in the Mud

The outcrop is quite large, and so are the number of tracks and the types of structures. Features of the site are therefore laborious to summarize. There are both deep and shallow tracks of all types, indicating that the viscosity or ability to penetrate the mud was variable. Dinosaurs passed across Era del Peladillo at different times, and the physical properties of the mud changed over time. On the same site there are both deep and shallow tracks, depending on when the mud was walked upon. The mud also behaved differently from one point to another. There are dinosaurs that passed at the same time but left tracks of a different depth. There are synsedimentary deformation structures that sometimes affect the dinosaur footprints, depending on whether the prints were made before or after the deformation. The Era del Peladillo surface with tracks is apparently the same tracking surface across the whole site. There are numerous W- and K-phase structures demonstrating this.

According to the behavior of the mud, the types of tracks can be classified (Table 3.3) as follows:

A. Deep tracks with vertical walls are manifest in sauropod footprints. This indicates the mud was of low viscosity but high cohesiveness.

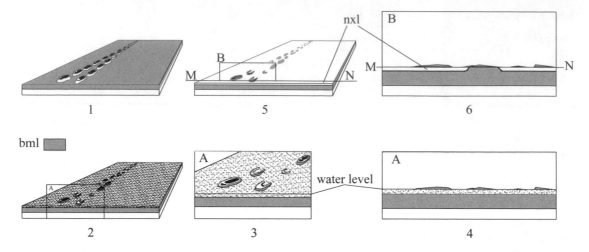

3.25. Formation of extrusion rims of dinosaur footprints at outcrops 2 to 4 of Era del Peladillo. (1) Dinosaur walked across soft, muddy (today black micritic limestone, bml) areas forming deep tracks with large extrusion rims. (2–4) Upper surface of sheet of water after dinosaur passed. It does not completely cover extrusion rims of footprints. (3) Close-up of portion of 2 enclosed by rectangle A. (4) Vertical section and lateral view of mounds formed by extrusion rims that protruded above surface of water. (5) Aspect of zone after sedimentation of upper bed (next sedimentary layer, nxl) of marly limestone. (6) Vertical section along MN line and side view of extrusion mounds. Section shows deformed pretrack bed (bml) and upper posttrack bed (nxl). Upper bed (nxl) molds to lower part of extrusion rims. Upper part of extrusion rims (bml) protrudes above upper bed (nxl) because there was no deposit of marly limestone at top.

B. Sometimes the extrusion rim height is such that it probably emerged above the water surface of the lake, because the next sedimentary layer (nxl in Figs. 3.25, 3.26) does not cover the top part of the extrusion rims (Figs. 3.25, 3.26, 3.27).

C. There are footprints in mud of low viscosity and cohesiveness where the walls have collapsed. If the track is wide (Fig. 3.28), the part that fall inward leaves a small mudflow.

D. If the tracks are narrow (theropod), the feet sometimes sank into the mud, leaving a metatarsus mark (Fig. 3.29). Alternatively, they sank so much that the mud covered them and the three toes exited together through the front (Gatesy et al., 1999) of the mark left by digit III during the K phase (Fig. 3.30). In some footprints, the mud slipped back into the gaps, resulting in toe and metatarsus marks that are so thin that they look like grooves on the surface of the rock.

E. There are superficial footprints next to or in the same sectors as deep footprints. This is attributed to the changes in the viscosity of the rock over time (some dinosaurs crossed before others), and from one point to another even at the same time (the depth of the footsteps varies within the same trackway). Both theropods and ornithopods are responsible for the superficial tracks. This is easily seen from the structure of the digital pads (Fig. 3.31).

F. The footsteps across surfaces at maximum rigidity leave no toe or pad marks. These tracks are oval depressions with concentric cracks due to tensional stresses in the surrounding area. An algal (cyanophyte) mat is thought to have formed a flexible mesh on the surface (Figs. 3.24, 3.32). Depending on the flexibility of the ground and the viscosity of the lower level, the surface folded to a greater or lesser extent, and in some cases split.

G. In several areas of Era del Peladillo there are footprints of variable depth of different ichnotypes. For example, in 5PL there

3.26. Wedging of upper layer (nxl, Fig. 3.25) on contact with black limestone (bml, Fig. 3.25) extrusion rim near edge of subsite 3PL. Lower photograph shows enlargement of upper image.

3.27. Upper part of an extrusion rim (bml, Fig. 3.25) which protrudes through top of nxl (Fig. 3.25) in 4PL outcrop. Footprint not covered by nxl bed.

3.28. Mudflow example in sauropod footprint 4PL44.2p. Other footprints are 4PL46.1 (uppermost part) and 4PL44.2m (to right). Mud movement results from combination of vertical footprint walls, deep footprint, and low coherence of mud. Arrow indicates mudflow direction. Drawn squares are 30 cm on each side.

3.29. 2PL outcrop. Semiplantigrade footprint 2PL169.1, right print that has completely collapsed. Metatarsus and digit IV preserved as linear traces. Hallux trace is possibly obliterated by tectonic fractures.

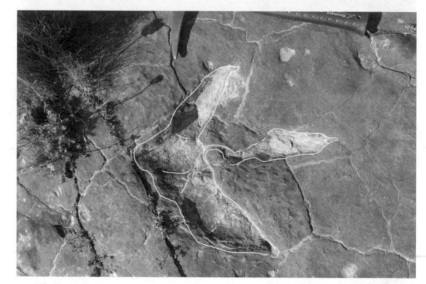

3.30. Collapse structures in left footprint 2PL162.6 similar to footprints illustrated by Gatesy et al. (1999). Pes sank through mud (low viscosity and low coherence), and exit of three digits was through distal digit III depression. It is interesting to note sigmoid striation (curved groove) between proximal ends of IV and III marks, produced by metatarsus movement.

are deep sauropod footprints crossing the site and shallow theropod and ornithopod prints (only on the western part).

H. In tracks at 1PL thought to have been made by a herd of ornithopods (Fig. 3.33) there are footprints scarcely 3 cm deep and others in excess of 15 cm.

I. At 3PL, there is a portion of the site ("stretch zone") where tracks were erased by the slumping of the mud. Some footprints have been erased (inside the stretch) while others (outside it) have not, so while one part remains, another has been obliterated (Fig. 3.34). This effect mud movement erasing the tracks is also found at La Virgen del Campo. At Era del Peladillo there are two parallel stretches of synsedimentary sliding of the mud

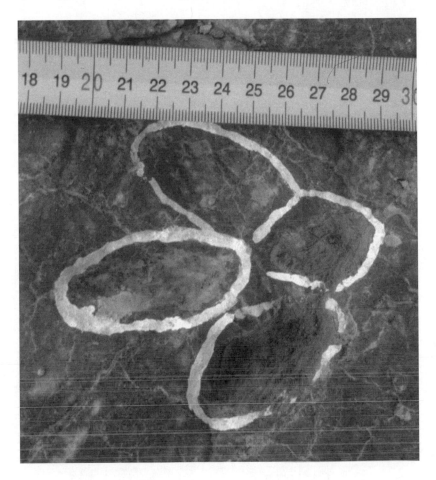

3.31. Right ornithopod footprint 5PL2.4 shows distinctive characteristics among groups of footprints in La Rioja. It is the smallest tridactyl ornithopod footprint in La Rioja (length < 10 cm). Four rounded pads (one for each of the digits II, III, and IV, and one for the "heel") result in superficial print on ground of high-viscosity mud.

3.32. Interpretation of formation of poorly preserved footprints that occur near edge of subsite 1PL. Flexible surface is thought to have been created by an algal (cyanophyte) mat. As dinosaur's weight passed over foot, algal surface variously bent and wrinkled, and in some instances tore to create cracks surrounding oval depression made by foot.

(Meléndez and Pérez-Lorente, 1996) that leave linear depressions, later filled by calcareous mud. The thickness of the filling layer is about 30 cm in the deepest part and about 15 cm or less at the sides.

Track Features (Ichnotypes)

At the site are theropod, ornithopod, and sauropod footprints showing different varieties with well-defined features (Fig. 3.35).

ORNITHOPOD FOOTPRINTS

Ornithopod footprints can be divided into three groups: ornithopod prints from subsite 5PL; tracks assigned to *Hadrosaurichnoides igeensis* Casanovas et al., 1993b; and isolated prints distributed throughout the site. The 5PL ornithopod footprints are grouped in the northwest end of the outcrop and are of variable size (Fig. 3.36). They are characterized by rounded pads in the toes and the heel. The largest of the group (5PL123) measures 46 cm in length, but the site also contains the smallest ornithopod footprint in La Rioja (5PL2.1), which is only 9 cm long. In the same

3.33. Footprints made by an ornithopod herd at 1PL outcrop. In this zone footprints are of variable depth but all are of the same ichnotype (*Hadrosaurichnoides*) and presumably made at the same time.

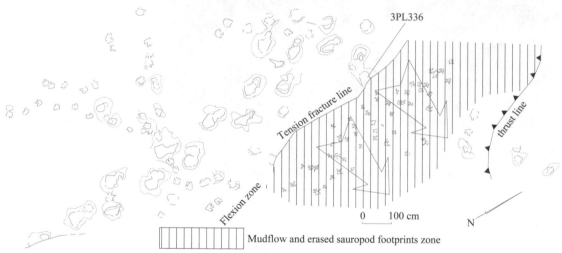

3.34. "Stretch zone" with theropod prints at subsite 3PL. Sector of black micritic limestone experienced synsedimentary sliding, as indicated by arrows. Previously made sauropod prints were destroyed when sediment moved and a linear depression was formed. A group of small-size theropod dinosaurs subsequently passed over this depression.

trackway is another print measuring 10.5 cm (Figs. 3.31, 3.36). Gierlinski, Niedzwiedzki, and Nowacki (2009) assigned this trackway to the ichnogenus *Dineichnus* (Lockley, Santos, Meyer, and Hunt, 1998). In those authors' drawings of the ichnogenus, however, the toes are constricted at the sides with tapered ends–features that are not visible in the small footprints of 5PL.

In 1PL and the lower part of 2PL, there is a group of more than 200 footprints with ornithopod features different from these of the subsite 5PL. Casanovas et al. (1993b) attributed these footprints to passage of a herd. The distinctive features of these tracks (Fig. 3.37) are as follows: the toes are closer together; the height of the mud in the interdigital spaces is less than that of the ground surrounding the footprints; and in the same interdigital areas there are striations or wrinkles parallel to the contour of the toe pads. Casanovas et al. (1993b) interpreted these features as indicating the presence of an interdigital web preventing the rise of the

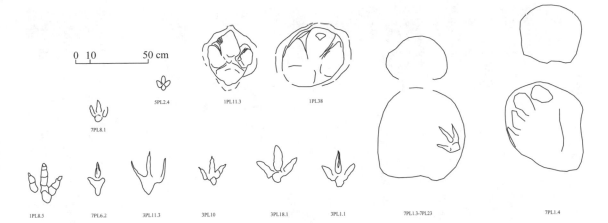

3.35. Footprints from PL track site. Considerable variability of footprint shapes is a consequence of the great extension of tracksite (Casanovas et al., 1999).

mud. The footprints were named as *Hadrosaurichnoides igeensis* Casanovas, Ezquerra, Fernández, Pérez-Lorente, Santafé, and Torcida, 1993b. Lockley and Wright (2001:439), on the other hand, stated, "These tracks may also be of theropod origin because they are longer than they are wide and the trackway is narrow." In the rest of the tracksite, and especially at 3PL, the ornithopod footprints are generally deep and large. They have not been attributed to any ichnogroup because the toe structures (types of pads, relative size and position) are not clear.

3.36. Map of Era del Peladillo 5 (5PL) subsite showing three sauropod trackways (5PL1, 5PL9, 5PL11), two theropod (5PL8, 5PL10) trackways, and a group of ornithopod trackways and isolated footprints. 5PL2 trackway contains the smallest ornithopod prints in La Rioja.

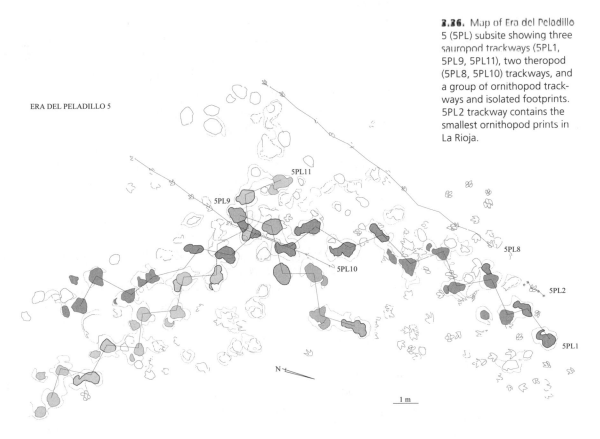

ERA DEL PELADILLO 5

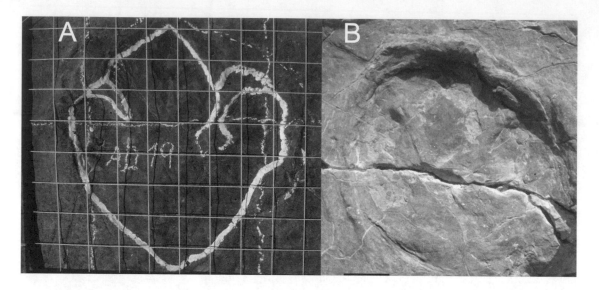

3.37. (A) Holotype (1PL11.3) and (B) paratype (1PL38) of *Hadrosaurichnoides igeensis* ornithopod footprint ichnospecies. Smaller squares are 5 cm on each side.

THEROPOD FOOTPRINTS

For theropod footprints there are several groups with different sizes and shapes.

1PL Theropod footprints have long, well-separated, and tapered toes. Various toe pad marks are visible (Fig. 3.35). Some of the theropod footprints are in a flexible algal surface. The average length of the tracks in each trackway ranges 26 to 36 cm.

2PL Theropod tracks have structures typical of low mud viscosity with collapsed walls. In some cases, the collapse occurs between the hypex of the three digits, while in others, which also have a mark due to the metatarsus, the collapse is total, leaving a track with three lines for toes and one for the metatarsus (Fig. 3.29).

3PL Theropod footprints are smaller than those just described. They show collapse structures and base incisions. There are nine trackways with footprints, the lengths of which range from 14 to 43 cm, with an average length for all tracks of 27 cm (Figs. 3.34, 3.35).

6PL Theropod footprints are visible in the slightly deformed part of the site. Although footprints in the eastern part of the exposure are deep and rounded and probably sauropod, those in the western part are shallow theropod footprints of varying sizes and forms.

7PL There are two types of theropod footprints (Fig. 3.35). One type is small (footprint length varying between 16 to 25 cm, average 20 cm), made by 12 dinosaurs with parallel trackways showing the same direction of travel. The other type is larger (length 31 to 35 cm, average 34 cm), with wider toes, and occurs in trackways in both directions of travel (Fig. 3.38). The small footprints have thin digits, with the rear of the track well defined. Sometimes a metatarsal impression is seen, but it is not clear whether this is due to slippage of the foot in the T phase. In less deformed footprints, a wide proximal pad, with digit II positioned forward and slightly separated from the rest of the foot, is seen.

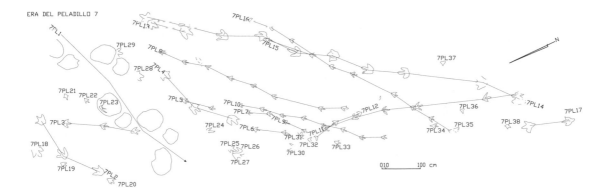

ERA DEL PELADILLO 7

3.38. Map of 7PL (Era del Peladillo 7) showing two types of theropod footprints and sauropod trackway. Small theropod trackways show similar orientation.

SAUROPOD FOOTPRINTS

Only 1PL has no sauropod footprints. The rest of the outcrops have many rounded or oval footprints chaotically arranged, with six trackways able to be distinguished. In 3PL there is a trackway (Fig. 3.34); at 4PL there is a second trackway going in a different direction from the previous one; at 5PL there are three trackways that cross at the same point but with different travel (Fig. 3.36); and there is a final trackway at 7PL (Fig. 3.38). In addition, Era del Peladillo contains isolated sauropod forefoot–hind foot pairs. With one possible exception, all sauropod ichnites (cf. Farlow, 1992) are of *Brontopodus*. The sauropod trackway at 7PL has been considered (cf. Farlow, 1992) as possibly *Breviparopus* (*Parabrontopodus* Lockley, Farlow, and Meyer, 1994). However, trackway 7PL1 is too short to conclude whether this is a narrow-gauge trackway or instead is a wide-gauge trackway with a change of direction that caused some of the prints to cross the trackway at its midline.

The footprints are usually distorted as a result of the behavior of the mud. No forefoot or hind foot structures can be discerned in any trackway except 7PL1. The forefeet are large and rounded, perhaps with marks of two rear structures due to digits. 7PL1.4 has pes prints with similar rounded depressions like those the first four digits would make (Fig. 3.35). They seem to be footprints from Titanosauria (Wilson, 2005), which are not consistent with narrow-gauge trackways.

Track Distribution in Relation to Bed Deformation Structures

ORNITHOPOD FOOTPRINTS

At 1PL (Fig. 3.39) there are 212 tracks from *Hadrosaurichnoides*, both deep and shallow. The deep ones are concentrated in the northern part of the site, while the shallow ones are to the south. All these dinosaurs crossed at the same time and before parts of the layer collapsed (Fig. 3.40). It is assumed that some of the tracks were erased by a mudslide because the group of tracks (including some oblique trackways) does not extend beyond the southern boundary of 1PL. There is just one trackway parallel to the predominant direction of the group. The ornithopod footprints at

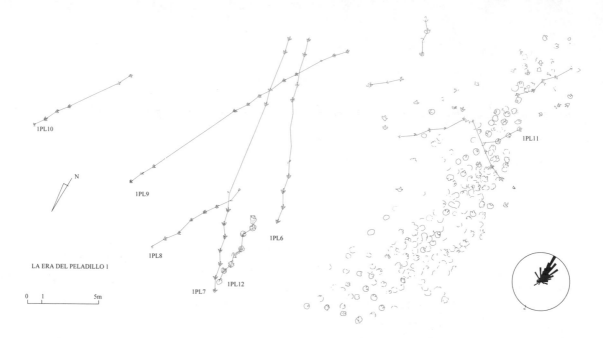

3.39. Map of subsite 1PL (Era del Peladillo 1) showing two groups of parallel theropod trackways (1PL6, 1PL7 and 1PL8, 1PL9, 1PL10) as well as footprints of ornithopod herd (*Hadrosaurichnoides igeensis*, IPL11). Rose diagram (circled) shows preferred direction of travel of ornithopods.

sites 2PL, 3PL, and 4PL are deep (up to 25 cm), with high extrusion rims. In general, the mud in the walls of the tracks held up well, and there was no collapse. At 5PL the ornithopod marks are shallow (less than 3 cm). At 6PL and 7PL there are no ornithopod footprints. The pattern of ornithopod footprints indicates that the sediment conditions were variable when the dinosaurs crossed. At 1PL, the soil was harder or softer in certain places; at 2PL, 3PL, and 4PL the mud was soft with a high consistency, and the ornithopod tracks in these outcrops are more than 15 cm deep; finally, at 5PL the ground was hard, so the feet penetrated only a few centimeters.

THEROPOD FOOTPRINTS

In 1PL there are two types of structures associated with these footprints. Some show the contours, pads, and nails clearly (Fig. 3.35), while others are only oval shafts (Fig. 3.24). The former are readily interpreted as marks left when the soil surface was soft. However, the others presumably formed when a flexible algal mat covered the area. The theropod footprints in the northern part of 2PL are structures typical of mud of low viscosity and cohesiveness. The feet sank in the mud; several types of collapse structures are evident. In some tracks, the walls have fallen in; others show a gap at the level of the hypexes separating the impression of digit III from the proximal part of the print (Figs. 3.30, 3.35); in still others, there is complete collapse of both the digits and the shaft of the metatarsus (Fig. 3.29). At 3PL the clearest theropod footprints occur in a sort of corridor bounded by two lines parallel to the direction of movement (Figs. 3.22, 3.34). The footprints must be shallow because there are no hallux or metatarsal marks, but they are deep enough to show some mud collapse structures, with some digits much too narrow for their length

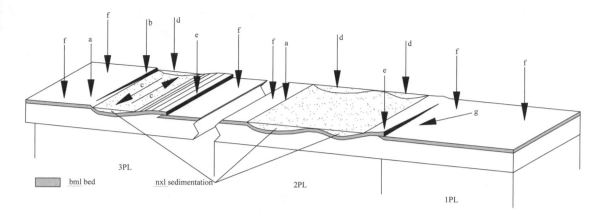

(such as print 3PL11.3; Fig. 3.35). The acetabulum height of the trackmakers is estimated to range between 63 and 140 cm. The distance from the abdomen to the ground would certainly be less (probably less than a meter). At 5PL are few theropod footprints; all are shallow and lack details worthy of mention. At 6PL the theropod footprints are concentrated in the western part of the site; subsequent trampling (dinoturbation) in the eastern part probably wiped them out. Finally, at 7PL there are both large and small theropod footprints, all of which are shallow (less than 3 cm). Both types of footprints show foot-sliding structures in the mud and collapse of the walls into the tracks.

SAUROPOD FOOTPRINTS

The southern part of 2PL and the aggregate unit consisting of 3PL, 4PL, 5PL, and 6PL have deep sauropod footprints (more than 15 cm), generally with high extrusion rims. The track distribution is chaotic in 2PL, 3PL, 4PL, and 6PL, and only two (wide gauge) sauropod trackways can be discerned. At 5PL there are three deep wide-gauge sauropod trackways that cross over each other, while in 7PL there is a relatively shallow (10 cm) narrow-gauge trackway of nine sauropod footprints (five incomplete manus/pes pairs). In 6PL the eastern part of the site is completely covered by footprints attributed to this ichnotype, with tracks superimposed on tridactyl tracks. Many of the rounded footprints go through the limestone layer into the marl below. The tridactyl footprints, erased by the previous ones, are shallower and easily preserved in the western part of the site, which was not crossed by the supposed sauropods.

Behavior and Barriers

This site has tracks distributed by ichnogroups as detailed below.

TWO GROUPS OF THEROPODS

At 1PL (Fig. 3.39) there are two groups of theropod footprints with parallel trackways. One shows three dinosaurs (1PL8, 1PL9, 1PL10) whose prints

3.40. At Era del Peladillo tracksite, bed bml (Fig. 3.35) experienced soft sediment deformation at two subsites (upper part of 2PL, and 3PL; Fig. 3.34) before deposition of upper bed (nxl). Distinctive features in bed bml include (a) folding; (b) tensional fracture; (c) theropod trackways showing two directions of travel; (d) "channels" or deep areas; (e) overthrusting and duplication of bml bed; (f) undeformed zones; (g) direction of herd of *Hadrosaurichnoides* markers.

have long, tapered toes and well-defined digital pads, which are parallel, showing the same direction of movement for about 25 m. The separation between the trackways is between 3 and 7 m, and the tracks are shallow (less than 1 cm). There may be more trackways parallel to these in the eastern part of 1PL under sediment layers yet to be investigated. Presumably these trackways were made by a group of at least three dinosaurs because of the features common to the three (morphotype, direction of motion, similar preservation). In an oblique direction to these three are two dinosaur trackways (1PL6, 1PL7) with the same features (morphotype, parallel nature, response of sediment). It is not known how many dinosaurs were in the two groups. In one instance (1PL6 and 1PL7), ornithopod footprints are superimposed on other possible parallel theropod trackways, while in the other (1PL8, 1PL9, 1PL10) there are probably parallel tracks under the covered area (east of 1PL). These sets of trackways would be appropriate for groups of few separated dinosaurs—very different from the accumulated prints of herbivore herds. If the outcrop had been small, only one sector of one of the five tracks would have been seen. The conclusion that the dinosaur walked alone would not have been correct.

HERD OF ORNITHOPODS

At the western part of 1PL, there is an apparent chaotic mixture of *Hadrosaurichnoides* ornithopod tracks. If the direction of travel of the footprints is summarized, a rose diagram showing a clearly preferred orientation is obtained (Fig. 3.39). The concentration of footprints is flanked by two trackways (1PL12, 2PL173) of the same type showing the same direction of travel as the preferred orientation. The trackway set is bounded at the eastern part of the group by 1PL12, with eight footprints that are linked by the mark of the tail. The trackway marking the western boundary of the herd is 2PL173, with four tracks. Many of the chaotic footprints overlap in such a way that they cannot be assigned to trackways. The orientation of the imprints and their overlapping in different directions (even though there is a preferred orientation) has been interpreted as having been made by a herd of herbivores that was not marching steadily in one direction but was instead slowly drifting along.

RETURN PASSAGE

At 3PL there is a stretch of the subsite showing parallel trackways of small theropod footprints that go in opposite directions: at least four trackways going in one direction and four in the opposite direction. These footprints are surrounded by different and deeper footprints. This arrangement of tracks does not occur at other subsites in the same layer (1PL, 2PL, 4PL–6PL). The stretch is bounded on the west by a tensional fracture and to the east by an overthrust of the layer on itself (Figs. 3.34, 3.40). The deep footprints on both sides of the corridor have disappeared from the interior as a result of the flow of mud. Some footprints are split by the fracture, so the side with no movement has footprints preserved while the other side does not. The interpretation is as follows:

- Dinosaurs walked over a muddy surface, and their feet penetrated more than 25 cm into the mud.

- A synsedimentary slip occurred on part of the layer. The west zone underwent tension and ruptured, causing a double layer in the east by the overlapping of the layer upon itself.

- A depressed corridor formed between the tensional rupture and the overlapping duplication areas. The tracks that were in the corridor were erased by the movement of the mud. The tracks in the ruptured area are preserved without removing the part situated in the sector which does not slip. There are footprints (3PL336) in which only a portion is preserved (on one side of the rupture) because the other portion has been deleted (on the other side of the rupture).

- The corridor was like an elongated furrow with a west wall over 60 cm high. It is thought that calcareous mud can sometimes lose up to 40% of its thickness after diagenesis (Meléndez and Pérez-Lorente, 1996).

- Dinosaurs passed along the corridor. The height of the dinosaurs' limbs is estimated at between 63 and 140 cm. The wall means the dinosaurs were more likely to walk along the corridor than perpendicular to it.

- Sedimentation in the next layer buried the depression of the corridor. This filling is thicker in the middle of the stretch than on the sides.

The trackways in the stretch (3PL10 to 3PL18) show that the animals moved along the same route but in opposite directions (Fig. 3.34). It even may be that the same dinosaurs went first in one direction, then the other.

CAUTIOUS DINOSAURS

Subsites 2PL, 3PL, 4PL, 5PL, and 6PL have abundant deep sauropod tracks, and some of the ornithopod tracks are likewise deep. Theropod footprints are generally shallow. This might indicate the preference of herbivorous dinosaurs for slurried surfaces where their legs penetrated deep into the mud. This would be consistent with the hypothesis of cautious dinosaurs of García-Ramos, Piñuela, and Lires (2002a), formulated after examining sites in Asturias: theropod dinosaurs are presumed to have had difficulties in soft ground, while the herbivorous dinosaurs had the same agility in both soft and solid mud environments.

The site at 6PL is divided into two sectors. The eastern part is completely covered with deep, rounded tracks, many going through to the layer beneath the one stepped on. The western sector has superficial tridactyl footprints with tapered digits. Some of the tridactyl trackways come from the heavily trodden area. The boundary between the two sectors is well defined. It is likely that after the tridactyl dinosaurs passed, a herd of sauropods walked through. It must also be remembered that the sediment was more resistant in the first stage (which saw creation of the tridactyl footprints) than in the second, when the sauropod tracks were made. It may be that algal mats were preserved during the first stage, then later destroyed. Alternatively, the sediment may have been softened before the second stage by a fresh input of water. The Era del Peladillo part occupied by the deep sauropod tracks is about 250 m in length. It is interpreted as showing passage a herd of these dinosaurs. There are, however, no criteria for determining the direction of travel of the group.

HETEROGENEOUS GROUP OF ORNITHOPOD FOOTPRINTS

In addition to the three deep sauropod trackways, isolated sauropod footprints, and three shallow theropod trackway, at 5PL there are shallow ornithopod footprints in one part of the site (Fig. 3.36). To the northwest are ornithopod tracks of various sizes. Among these is the smallest ornithopod footprint found in La Rioja (Fig. 3.31). This concentration of different-size ornithopod footprints is only found in this part of Era del Peladillo. It can be interpreted as a heterogeneous group of dinosaurs of the same ichnotype.

GROUP OF SMALL THEROPOD FOOTPRINTS

At 7PL (Fig. 3.45) there are eight trackways of small theropod ichnites that go in the same direction as isolated footprints of the same type. The trackways of large theropod prints have the same orientation but go in both directions. This parallelism is repeated at many of the outcrops studied. 7PL is a small site, so it cannot be inferred that some kind of barrier determined the orientation and direction of the trackway, as at 3PL. The sauropod trackway indicates nothing about the dinosaur behavior because it only contains nine tracks. The information does not indicate whether this was a group of juveniles or of small adult dinosaurs, as in Barranco de Valdebrajes. The trackmakers responsible for such small theropod footprints are often thought to have been small ornithopod dinosaurs rather than theropods. If so, the trackmaker candidate may have been the same kind of dinosaur responsible for the ichnogenus *Dineichnus*.

General Information

A general scenario can be proposed for the sauropod dinosaurs' simultaneous crossing of the six outcrops of the upper layer (1PL to 6PL). First, theropod dinosaurs walked through the area before the sauropod dinosaurs. The traces of these dinosaurs are the theropod footprints of 1PL, the north part of 2PL (distorted during synsedimentary sliding), and 6PL. The viscosity of the sediment was high in 1PL and 6PL, but low in 2PL. Next, the deep sauropod and ornithopod dinosaur footprints (2PL, 3PL, 4PL, 5PL and 6PL), the *Hadrosaurichnoides* marks at 1PL, and the north part of 2PL (trackway 2PL173) were made before the sedimentary slump. Then there was a synsedimentary slump, and small dinosaurs passed through 3PL. It has not yet been possible to establish the relative time of passage for the group of ornithopod dinosaurs at 5PL. The average density of footprints is about one per square meter.

The diversity of trackway orientations, reflecting the passing of groups and herds, changes in the physical properties of the ground, the presence of desiccation cracks in some places, and the flexible ground, indicate that the history of the site is complex. To this must be added the slump of the stratum with tracks, which was capable of erasing some of these tracks and preserving others. Also worth noting is the presence of extrusion rims that were higher than the water height of the next limestone layer, and that were not eroded after its deposition. Sauropod footprints with large extrusion rims seem to have been made after at least one set of desiccation cracks.

Work on recovering and maintaining the site in Era del Peladillo is ongoing. The procedure is similar to that used in Las Navillas and includes the preliminary analysis of samples of fresh and modified rock and its examination after accelerated aging tests. In addition to the destructive effects of frost and daily temperature changes (night–day), a large part of the deterioration is caused by the action of roots penetrating between the two hard layers and lifting the upper one. This area is home to woody scrub plants (gorse, rosemary, and thyme) that survive well in dry areas and have deep roots. The shafts of the sauropod and ornithopod footprints in the soft sectors are like huge plant pots, where clay accumulates and absorbs rainwater, promoting the colonization and growth of plants inside.

Main Features of the Site

Interest in the site is likely the result of a new ornithopod ichnotype with interdigital web, the presence of the smallest ornithopod footprint in La Rioja, the coexistence of the three types of dinosaur footprints common to La Rioja, and one of the three outcrops (7PL) having possible footprints of the type *Breviparopus* or cf. *Parabrontopodus* in La Rioja. In the upper layers (1PL to 6PL), the footprints are *Brontopodus* or wide gauge; coexistence on the same site of two herds and a group of herbivorous dinosaurs, probably with little difference in time (it may be deduced that there are three successive stages in the occupation history, theropod–herbivore–theropod, taking into account the sites 1PL to 6PL); variation of the viscosity of the sediment according to the outcrop and footprint stages; and the first dinosaur footprints from La Rioja interpreted as having been made on a flexible surface, with rupture of the elastic layer; and extrusion rims on some prints that are higher than the sedimentation top of the following layer.

History

This section describes individual footprints in isolated occurrences from La Rioja. These tracks are found as impressions or as natural casts in blocks of rock fallen or separated from their place of origin, in some cases having been incorporated into base layers on walls. Of these, the only published descriptions are from the study of the Navajún mine quarry (Mina Victoria site) and those of Valdeperillo (Ansorena, Díaz-Martínez, and Pérez-Lorente, 2007–2008).

Some inhabitants in the region have tracks in their homes (shafts and casts) that were collected by them or by their ancestors in the countryside. Some people leave them outside the entrance to their homes or display them inside. Probably the first referenced footprints in blocks date from 1989, when several footprints near El Villar de Poyales, discovered by a schoolteacher in the village, were collected. They were deposited that same year in the Enciso Paleontological Center. Subsequently, casts and undertracks were found in digs at Peñaportillo (Munilla) and Los Cayos A. Later still, more isolated tracks were found in the outlying areas of Enciso and Igea, of which the most striking tracks and casts are deposited in the collections of the paleontological centers of those two places. These are not the only known examples. There is an oral report of at least one wall located near Enciso, separating two fields and containing a large number of isolated footprints. The blocks have not yet been removed because sufficient facilities for recording and storage have yet to be found.

The site of Navajún was created as a result of explosions at the quarry face of the pyrite mine. The owner of the mine separated the blocks with footprints and communicated his findings to the heritage commission of the La Rioja government so they could be studied. During our visit, the employee who accompanied us showed the location in Valdeperillo, where we found dozens of pieces of rock with natural casts and footprints.

Many of the tracks collected in the Enciso Paleontological Center come from large construction projects (roads, reservoirs, quarries, wind farms), or forest paths and cattle tracks. These specimens are identified during the paleontological monitoring of the works. In La Rioja, an artistic and historical heritage law protects the entire area with Jurassic–Cretaceous continental outcrops (Cameros Basin). Any work with material removed in these areas must be checked by a geologist to monitor whether outcrop fossils of patrimonial interest come to light.

Description of Location

MINA VICTORIA AND VALDEPERILLO OUTCROPS

The footprints discussed in this section occur at the sites (Fig. 3.41) of Mina Victoria (MIV), Navajún, and Valdeperillo (VLP). There are also isolated footprints in various separate locations across the Cameros Basin (e.g., Igea, Cornago, Préjano, Enciso, Munilla, Hornillos de Cameros,

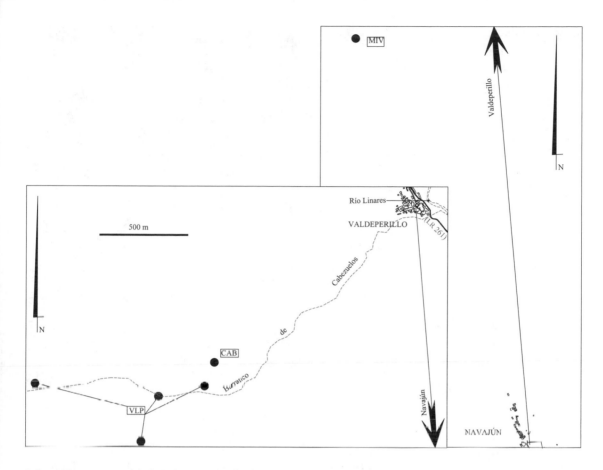

3.41. Location of outcrops of Mina Victoria (MIV), Valdeperillo (VLP), Cabezuelos (CAB), and end segments of line joining Valdeperillo and Navajún. Line measures 12.5 km.

Jalón, Torremuña, and Cabezón de Cameros). The geologic groups in which they have been found are those of Oncala, Urbión, and Enciso. Mina Victoria and Valdeperillo are about 10 km apart, while the distance between the towns of Valdeperillo and Navajún is 12.5 km (Fig. 3.41). At these sites, fossils of semionotiform fishes (Actinopterygii, Neopterygii), gastropods (Viviparidae and "hydrobids"), and bivalves (Unionida) were found (Bermúdez-Rochas, Delvene, and Hernán, 2008). The site closest to Valdeperillo is Cabezuelos, which is a few meters from occurrences with isolated footprints (Moratalla, 1993). The following are some distance away: Barranco de Acrijos (2.5 km), Valdenocerillo (3.8 km), and Los Cayos (4 km) (Casanovas, Pérez-Lorente, and Santafé, 1989; Casanovas et al., 1995c; Moratalla, 1993). One isolated track has been cited in Valdemadera, which is 3 km from Mina Victoria. Access to the two places is by a main road. At Mina Victoria, the road leads to a pyrite mine, where the blocks are, and at Valdeperillo, the first track is at the roadside. To get to the rest of the Valdeperillo footprints, one must walk through woodland.

The footprints at the Mina Victoria can only be seen in blocks that have fallen from the vertical wall of the quarry face of the mine (Figs. 3.42, 3.43). In this area alone, there is more than 1 m^2 of stratification surface in the mine blocks. The dip of the strata is 10° toward the interior of the mountain, and the rocks crack easily. In the vicinity of the mine,

3.42. Vertical quarry face of Victoria mine. Face shows sandy and silty layers. Sandy layers contain footprints. People in photograph are geologists taking pyrite mineral in lower slates.

no other area with footprints has been found. Although the outcrops in the strata are continuous and long, no large stratification surfaces are seen. The layers can be followed across the terrain for more than 10 km without interruption. The tectonic structure is simple, so the layers are like huge, flat boards at a slight angle. The nearest footprint sites studied are in the same Urbión Group of rocks but far from the Mina Victoria site (Barranco de Acrijos at 7 km away [Casanovas et al., 1995e], Cabezuelos at 10 km away [Moratalla, 1993]). Both sites are relatively poor, with only 11 footprints in the former and 13 in the latter. At the same time ichnological data were taken from the Mina Victoria, the area between the Cabezuelos and Acrijos was prospected. At the Valdeperillo site, we found abundant casts and both large and small blocks of rock removed from their original position, with natural casts and footprints whose condition was generally poor. In addition, natural casts were seen at the base of many layers in the cliffs of the Cabezuelos ravine (Fig. 3.44).

OTHER ISOLATED FOOTPRINTS IN LA RIOJA

A reservoir is currently being built in Enciso that sits at the base of the Enciso Group. The affected rocks, which are predominantly siliceous, are characterized by not having large outcrops of the tops of layers. Because the work is being carried out on land in the Weald Facies (a protected zone of the Cameros Basin because of its paleoichnological heritage), a prospecting program is scheduled in the area affected by the reservoir. In the flooded area, no site has been detected, and in the boundary zone

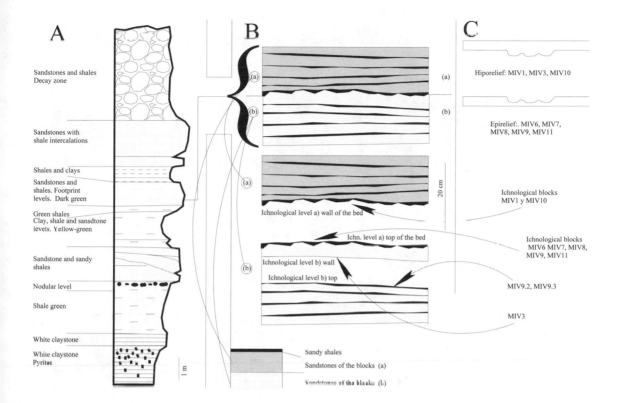

A

Sandstones and shales
Decay zone

Sandstones with
shale intercalations

Shales and clays

Sandstones and
shales. Footprint
levels. Dark green

Green shales
Clay, shale and sansdtone
levels. Yellow-green

Sandstone and sandy
shales

Nodular level

Shale green

White claystone

White claystone
Pyrites

1 m

B

(a)

(b)

(a)

Ichnological level a) wall of the bed

Ichn. level a) top of the bed

(b)

Ichnological level b) wall

Ichnological level b) top

20 cm

Sandy shales

Sandstones of the blocks (a)

Sandstones of the blocks (b)

C

Hiporelief: MIV1, MIV3, MIV10

Epirelief:. MIV6, MIV7,
MIV8, MIV9, MIV11

Ichnological blocks
MIV1 y MIV10

Ichnological blocks
MIV6 MIV7, MIV8,
MIV9, MIV11

MIV9.2, MIV9.3

MIV3

3.43. (A) Partial working front (quarry face) stratigraphic log of Victoria mine and location of member with footprints (B) Detail of sedimentary levels that blocks come from and location of studied blocks. (C) Types of footprint structures in each block. 3.47 shows portion of wall surface of (a).

above the maximum stored water height there are two sites (La Muga [Caro et al., 1995]; Carretera de Soria [Casanovas et al., 1995d]). Of the relatively large blocks of rock that have been released by blasting, 12 have been found with tracks. Some, because of the huge volume of rock they are contained in, have been dumped after constructing the road, while others were able to be rescued and transported to places accessible to visitors. Melero and Pérez-Lorente (2011) have studied the footprints of the block and those that cropped up in the dam abutments that could not be recovered.

To supply aggregate for the dam, a limestone quarry was excavated in the Oncala Group. One of the walls has gaps that could be footprints. Work was done on the access road to the quarry; blasting revealed parts of footprints visible in the blocks. Work to improve or widen the roads in the Cameros Basin rocks of La Rioja and jobs that involve widening or settlement of slopes of rural roads (for forestry and farming) have also provided footprint casts deposited in various places. Large fragments are left on the sides of roads, some visible to visitors, while others await their final placement. The number of footprints destroyed in the blast work of quarries, reservoirs, and roads must be enormous. In addition, several blocks containing footprints (one per block), or consisting only of a single natural cast, have been located in walls surrounding agricultural fields. After plowing reveals them, farmers leave the rocks on the edges of fields, either to separate the fields from neighboring properties or to retard soil erosion. In some paleontological excavations, chunks of strata have been

3.44. Cast (sauropod manus?) at base of bed of limestone, Valdeperillo site.

lifted, under which footprints have been revealed. Certain fragments of rock have revealed casts of footprints (e.g., at the site of Peñaportillo). These fragments are deposited in the Enciso Paleontological Center.

Isolated tracks are found everywhere in the region except in Oliván Group outcrops. It has been postulated (Pérez-Lorente, 2007) that there are no large outcrops of rock in the Tera, Oncala, or Urbión groups because weathering has reduced the hard layers into small pieces as a result of their composition and degree of fracturing (Fig. 3.45). The density of footprints, at least in the Oncala and Urbión groups, should be similar to that of the Enciso Group, although there are fewer inventoried sites. The Tera Group is not in such a favorable position because of the nature of its sediments—usually conglomerate and clay.

Lithology

Most of the blocks of rock cited are sandstone. So far, natural casts in limestone have only been found after reservoir explosions and during road-widening work after breaking through thick calcareous sedimentary formations or units (in the Enciso Group and Leza Formation of the Oncala Group). Natural casts are isolated or attached to the base of strata. It is probably easier for casts to become separated from sandstone

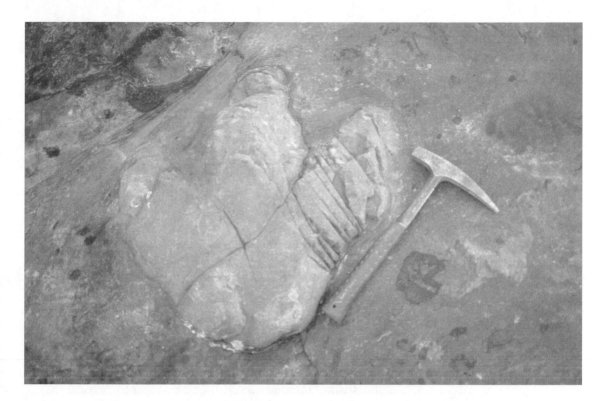

than from limestone because the formation conditions are different. The hole of many footprints is filled with sand left by flowing water, after which the water may remove the print's sedimentary material or deposit a thin or thick bed of sand over the print (Fig. 3.46). Such a stream may also erode a large section of the tracking surface. In contrast, carbonate sedimentation usually covers the entire surface so that casts are strongly cemented to limestone deposits.

Track Features in Mud

In general, rocks with tracks cut away by human action or large detached boulders from cliffs and slopes have all sorts of structures, including deep and shallow natural casts and hollows. For casts to remain intact, however, they must be relatively deep or form thick enough isolated units (Fig. 3.46). Mina Victoria footprints show the first style of preservation and some Valdeperillo casts the second. These two preservation modes also characterize other isolated footprints of La Rioja. Specimens from public works and detached limestone blocks, such as those seen at a site not yet inventoried (Las Peñas Amarillas) in the north of the Las Losas site, belong to the first group, while some isolated footprints deposited in the paleontological centers, and some specimens from rural walls and removed by farmers, belong to the second. For footprints to be deep, the walking surface must have been mud of low viscosity (for the foot to penetrate) and high consistency (for the walls of the footprint not to collapse).

3.45. Broken sandstone natural cast (rigid) included in siltstone (more plastic). Siltstone sediment flows but sandstone is deformed by system of fractures. Undetermined isolated tridactyl footprint. Urbión Group, near Cornago village.

3.46. Process of forming sand casts. (Left) Separate casts as loose stone. (Right) Casts attached to upper sedimentary level. Possibility of preservation is greater if they are thick and if they are attached to upper layer.

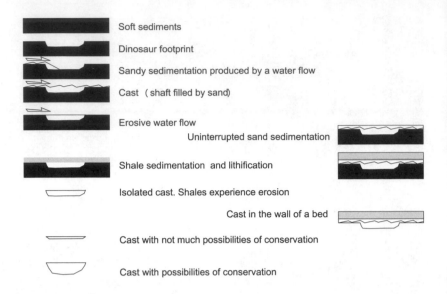

Soft sediments

Dinosaur footprint

Sandy sedimentation produced by a water flow

Cast (shaft filled by sand)

Erosive water flow

Uninterrupted sand sedimentation

Shale sedimentation and lithification

Isolated cast. Shales experience erosion

Cast in the wall of a bed

Cast with not much possibilities of conservation

Cast with possibilities of conservation

The large blocks and most isolated footprints are formed by later sand filling. The sand must be swept away by a stream without excessive erosive power for the hole to maintain its depth (Fig. 3.46).

Footprint Distribution According to Their Structures

MINA VICTORIA FOOTPRINTS

At Mina Victoria there are footprints on the tops and in the walls of layers (Fig. 3.43), showing a truly comprehensive range of direct and indirect structures. On the top, tracks occur as undertracks, subtracks, and stamps, while natural casts and undertracks are seen in the walls. The Mina Victoria detached blocks have tracks from both walls and the tops of layers (Fig. 3.47, 3.48), as sometimes the two study surfaces are preserved. The footprints are located in several loose blocks that can be correlated with rocks from the working quarry face. The stratigraphic column (Fig. 3.47A) shows that footprints come from a single horizon. This horizon consist of two units (Fig. 3.43B; units or levels a and b) that can be seen in the loose blocks. Particular blocks (Fig. 3.43C) contain tracks preserved in a different manner in each of the two units.

The stratigraphic wall surfaces (units a and b) have projections left made by the foot descending in the lower stratigraphic levels (Fig. 3.43B, surfaces MIV1 and MIV10, lower surface of a, and surface MIV3, lower surface of b). The footprints are preserved as hyporeliefs (casts; Fig. 3.47). The surface also show prominent sedimentary (prod and bounce) marks. The upper stratigraphic surfaces (units a and b) have footstep shafts, and in the detached blocks, their filling is also sometimes preserved. The sedimentary bounce and prod marks are like hollows in the stratification surface (Fig. 3.43B, surfaces MIV6, MIV7, MIV8, MIV9, MIV11, upper surface of a, and MIV3, upper surface of b). The upper and lower surfaces

3.47. Photograph showing partial face of MIV10 block. Mold of sedimentary structures (ripples, prods, bounces) and some footprint casts can be seen. Mina Victoria site. Block is rotated so that surface layering in photograph is not the top but a wall of the layer.

are part and counterpart of the same surface. Thus, wave ripples occur in the upper surface of units, while the lower surface preserves a copy of the ripples made in the sediment that covered the units (Fig. 3.47). However, neither of the two surfaces is the tracking surface. The footprints of the MIV10 block are not real tracks but undertracks and subtracks. The selected footprints were formed as follows (Fig. 3.46). First, several sandy silt laminites were deposited on a layer with straight and parallel ripples and other flow structures (prod and bounce casts). Next, the dinosaurs

3.48. Photograph and drawing of III toe distal end of MIV7. Subtrack and undertrack (silt and sand levels) and cast (sand inside marks of toe and nail) in footprint of Mina Victoria tracksite. Tracking surface is placed on levels of silt. Horizontal line, silt; dotted line, sandstone. Size of grid squares is 5 cm.

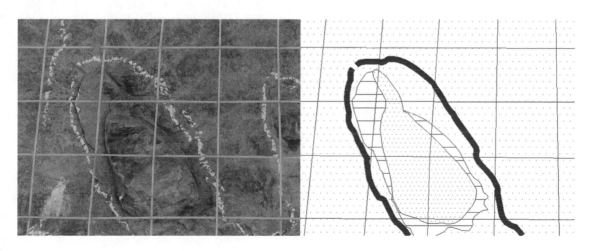

3.49. Valdeperillo, 2VLP6 footprint. Sauropod footprint cast in which claw has been well marked. Limestone of Urbión Group. Scale bar = 10 cm.

passed, deforming the substrate and leaving undertracks on the surface with ripples. Some dinosaur claws cut through the laminites into the layer with ripples. Finally, the holes and claw tracks were filled with a new supply of sand (Fig. 3.48).

VALDEPERILLO FOOTPRINTS

Valdeperillo has hollows and casts of footprints that can be shallow or deep. Several isolated sandstone casts have been found, including three well-preserved shaft specimens and six casts at the base of in situ strata (i.e., not removed from their original position). Isolated casts have been detected both in blocks displaced left by farmers on the edges of the fields and isolated on slopes and noncultivated places. They are usually large with rounded edges. Weathering may have played an important role in the appearance of the footprints. There are footprint shafts in at least three blocks detached from their original location. The shafts have less rounded footprint outlines, possibly because they have spent less time exposed to the elements. In the natural casts seen in the wall of the bed (Figs. 3.44, 3.49) on the edges of cliffs and slopes, footprint structures can be seen very well. No skin impressions have yet been found in these types of tracks in La Rioja. However, claw structures have been found,

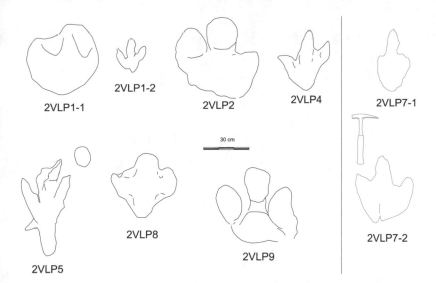

the surface of which is smooth and different from the rough skin of the feet (Fig. 3.49).

Track Features (Ichnotypes)

In both isolated casts and blocks created by human activity, sauropod, theropod, and ornithopod tracks have been observed (Table 2.4). The large ornithopod footprints are of the *Brachyiguanodonipus* type Díaz-Martínez et al. (2009) in the sense that they are massive, wider than they are long, and have rounded pads on each digit. Small ornithopod footprints have the same features and are only distinguished from the former prints by their size. Some rounded and wide ornithopod digit marks may be either stamps or undertracks (Fig. 3.50, 2VLP1–1, 2VLP8). The theropod footprints are both large and small (more or less than 25 cm in length, respectively), with wide or narrow digits. Some of these footprints are casts and others are hollows or shafts. At the site of Valdeperillo, a theropod cast was also found with both metatarsal (semiplantigrade) and hallux (Figs. 3.50, 3.51, 2VLP5) impressions, and possibly a manus print as well. Sauropod footprints are less common. They are only distinguished clearly in casts at the base of strata in rocky outcrops, as forefoot crescentic marks or as hind foot impressions with claws (Fig. 3.49).

Distribution of Isolated Footprint Ichnotypes in La Rioja

The number of isolated sauropod footprint casts is small. Many have been discovered at the base of sedimentary layers that are located in situ. They have only been seen at the Valdeperillo site in a block close to a rocky outcrop. This cast (Figs. 3.44, 3.49) is in a limestone of the Urbión Group. There may be isolated specimens as well as some in blocks artificially torn loose by blasting, during which these types of footprints go completely

3.51. Photograph of footprint 2VLP5, showing possible manus print. Scale bar = 10 cm.

unnoticed. Many of them must be rounded, complete, or incomplete depressions and casts that would be difficult to identify. The marks of digits or claws are never seen and are easily eroded. Isolated theropod footprints are distributed throughout the Oncala, Urbión, and Enciso groups. Usually they are more than 25 cm long, although some are smaller. Almost all such tracks are in sandstone, although some are in limestone. Ornithopod footprints are also found in all three stratigraphic groups. Most notable are the large footprints of this kind (over 60 cm long), one of which was found in the Oncala Group (Leza limestone), another (in sandstone) in the Urbión Group (Figs. 2.2, 3.52), and three more in the Enciso Group (in sandstone). The total number of isolated footprints (in blocks of rock or as isolated natural casts) collected in the field and described in one or other study are as follows: 32 theropod (mostly hollows), 17 ornithopod, one sauropod, and 14 unidentified footprints. The large proportion of ornithopod tracks found is surprising considering that the amount of theropod footprints is usually more than twice that of ornithopod prints in the Cameros Basin.

General Information

The chances of finding sites with footprints depend on the position of the strata with respect to the slopes and the lithostratigraphic group (Fig. 3.53). Although on some slopes there are large outcrops from the upper surface of the layers, on opposite slopes the cross section of the strata is

3.52. Ornithopod footprint (sandstone cast) from Urbión Group deposited in Enciso Paleontological Center. Skim marks are produced in K phase (from exit of foot). To compare scale, see Fig. 2.2.

prominent. If the rocks are slippery, sites are not easily found on either of these types of slopes. The best prospect for natural casts and other isolated footprints is in the places where the profile of the strata is prominent (Fig. 3.53, center).

Outcropping rocks in Mina Victoria and Valdeperillo are in the Urbión Group and have the common properties of being fragile and slippery. The fragility (Pérez-Lorente, 2002a) is due to a combination of three factors: lithological composition (silty quartzite or carbonate), metamorphism effects, and effects of tectonic deformation. Many of the Urbión Group rocks also have disseminated pyrite, which makes the sandstone disintegrate easily as a result of natural alteration induced by the iron sulfide breakdown products they contain. In the Mina Victoria and Valdeperillo areas, there is a large amount of disseminated pyrite. Siliceous rocks in the Oncala Group have similar properties to the Urbión Group. The carbonate rocks, other than those of the Leza Formation (Oncala Group), occur as slabs, and others are ocherous (with many voids and cavities). It is difficult for these to be preserved or for any footprints they contain to be recognizable.

Assuming that the great number of loose blocks containing tracks is in the same proportion in all three groups (Oncala, Urbión, Enciso), then this must mean that the number of footprints impressed into the rock is similar in all the groups. This indicates that the reason why the largest proportion of tracksites occurs in the Enciso Group is because large rock outcrops there are better preserved and do not disintegrate, compared with the Oncala and Urbión groups, which are unable to offer large, continuous rock outcrops (Pérez-Lorente, 2002a). The general map of the Cameros Basin (Pérez-Lorente, 2007) illustrates this distribution. Although the density of sites in the Enciso Group is high, it is less so in the Urbión and Oncala groups. The data gathered and the collection of isolated footprints indicate that this distribution is due to two factors. First, in the siliceous rocks of the Urbión and Oncala groups, the exposed rocks are broken into blocks and small pieces. Second, when carrying out

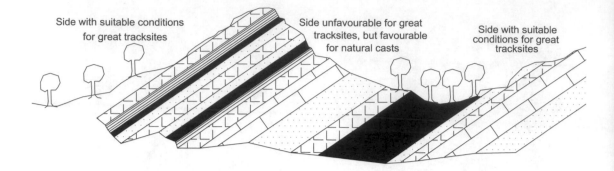

Side with suitable conditions for great tracksites

Side unfavourable for great tracksites, but favourable for natural casts

Side with suitable conditions for great tracksites

3.53. Favorable and unfavorable places for footprint sites. If dip angle of stratification is equal to topographic slope angle, location is favorable for tracksite. If stratification dip is cross-slope, then location is favorable to find loose rocks (with traces and casts) and casts at base of strata.

surveys and other work, complete or partial tracks (casts, undertracks, shafts) are found in isolated, displaced blocks in the Oncala, Urbión, and Enciso groups. Also, many footprints are found in natural cuts in prominent stratification areas.

Many layers in the Cameros Basin contain footprints that are only seen in blocks removed by nonnatural processes. In general, fragments removed by blasting or by removal of rocks with heavy machinery show surfaces that would never have emerged complete. The natural fracturing of the rocks and erosion are responsible for their disintegration. The size of natural blocks with footprints is highly variable – typically on the order of several cubic meters (Fig. 3.54) up to casts the size of a hand. Examples of the first kind (in layers resistant to dense fracturing), containing many shallow and deep footprints, are located in blocks north of Las Losas (Las Peñas Amarillas tracksite). If the fragments are on the order of several square meters, the number of footprints is normally such that the site has been identified by local people (shepherds or hunters) or by paleontological prospecting. The density of tracks cannot be determined in the absence of outcrops exposing track surfaces. The absence of sites with footprints detected in some areas of the Cameros Basin is not indicative of the actual content of footprints in the rocks but rather the degree of fracturing and weathering. If there were enough time and resources to search for and collect specimens recognized outdoors, the catalog of fossil footprints in La Rioja would increase by several hundred.

Main Features of Sites

In large areas without recognized sites in the Cameros Basin there are significant numbers of outcrops with dinosaur footprints. The stratigraphic group where this survey was carried out was the Urbión Group. There are also similar sites with dinosaur footprints in the Oncala Group. The main type of natural casts found is ornithopod; the rounded and massive forms and the size of the footprints are probably a positive factor in their conservation. Ornithopod footprints probably accumulated in places with soft mud where the feet penetrated further into it, while the theropods frequented areas with harder ground. Sauropod footprints are not seen, except those in the base layers of walls and slopes where the rocks are

3.54. Example of large detached blocks of rock. Las Peñas Amarillas tracksite (north of Las Losas site). Limestones of Enciso Group. In the blocks are several trackways that are being studied.

more prominent. I do not know why isolated sauropod manus casts or subtracks have not been found in La Rioja.

History

The site of Los Cayos (LC) was discovered in 1985, and additional outcrops were added until the current number of six cataloged outcrops was reached. The first outcrop was Los Cayos A and the last was Los Cayos S, first described in 2003. There were several digs at Los Cayos until 1997, when all excavations at footprint sites were banned in La Rioja unless they were accompanied by a restoration and conservation program. All the outcrops are mainly in siliciclastic strata in the Barranco de Los Cayos, from which the site gets its name. There are still a few tracks in limestone left undescribed. According to the mapping of the area, the sites are in the upper part of the Enciso Group. This site was dug and studied by a joint team from Universidad Autónoma de Madrid and Iberdrola; the team is currently part of the Instituto Geológico y Minero de España. In 1989, a roof and fencing were placed around part of Los Cayos A, an action promoted by the two cited institutions: the electric company Iberduero (now Iberdrola) and the Universidad Autónoma de Madrid.

Description of Tracksite

As in El Peladillo, Los Cayos is a set of scattered outcrops spread over a wide area. The outcrops are assigned the identifying letters LC, followed

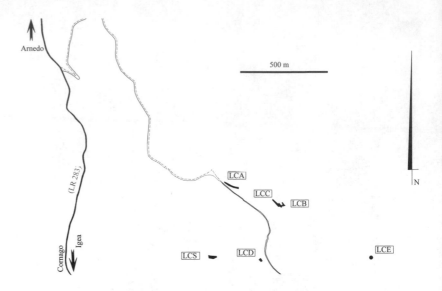

3.55. Location of outcrops (LCA, LCB, LCC, LCD, LCE, LCS) of Los Cayos tracksite. Local road is labeled LR 283.

by letters from A to E in the order in which they were worked. LCA, LCB, and LCC were the first discovered, in that order. After them were LCD, LCE, and finally LCS (Fig. 3.55). Although the outcrops of LCA, LCB, LCC, and LCD are relatively large, with many tracks, LCE and LCS are small and not relevant for study. In the literature on the site, the symbols referring to sites may be followed by others identifying them: the outcrop sector, the layer with tracks (if there are several in the outcrop), and finally the trackway and the number of the track.

The Los Cayos outcrops are in multiple sedimentary layers, most of which are sandstone. Some included more than one sandstone layer. There are four groups of superposed outcrops. The stratigraphically lowest is LCS, which is located furthest southwest, with tracks in 10 strata. Some meters east of LCS is LCD, much of which was destroyed during the construction of a road. The upper outcrops are located on sandstone that stands out in the terrain, with LCA, LCB, and LCC occupying at least five layers of the same stratigraphic sandstone member. Finally, LCE is the stratigraphically highest and is separate from the others. Adjacent to LCB and LCC outcrops are several locations where digging was stopped. There are also unaccounted footprints in small stretches of rock. The nearest tracksites are those of Valdeperillo (Cabezuelos), some 3.5 km away, and El Chorrón del Saltadero and La Era del Peladillo 6 (6PL), 3 km away.

Los Cayos is one of the tracksites most often visited by tourists, along with other sites in La Rioja (La Virgen del Campo, Barranco de Valdecevillo, and Munilla). Access to one part of Los Cayos is easy: the area near LCA is reached by a cart track, followed by a few meters' walk along a path that is part of the same track. This is the only subsite whose location is well indicated, but access to LCB and LCC is simple. At these latter two subsites, the land is terraced so that getting to the tracks is easy. Overall, it is a suitable environment for any kind of scientific learning activity.

Lithology

All the Los Cayos outcrops are associated with fluvial–lacustrine sequences of sandstone (Moratalla and Hernán, 2008; Mulas et al., 1988) in the Enciso Group. Freshwater molluscs (Unionoidea, Trigonioidoidea, Neomiodontidae, Cassiopidae) have been described from the site (Delvene and Munt, 2011). The thickness and structures of the strata change from one subsite to another. Those with larger outcrops (LCA, LCB, LCC, LCE) show repeated alternating layers of shale and sandstone. This stratigraphic interval is considered to be lacustrine littoral (Moratalla, Hernán, and Jiménez, 2003). The sandstone levels are usually alternating, thin, weakly cemented parallel plates with sandstone of variable silt content. Some have sedimentary structures in the top layers, such as ripples of parallel ridges that cross the entire outcrop. The sedimentation of the shale levels is so fine that small footprints (a centimeter in size or less) of turtles and birds are preserved.

Track Structures

Almost all the described tracks are in sandstone (whether laminated or not) and can be casts, subtracks, real tracks, or undertracks (Figs. 3.56, 3.57). The undertrack shafts (LCA) are filled with the deformed material of the subtracks, some with a neutral level or dead zone as described by Allen (1997). There are also casts from real tracks, both in outcrops (LCD) and in the detached stones surrounding the outcrops. Among the layers

3.56. Los Cayos A. Ripples in study surface that pass through footprint. Study surface is not tracking surface, and footprint is an undertrack.

of limolitic sandstone are layers of fine sediment that can show small footprints.

UNDERTRACKS WITH RIPPLES AND SUBTRACKS

The study surface at LCA has parallel ripples with rectilinear crests. The ripples occur both in the pretrack surface and at the posttrack surface, crossing the footprint interior (Fig. 3.56). The track shafts may either be empty, or partially or completely filled with sediment. Some of the shafts show signs of having been artificially excavated, in which case the filling material has been removed. The track filling material is deformed (Fig. 3.57). Many subtracks are indistinguishable from casts because they have

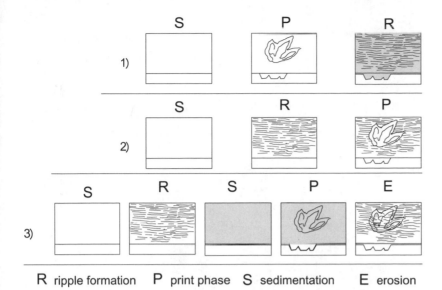

1)   S   P   R

2)   S   R   P

3)   S   R   S   P   E

R ripple formation   P print phase   S sedimentation   E erosion

3.58. Temporal relationship between ripple marks and footprints al LCA. Three possible scenarios are presented. Each is illustrated by a temporal sequence (illustrated from left to right) of boxes; horizontal line across each box separates plane view of sediment surface (line above) from cross section through sediment layer (line below). In scenario 1, sediment layer is deposited (S) after dinosaur walks across surface to create print (P). Sand layer moving in ripples (R) fills track to create natural cast, and ripples are undeformed. In scenario 2, dinosaur steps across rippled layer, squashing ripples. In scenario 3, ripples are buried by sediment layer, somewhat protecting them. The dinosaur walks across this new layer, creating undertracks that deform, but do not crush, ripples. Preservation of ripples with respect to tracks at LCA is consistent with scenario 3.

apparently been filled by massive (homogeneous) sediments. Thus, the structures produced by the deformation of the sediments beneath the feet cannot be discerned.

The ripples were formed before the formation of the tracks and deposition of the level on which the dinosaurs walked. The top of the rippled bed was beneath the tracking surface when these prints were made. Determining the temporal relationship between ripples and footprints (Fig. 3.58) involves consideration of several possible scenarios.

- *Scenario 1.* If the ripples were made after the formation of the tracks, the track shafts would be filled, and the ripple crests and troughs would pass over them. The sequence of events would be as follows. First, a dinosaur walks across a soft substrate and leaves footprints. Then a stream of water moves the sand over the ground. Finally, the sand fills the track shafts, and ripples pass over them without being deformed from one part of the ripple to another, and over the footprints.

- *Scenario 2.* If footprints were made directly on the ripples, then they should sink the surface stratification and crush the crests, leaving them flattened within the footprint. First, a stream of water moves the sand, leaving ripples on the surface. Then a dinosaur walks over it, flattening the tracking and ripple surfaces.

- *Scenario 3.* For the crests to be deformed without crushing, they need to be protected by an overlying sediment level. First, a stream of water moves across the sand, leaving ripples on the surface. Then several sedimentary layers are deposited above the ripples, protecting them. A passing dinosaur next depresses the original surface and leaves undertracks in the sedimentary layers below while deforming the ripples inside the still-soft sediment.

3.59. Right footprint, Los Cayos C (LCC). Several digital pads can be seen in digit IV (rightmost toe mark) and claw mark on digit II (leftmost toe mark). Deeper parts of footprint are wrinkled by fine ridges and microfractures. Wrinkles converge at tips of digits, at footprint "heel," and around hypexes between digits.

It is clear that the LCA study surface matches the third scenario, and so it is the tracking surface. The tracks in this surface are not true footprints but undertracks.

UNDERTRACKS WITH WRINKLES

Most of the dinosaur footprints at LCB and LCC have small, wavelike structures (similar to those cited by Carvalho, 2001) involving ductile (Figs. 3.59, 3.60) or fragile (Fig. 3.59) deformation of the mud (Milan, Clemensen, and Bonde, 2004). In this case, neither the waves nor the ruptures indicate that this area was in direct contact with the skin of the dinosaur's foot. The formation of these tracks involves a fine lamination that is not destroyed beneath the tracking surface but rather wrinkled. Wrinkled surfaces inside tracks (Fig. 3.61) are placed several layers below direct structures. These structures can be seen in side cuts made of footprints within the thin laminated layers. Many of the footprints in LCA, LCB, LCC, and LCE were made in layers of finely laminated sandstone.

3.60. Footprint at Los Cayos B (LCB). Distal end of digit III is at center, pointed toward bottom. Note wrinkles near apex of digit III, as well as digital pads and claw marks.

OTHER STRUCTURES

The footprint shafts of LCA and LCC are usually partly or wholly filled (casts or subtracks). Some footprints are filled with deformed sediment (Fig. 3.62) similar to those illustrated by Gatesy et al. (1999) and also to others seen in other outcrops in La Rioja, such as La Era del Peladillo or Santisol (Pérez-Lorente, 2003c). Figure 3.62 shows the following features:

1. The exit area of digit III is a small hole, the bottom of which contains severely deformed material. The contact between the walls and the bottom is a clear line.
2. Mud collapsed between the hypexes, which closes the shaft at the proximal part of the digit III mark. An incision or skim mark (Allen, 1997) occurs along the collapse line.
3. Associated with feature 2, lobes of mud are evident between digit III and the two side digits.
4. The rear central area of the print is a flat base, with no pad marks.

This structure can be interpreted as being produced by a foot completely penetrating the mud. On exiting, the three digits likely folded together and left the mud through the central hole at the front of the print. Alternatively, digits II and IV exited separately and vertically at the side of the tracks, but there are no structures to support this latter hypothesis. The mud collapse features curve the shaft walls of the footprint inward.

SUBTRACKS AND NATURAL CASTS

The undertracks are formed below the tracking surface. The tracking surface itself has eroded away at many of the outcrops in Los Cayos. Sometimes part of the tracking surface is deformed (posttrack surface; Gatesy, 2003) and/or displaced (dead zone; Allen, 1997) in the interior

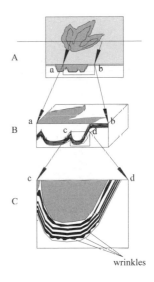

wrinkles

3.61. Structure of footprint wrinkles. (A) Tridactyl footprint (dark gray) seen in surface view (light gray region). Line cuts across print; bottom portion illustrates section across print (a–b). (B) Enlargement of section (a–b); note depressions in print surface corresponding to toes. Segment (c–d) shows section across one toe. (C) Enlargement of segment (c–d). Black-and-white layers correspond to sedimentary laminations beneath toe. Wrinkles in lowermost layer represent these shown in Figs. 3.59 and 3.60. Several wrinkle layers are commonly superimposed in same section.

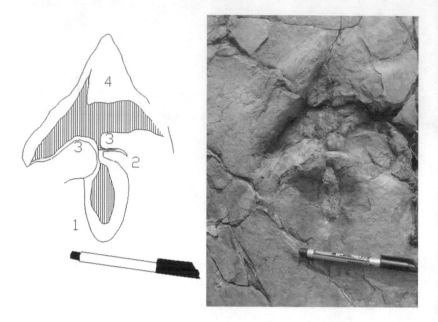

3.62. Interpretive drawing and photograph of footprint from Los Cayos B, showing features of interaction of foot of dinosaur with mud. Vertical lines in drawing indicate deformed area and base of track. Key to features: (1) exit scar of three digits; (2) collapse of mud at base of digit III; (3) interdigital lobes produced by collapse of mud at hypeses and base of digit III; (4) smooth proximal sole zone of print (lacking pad marks). Marker (14.5 cm long) provides scale.

of all the material forming the footprint. There are footprints (Fig. 3.63) where the foot deformed several layers of sediment, which were then partly pierced by the claw. At the tip of some digits with wrinkles on the soles, the claw mark can be seen as a result of its greater penetration in the mud. In one of the profiles (Fig. 3.63), the filling of the area pierced by the claw remains, cutting through layers in a manner similar to that described by Boutakiout et al. (2006), recording the following sequence of events: First there is deposition of thin layers of fine sand. Next, dinosaurs pass, whose feet deform or penetrate through the laminite layers, leaving side drag folds, wrinkles, and incisions or slide marks on the print base. Undertracks (indirect structures) are formed, as are fissures produced by the claws that cut part of the ground (direct structures). Finally, sediment fills the gaps.

Although the footprint section may be complete with all its parts (cast, direct structures, undertracks), what is less noticeable is the stamp or direct structures of the pes sole. The parts most easily seen in the track (Fig. 3.63) are the undertrack and the filling of the shaft in the track (natural cast). The sedimentary layers that were between the study surface and the tracking surface have eroded. The filling of the part pierced by the claw is like a narrow vertical sheet. The casts seen at LCC are filled with sand with less clay and have a massive structure. In cases where the sedimentation that followed after the formation of casts comprised silt layers, ridges formed on top of the footprints above the sandstone casts as a result of the lower compressibility of the sand filling. The shale layers lost more volume by compression than the sand and adapted to the shape of casts with projections, which are overtracks (Fig. 3.64).

Footprint casts occur at various levels and outcrops. At LCD there are ornithopod casts embedded in the top of the strata forming the outcrop (Moratalla and Hernán, 2008). Several levels below LCA, but in the same

3.63. Natural cast of footprint at Los Cayos C (LCC). Sand filling of cleft produced by nail going through substrate below tracking surface can be seen.

sandy stratigraphic member, theropod casts are seen. These structures involve the filling of the shafts by sand or silt carried by a stream that was unable to erode the base of the track shafts.

DIRECT STRUCTURES

Not all footprints seen in the study areas LCA, LCB, LCC, and LCE were produced on the same tracking surface. Some impressions, such as tortoise footprints and trackways with tail markings, must be direct structures. The rest of the outcrops are different. There are tracks that must be direct structures at LCD, some of them filled with sediment, and there are some direct structures in other layers at LCS, but observation conditions are not good at either site.

3.64. Sequence of events resulting in formation of overtracks due to differences in compressibility of sand and mud. (A) Formation of footprint. (B) Sand fills footprint shaft. (C) Several mud layers cover footprint and tracking surface (cf. Paik, Kim, and Lee, 2001). (D) Compaction of sediments during diagenesis and lithification. (E) Erosion to create present state.

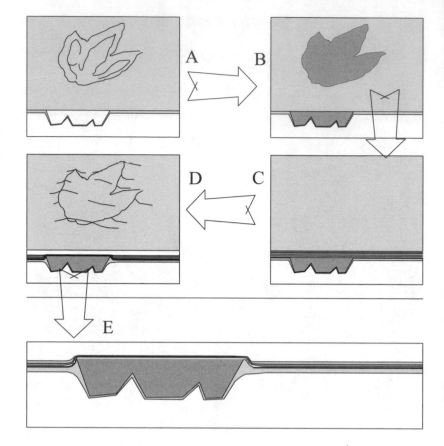

Ichnotypes

THEROPOD FOOTPRINTS

Bueckeburgichnus The three basic types of dinosaur footprints found in La Rioja occur at Los Cayos, as well as ichnites from other vertebrates (birds, pterosaurs, turtles). In some cases, the ichnogenus for some of the footprints has been determined as described below. The theropod footprints in LCA are attributed to *Bückeburgichnus maximus* Abel, 1935 (cf. Moratalla, 1993). This ichnogenus and its successive adaptations (*Bueckeburgichnus*, Lockley, 2000) are problematic for several reasons, explained by Thulborn (2001). The ichnogenus (Kuhn, 1958) is based on a cast with the marks of the metatarsus and hallux. Of the dozen footprints found at the site (Bückeburg), there is only a photograph of two of them, one with the marks of the metatarsus and hallux (from which Kuhn [1958], but not Abel, defined the ichnogenus) and another tridactyl footprint without a hallux mark, which is what Lockley (2000) used (cf. Thulborn, 2001). Finally, the footprints with the metatarsus and hallux marks on the site are completely different from those used by Kuhn (1958). Without going into further problems associated with this name, we must add that the footprints (shafts) visible at LCA are all undertracks (Figs. 3.56, 3.57, 3.59) and not true footprints, and that the allocation of an indirect impression to an ichnogenus is not recommended.

Some of the undertracks (Fig. 3.57) in the western part of the LCA outcrop also have a hallux mark. The tracks with this impression alternate with tridactyl footprints in the same trackways. Moratalla (1993) interprets this alternation by variation in the foot sinking into the mud. More recently, Moratalla and Hernán (2008:169) assign the footprints of the middle–great size of Los Cayos to "conjunto *Buckeburgichnus–Megalosauripus.*"

THERANGOSPODUS The footprints in LCC and LCE have been attributed to *Therangospodus* Moratalla, 1993. The indicative features are low interdigital angle, elongate heel, symmetrical hypexes, subparallel digit contour line, pronounced medial notch (between digits II and III), and digits slightly tapered. Moratalla stresses that at the second outcrop (LCE), the notch is not very pronounced.

The *Therangospodus* prints at LCC and LCE are undertracks, at least in most cases. In almost all the footprints there are striae and/or a parallel series of wrinkles or corrugations on the inside. The striae and wrinkles are elongated, and along the digit, marks are almost rectilinear and parallel to the edges throughout. However, they form an angle at the apex of the toes with the vertex at the distal end (Fig. 3.50). In the area where the toes meet, they are curved and surround the hypexes, which appear to be the center of the described curves (Fig. 3.59), joining the most external curves in the heel of the track. In some of these footprints (Fig. 3.63), the sand filling the space left by the claw penetrating the mud is preserved.

THEROPOD FOOTPRINTS WITH TAIL DRAG MARKS Because of the scarcity of tail tracks described to date, two theropod trackways, LCB2-R1 and LCB2-R5 (Moratalla, 1993) with associated narrow and sinusoidal marks (Figs. 3.65, 3.66), are of some interest, as these latter marks were supposedly made by the tails of dinosaurs. They are narrow, linear depressions in the midline of two theropod trackways, sigmoid with a V-shaped cross section, with parallel grooves and edges caused by extruded mud. These characteristics are similar to those suggested by Hunt and Lucas (2004) and Dalla Vecchia et al. (2000) to identify dinosaur tail marks. The narrow marks with a V-shaped section are typical of theropods (cf. Jiménez Vela and Pérez-Lorente, 2006–2007). According to Torcida et al. (2003), most trackways with tail markings have peculiar features. They were made by normally bipedal dinosaurs with tail marks associated with trackway segments where the dinosaurs adopted a quadrupedal position, or they are generally associated with semiplantigrade footprints and/or prints with hallux marks. Torcida et al. (2003) indicate that the dinosaurs had to adopt special postures to enable the tail to touch the ground. The tail mark is clear in both trackways (Fig. 3.66). Where an interaction with the footprints can be distinguished, the tail markings occur after the footprints are laid down. In LCB2-R5, the tail mark occurs with footprints showing the hallux, and in LCB2-R1 with semiplantigrade footprints.

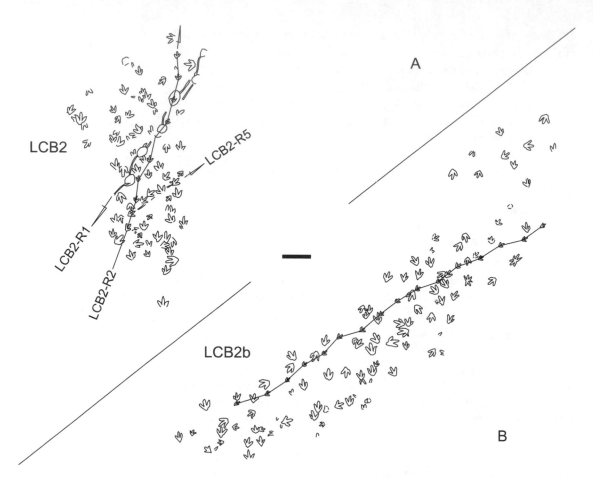

3.65. Two parts of Los Cayos. (A) Segment of LCB2 showing two trackways with tail drag marks, LCB2.R5 (footprint with hallux), and LCB2-R1 (semiplantigrade). Arrows indicate direction of travel. (B) LCB2b sector, in which wrinkles and bases of footprints are clear. Noted too is longest trackway of Los Cayos tracksite (16 footprints over about 12 m). Modified from Moratalla, Sanz, and Jiménez (1999). Scale bar = 1 m.

ORNITHOPOD FOOTPRINTS

Moratalla and Hernán (2008) cite the LCD ornithopod footprints as large ornithopods of the Iguanodontoidea clade. They are considered similar to ornithopod prints at sites in La Rioja described in this book (Valdemayor, 1PL, La Cuesta de Andorra) and other European sites of the same age. The features of these tracks are similar to those of *Iguanodontipus* Sarjeant, Delair, and Lockley, 1998, typical of the Upper Jurassic–Lower Cretaceous. Moratalla and Hernán (2008) say that because phylogenetic analyses of derived ornithopods are in a state of flux, for now, these prints can only be assigned to ornithopod dinosaurs.

SAUROPOD FOOTPRINTS

Moratalla and Hernán (2008) describe the sauropod tracks at LCS. The most notable aspect of the tracks is the presence of the impression of digits I and V in a left manus print. The forefoot has the impression of a wide central digit, which corresponds with titanosaur or titanosauriform dinosaurs. These authors identified the footprints as *Titanosaurimanus* ichnospecies. The trackway is short, and it cannot be established whether it is narrow or wide gauge.

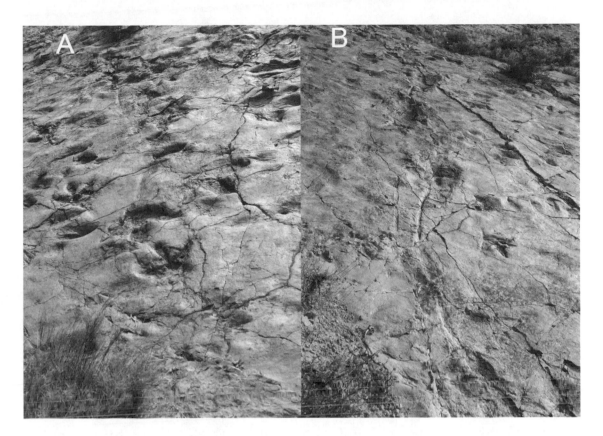

FOOTPRINTS OF OTHER VERTEBRATES

In addition to the footprints of dinosaurs, at LCC1, 58 small tridactyl turtle footprints, in what seem to be a number of trackways, have been cited (Moratalla, 1993). Each track is a set of three separate parallel slits barely 1 cm in length. The slits are tapered in what we interpret as the direction of travel (Fig. 3.67). They have been attributed to turtles and are currently the only place with this kind of footprint discovered in La Rioja. There are, however, plenty of places where shell fragments and other skeletal remains of these animals have been found. This same outcrop includes marks supposedly caused by birds (Moratalla and Sanz, 1992; Moratalla, Sanz, and Jiménez, 1992). Consisting of three small, thin, open, and tapered digits, some of which also have hallux marks, they are between 2 and 8 cm long, including the hallux. Finally, on one side of LCA, in one of the lower layers, is a set of fewer than five pterosaur footprints. This is the only site with pterosaur tracks described in La Rioja.

Track Distribution

In Los Cayos there are three types of tracks common to La Rioja (theropod, sauropod, ornithopod), although they do not occur together in the same outcrop. At the LCA, LCB, LCC, and LCE outcrops there are no ornithopod or sauropod footprints; all are theropod footprints. There are

3.66. Photograph of two tracksites with tail drag. (A) LCB2-R5 with distinct toes and hallux mark; trackway direction toward observer. (B) LCB2-R1 with indistinguishable anatomical details (pads, claws, etc.). Trackway from observer's point of view. Dinosaur goes from bottom to top.

3.67. Marks of three toenails of turtle. On same study surface are dinosaur footprints.

mostly undertracks and subtracks. This is not to say there are no direct tracks or natural casts, but that these latter structures are abundant. There are incisions that result from digits that sank through to lower levels and collapsed the structures of the walls, some of which may pinch together. There are some undertracks with tapered ends of claws and digital pad structures (the Los Cayos tracksite is not the only place where this structure association is cited). There are localities in outcrops with algal laminites with a series of superimposed undertracks with pad marks.

The LCD outcrop consists mostly of ornithopod footprints but includes one known theropod footprint. More cleaning and digging in the surrounding area may reveal more carnivore footprints in the same layer. Presently known ornithopod footprints are deeper, and they probably all contain a cast of the footprint. The theropod footprint is shallow (less than 3 cm deep). Both types are on the top of the same layer.

LCS is a set of separate outcrops along a ravine. It contains about 10 layers with sauropod, theropod, and ornithopod footprints, but none with more than one ichnotype in the same layer. No conclusions can be drawn from this apparent distribution because small outcrops usually have a reduced number of ichnotypes (Casanovas et al., 1999).

Relation between Footprints and Lithology

LCA, LCB, LCC, and LCE are in a sedimentary sequence with an obvious pattern. LCE is the highest outcrop, about 10 cycles higher than LCB. Each

3.68. Partial view of interior of area protected by shelter and fencing at Los Cayos A. In top left-hand corner is a photograph of shelter. Outcrop continues to right of shelter. Shelter and fencing work was carried out by an agreement between the Iberduero (now Iberdrola) electric company and the government of La Rioja.

cycle consists of a group of sandy levels, generally less than 1 m thick, and other shale levels (about 5 to 10 m thick). There are no significant limestone levels interspersed. The sand levels are predominantly fine lamination. According to Moratalla, Hernán, and Jiménez (2003), the sedimentary environment must have been at the margin of a shallow lake, but almost without emersion, because no soil developed there. The highest levels have no limestone deposits, so these are probably the highest part of the Enciso Group.

The lithological variation of the layers at LCD and LCS is greater, with the sandstone laminations sometimes being thinner. There are abundant limestone layers interspersed in the series, and the shale levels are much thinner than in previous outcrops. Limestone sedimentation is consistent with areas remote from, or protected from, the influence of fluvial siliciclastic input. At LCD, most of the tracks have rounded outlines, and a round pad discernible in the digit marks is usually found in these footprints. This is an accumulation of ornithopod footprints. Regrettably, LCD was destroyed during the construction of a road, although the area could be extended with some digging and clearing in the vicinity.

At Los Cayos, there is therefore a distribution of the types of footprints according to the lithology. In the higher outcrops (LCA, LCB, LCC, LCE)—with sandstone/shale cyclic sedimentation—there are footprints of theropods, pterosaurs, turtles, and birds. In the lower area—where the sediment is mixed with no apparent simple cyclic nature (LCD, LCS)—there are theropod, sauropod, and ornithopod footprints. Despite the distribution of footprint groups there are no criteria for discussing the preference for certain habitats or herd behavior of dinosaurs at the Los Cayos tracksite.

Conservation

The problem of destruction of the rocks by natural causes is apparent at this site. Most of the outcrops (LCA, LCB, LCC, LCE, part of LCS) are in compact sandstone levels that resist the action of atmospheric agents relatively well if they are not fractured. Plants are most responsible for destroying the rock, as the thickets of thyme, gorse, and rosemary can insert roots through the cracks. The cold that freezes the water that gets into cracks of the rock is another major destructive agent. A cover has been built over a portion of LCA, with apparent beneficial effects, because inside the cover, there is virtually no vegetation (Fig. 3.68). Another problem is human intervention. Much of the LCD site was destroyed during construction of a new road, when a digging machine broke through and destroyed most of the outcrop. Fortunately, several studies were done before this disaster (Moratalla, 1993; Moratalla and Hernán, 2008; Moratalla, Hernán, and Jiménez, 2003), with data published from the destroyed part.

Track Density

The density of tracks is similar to the rest of the sites in the zone. However, the total number of more than 1000 dinosaur footprints in uncovered parts is remarkable. At several outcrops (LCA, LCB, LCC, LCD, LCS) there are fronts where digging was abandoned and could be continued. Meanwhile, at some (LCD, LCS), simply cleaning the rock and removing weeds and loose debris would probably bring to light several hundred more footprints.

This is not the only footprint bonanza; sections throughout sedimentary levels (especially in the thin laminated sandstone) reveal many footprints trapped inside these hard levels. These are also important for study because they show the effects of deformation produced by the autopodia in the sedimentary laminations and layers. In the profile of LCA there are superimposed footprints in the upper part of the sandstone member. Inside the sandstone layers beneath the surface of the LCA study area there are different layers, each of which has been a tracking surface, i.e., there is an undetermined number of dinosaur crossings separated by sedimentation periods. In addition, at the base, casts and what are probably subtracks can be seen. The laminites may correspond to periods of climatic variations over time.

Los Cayos has reported a total of 17 layers with tracks in outcrops, not including those inside and on the wall of the sand levels. The potential for sedimentary patterns between LCB and LCE has also not been taken into account. The number of footprints in the environment of Los Cayos could certainly be increased by extending the levels already worked. The richness of dinosaur tracks in Los Cayos and its potential is enormous.

MAIN FEATURES OF THE TRACKSITE

The environment of Los Cayos is notable for its diversity of dinosaur footprints (theropod, ornithopod, sauropod) and other types of contemporary vertebrates (turtles, pterosaurs, birds). Although many of the tracks are not true footprints but undertracks, with very few stamps, this does not prevent them from being classified into ichnological groups to determine restricted ichnofacies. There are many mud deformation structures produced by the footprints at several depth levels, related to viscosity conditions of the varied layers. There are examples of physical behavior ranging from collapsed structures of mud in deep layers (low viscosity, low consistency) to possible fluid structures (wrinkles). There are also different types of filling of shafts. The fillings are casts (i.e., undeformed material filling the impressed footprint shafts on the tracking surface) and subtracks (i.e., shafts filled with sediment deformed by the footsteps). Various study surfaces show such structures accompanying stamps and overtracks.

Los Cayos also has particular trackways whose significance has not yet been studied. An example is the coexistence of trackways containing two types of tracks (tridactyl, tetradactyl) left by the same dinosaur, located among other tridactyl trackways. The reason for the existence of the two tail-mark trackways has also not been determined: one accompanies tetradactyl footprints and the other semiplantigrade footprints. The number of tracks and footprint types found at Los Cayos, the variety of structures associated with the footprints, and the sedimentary environment should provide information and results that cannot be obtained at smaller tracksites.

History

The Valdeté (VLD) site was discovered by Santiago Jiménez, a member of the research group from the Universidad Autónoma de Madrid, and the electric company Iberduero (now Iberdrola). A study was started in 1985. The first publication to mention it was a promotional book (Moratalla, Sanz, and Jiménez, 1988). Two subsequent studies were carried out in 1993 and 1994, which interpreted the single trackway at the site as having been made by a lame ornithopod dinosaur.

Description of Site

The site is located on a piece of open ground accessible by a path made for reforestation and can only be reached on foot. The outcrop is easy to see because the rock slopes smoothly and is well indicated by vertical

Valdeté

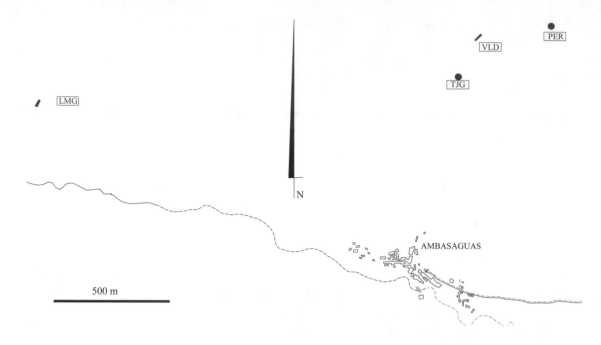

500 m

3.69. Location of site of Valdeté (VLD) and neighboring sites Perosancio (PER), Tajugueras (TJG), and La Magdalena (LMG).

signs. There are other sites nearby (Fig. 3.69): 1900 m away is La Magdalena, studied by Moratalla (1993); 180 and 336 m away, respectively, are Tajugueras and Perosancio, both with no known scientific work undertaken. Other sites (not shown in Fig. 3.69) are Santa Lucia, 1900 m to the east but not yet studied, and the sites of Corral del Totico (1TT and 2TT), 2100 and 2200 m to the west, described in this volume. No other fossil sites have been detected nearby, although it is easy to find rock containing fish scales and abundant ostracods in the surrounding area.

Lithology

The rock is a gray limestone that has turned a yellowish color on weathering. The tracks are found in the upper part of a unit of 160 cm thickness. Weathering has separated the layers into varying thicknesses. Sedimentary structures are only laminites, with possible interference ripples and desiccation cracks at the top. All authors who have worked in the area recognize that the limestone is of lacustrine origin.

Track Features in Mud

The depth of the tracks varies. The shallowest (VLD-R1.9, VLD-R1.10) are between 1 and 2 cm deep. The deepest (VLD-R1.11) reaches 6 cm. The extrusion rims are not very well defined. In general, the deepest part of the tracks varies between 3 and 4 cm beneath the top of the extrusion rims. All tracks have a depressed interdigital area, which is better defined for digits III to IV than digits II to III. This depression is probably due to the behavior of the mud rather than any other reason. The distal end of digit IV is more widely separated from digit III than is the distal point

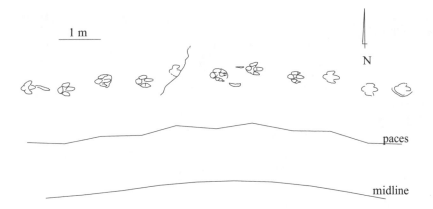

3.70. Trackway VLD-R1. (Top) Sequence of footprints. (Middle) Position of successive paces of trackway traced. (Bottom) Curve showing path of animal's midline along trackway, indicating that the dinosaur turned slightly to the right.

of digit II. The footprints were made after the desiccation cracks in the surface of the limestone.

Track Features (Ichnotypes)

These footprints have been attributed to ornithopods. Their features (Moratalla, 1993) are as follows (Fig. 3.70): they are mesaxonic, U-shaped, tridactyl footprints 41 cm long and 33 cm wide; the print outlines and distal ends of the digits are rounded; digit III is wide, robust, and relatively elongated, with a U-shaped contour; digit IV is longer than II with the two hypexes symmetrical (equal distance from the proximal part of the heel); one oval pad per digit, and a circular pad at the heel; notches in the print outline at the proximal ends of both digits II and IV. The digital divarication is low (23–26°), and the high value of the ratio (l – a)/a is in fact more typical of theropod dinosaurs in footprints of this size.

After examining the ornithopod foot skeletons known at that time, Moratalla (1993) concluded that the closest match to the Valdeté prints (as a result of the narrowness and length of the digits) is *Camptosaurus*. However, given the few tracks of this type found, the final assignation is Iguanodontidae indet. Contrary to Moratalla's interpretation, however, there are some observations that do not justify the allocation of these tracks to ornithopods. Moratalla (1993) acknowledged that the tracks are longer than they are wide (l = 41 cm, a = 33 cm) and that the angle between the digits is small. In addition, some of these tracks (Fig. 3.71) show features that can be interpreted as marks of pads (in the sole of the footprint and on the contour line), and the outer digits are asymmetrical with respect to the axis of the middle digit; that is, the outer digit (IV) makes a larger angle with respect to the middle digit (III) than the internal digit (II).

Distribution of Tracks

The Valdeté trackway has significantly longer (13%) steps that begin with the left foot [L] than it does steps that begin with the right foot [R] (Moratalla, 1993). The sequence of steps (in cm) is as follows: [L]96 (–)[R]88 (+)[L]99 (–)[R][86] (+)[L][97] (–)[R]89 (+)[L]101 (–)[R]90 (+)[L]99 (–)[R]85,

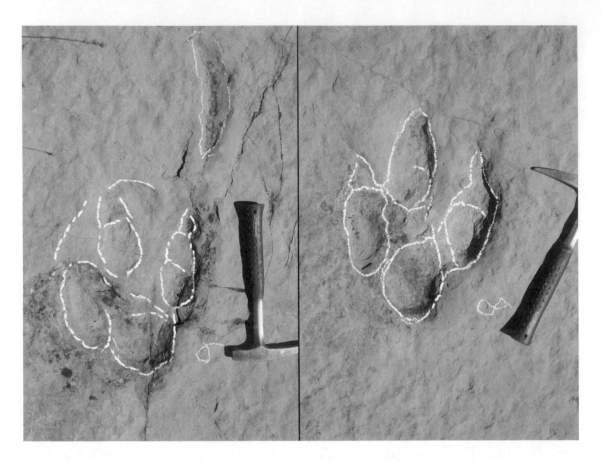

3.71. Two right footprints from trackway VLD. (A), VLD-R1.6. (B) VLD-R1.8. Boundary of digital pads, acuminate tips of toes, and asymmetric interdigital angles (II^III < III^IV) are visible. Print VLD-R1.6 also show slight interdigital depression that can variably be seen in other footprints of trackway.

which show an alternating increase (+) and decrease (−) along its length. The figures in square brackets were not provided by Moratalla (1993) because of a rupture across the site, which may have distorted the original value. Moratalla concluded that this must be due to some pathological factor. If it was a foot injury, it would have been the left foot that was affected. However, there is no difference in the footprints to support this possibility. The problem or defect could be in any part of the limb, including the hip. Lockley et al. (1994) suggested that this trackway is one example of many that exist around the world of footprints left by lame dinosaurs.

Examination of the trackway (Fig. 3.70) shows that the dinosaur turned to the right. The question arises of whether turning left was just as likely. Examination of other sites in La Rioja with long (sinusoidal) trackways shows that some of these have part of the trackway in which some of the steps are larger on one side than the other (Pérez-Lorente, 2003c). This is also seen, for example, at La Canal (Casanovas et al., 1995a), where the first 18 steps show a turn to the left and where the right–left steps are greater than the opposite: [L]96 (−)[R]79 (+)[L]87 (+)[R]96 (−)[L]83 (+) [R]102 (−)[L]96 (+)[R]98 (−)[L]90 (+)[R]99 (−)[L]91 (+)[R]95 (−)[L]90 (+) [R]105 (−)[L]92 (+)[R]97 (−)[L]91.

At the site of La Barguilla, there is a trackway that turns in two directions: first to the left (the right–left steps are larger) and then to the right (the left–right steps are larger). The final part of the track also turns to the left but cannot be analyzed as a result of the poor preservation of the footprint, LBG2.14: [L]105. (–)[R]70 (+)[L]106 (–)[R]91 (–)[L]77 (+)[R]95 (–)[L]69 (+)[R]100 (–)[L]82 (+)[R]92 (+)[L]95 (+)[R]100 (–)[L]102 (+)[R]111. The Valdeté trackmaker may well have been lame, but it is also possible that the differences in step lengths only reflect its changing direction of travel. Therefore, although possible, it is risky to assume that VLD-R1 is a trackway left by a lame dinosaur.

Jenny and Jossen (1982) reported limping theropod trackways at the town of Aït Blal, from the Lower Jurassic of the Atlas Mountains (Morocco). The Aït Blal limping theropod trackway (Ishigaki and Lockley, 2010; Lockley et al., 1994) is larger than that of Valdeté, with left and right turns. In this example, the difference in steps still occurs despite the animal's turns, and thus it was likely limping.

General Information

The track density in Valdeté is low, 0.5 per m^2, which means that on average you need to go 2 m^2 across the outcrop before finding a track. The nearby sites of Perosancio, Tajugueras, La Magdalena, and Santa Lucia also have very few tracks. A little further away are the sites of La Cuesta de Andorra, with an ornithopod trackway, and the Corral del Totico and Barranco de Valdegutierrez trackways, which are exceptions regarding the number (349) and type of footprints (rounded ornithopod, sauropod, and typically theropod footprints). The position of the levels in Corral del Totico is lower than the previous ones, although the sedimentary features are the same. The site is already quite changed by weathering. The layer is peeling, and there are lots of loose fragments, even at the edges of the footprints. The action of frost on the cracks, and probably the high diurnal temperature changes, are responsible for breaking the rock fragments into centimeter-size or even decimeter-size pieces. The rate of destruction of the site seems high.

Main Features of Site

The footprints of Valdeté have been used to define a type of ornithopod dinosaur different from types proposed in the Cameros Basin (Díaz-Martínez et al., 2009; Moratalla, 1993). The debate over whether it is actually a theropod or ornithopod footprint is still open. In any case, it is a footprint with mixed features, if it is a stamp. This was also the first site in La Rioja where a lame dinosaur was said to occur. Although the curvature of the trackway complicates interpretation of steps according to a long/short pattern, it may well have been made by an animal with such a defect. The abundance of sites and the number of footprints per site decrease toward this area. This is likely due to the orientation of the ravines, which are perpendicular to the direction of the strata, thus not allowing the development of large surface outcrops.

3.72. Location of sites Barranco de Valde-gutiérrez (BVG), Corral del Totico (1TT, 2TT), and other neighboring points with tracks: La Magdalena (LMG), La Cuesta de Andorra (AND), Barranco de la Sierra del Palo (BSP), Lastras 9A, 9C and Lastra 10A, Navalsaz (N∠).

Corral del Totico and Barranco de Valdegutiérrez

History

In 2001, S. Doublet, a French geologist studying for her Ph.D. by doing work in the Cameros Basin, showed me a new location with dinosaur footprints near the old bridle path between Navalsaz and Ambasaguas. The site is near where Brancas, Martínez, and Blaschke (1979:61–72) had described the "conjunto de Lastras 7-A, B, C y 9-A, B, C." The same year, 2001, a site covering an area of 170,000 m^2 was investigated between Barranco de Valdegutiérrez and Corral del Totico (Fig. 3.72). The site found by Doublet was identified and located, and it was given the name Corral del Totico 1 (1TT), followed by the "slabs 7-A, B, C and 9-A, B, C" included in the outcrops of Barranco de Valdegutiérrez and Corral del Totico 2 (2TT). The drawings made by Brancas, Martínez, and Blaschke (1979) have very few reference points (Fig. 3.73).

The exposure named Lastra 9-C, with two footprints, must be at the bottom of the ravine; exposure 9-A (41 footprints) is assumed to be high on one side of the gulley, extending over several yards along this channel; exposure 9-B, said to have 114 footprints ("on descending from the top of the mountain, a large number of tracks were found, although mostly eroded"; Brancas, Martínez, and Blaschke, 1979:72), is identified (Fig. 3.73) as 2TT.

It is assumed that the location of the "Lastras 7-A, B, C" is in the lower part of Barranco de Valdegutiérrez. The description of the site is similar: abundant bush, and within it, in the clearings with no debris, dinosaur footprints. The number of footprints in the exposures (19 in Lastra 7-A, 12 in Lastra 7-B, eight in Lastra 7-C) implies that the clean

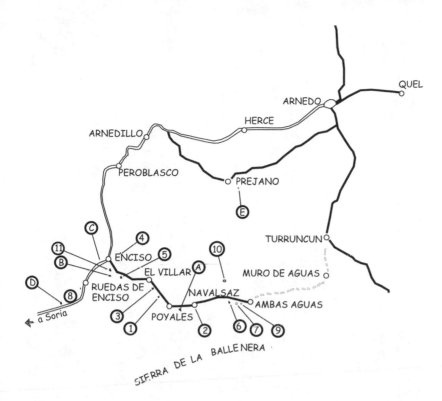

3.73. Old map of tracksites modified from Brancas, Martínez, and Blaschke (1979): Lastra 2 (NZ), Lastra 6 (AND), Lastra 10 (10A-BSP), Lastra 7 (part of BVG), Lastra 9B (2TT).

surface layer is relatively large, leading to the belief that the outcrops are on the same side of the ravine bed.

The first exploration stage (2001), in addition to locating the sites 1TT, 2TT, and Barranco de Valdegutiérrez (BVG), found a new one, 2BVG. In 2003, 1TT and 2TT were mapped and information taken. The second exploration phase (2003–2005) involved study of the Barranco de Valdegutiérrez area (Fig. 3.75), where 28 outcrops (1BVG to 28BVG) were located and described.

The exploration work also confirmed that the Corral del Totico outcrops (1TT and 2TT) have well-defined limits, but in Barranco de Valdegutiérrez, there is a great possibility of expanding the surface for tracks. In 1TT and 2TT, the strata are cut by erosion to their east side and are buried beneath the hillside to the west. The density of footprints is 0.3 per m^2 in 1TT and 0.4 per m^2 in 2TT – a very low number.

Description of Location

Barranco de Valdegutiérrez and Corral del Totico are in geographical alignment with a large number of sites. A line can be drawn from Valdeté in the west to Las Losas in the east, measuring 12 to 14 km, which passes through 18 studied sites. The largest concentration of dinosaur tracks on that line are in Barranco de Valdegutiérrez and 2TT.

The two sites have no roads that can easily be accessed by car or on foot. The setting is also mainly scrubland with thorny vegetation (gorse, dog rose), making it inhospitable. The 1TT footprints are difficult to find

3.74. Photograph of 2TT outcrop taken from opposite side of ravine with excellent viewing conditions of study surface. It is possible to see shadows of footprint shafts.

because the outcrop is small (60 m²) and the footprints' position is hidden, unless the traveler passes over them. The 2TT ones are, however, visible from the hill opposite the outcrop (Fig. 3.74). The 2TT surface area is about 600 m² and has always been clear. However, anyone seeing the footprint shafts and their shadows may not have recognized them for what they are. The site photograph (Fig. 3.74), taken before the study, shows how the site has always looked. The surface is clean because rainwater falls down hard enough to keep the rock clean of loose stones and dirt. Moreover, the rock has no gaps to allow plants to grow.

The only path providing access exits the road and crosses the site from west to east across the southern end of Barranco de Valdegutiérrez. It is an old bridle path for people and horses, stretching from the Valdegutiérrez spring to the Corral del Totico. This path has fallen into disuse and is now largely covered with vegetation. Today, the only accessible

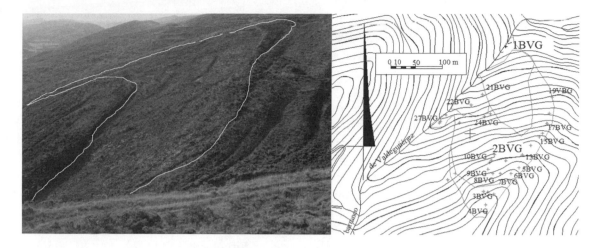

point is the path from Navalsaz to Ambasaguas (between the sites of the Cuesta de Andorra and Valdeté, and bordering it to the north) and the road from Enciso to Los Cayos (well south of the outcrop). To reach the outcrops, you have to walk through thickets and thornbushes, with occasional gaps left by cattle.

The site 1BVG (20 m²) is located at the north end of Barranco de Valdegutiérrez and did not need cleaning because it is at the bottom of the Barranco de Valdegutiérrez. At the 2BVG site, by the path of the Valdegutiérrez spring to the Corral del Totico, 4 m² were cleaned (from the total site area of 18 m²) for investigation. The two outcrops are in the eastern part of the Barranco de Valdegutiérrez. In the open spaces of the slope, free of brush and debris, black limestone outcrops occur. There are dinosaur footprints in all clean areas of the rock (Fig. 3.75).

At Barranco de Valdegutiérrez, the layers with tracks are parallel to the slope of the ravine (Fig. 3.75), and the footprints are exposed along the entire slope, between the loose stones and scrub. It is easy to remove the mantle of debris and brush and expose a large area of footprints. Barranco de Valdegutiérrez extends over 31,000 m², which makes it the second largest site after the Ardley Quarry (500,000 m²; Day et al., 2004). The surfaces with tracks at Barranco de Valdegutiérrez are at the top of three layers of hard, dark limestone with the same orientation (direction and slope) as the hillside. The areas free of vegetation and debris contain 0.3 (1BVG), 1 (2BVG), and 2.5 (21BVG) footprints per square meter. Applying these values, the Barranco de Valdegutiérrez site contains between 9000 and 77,000 footprints. In a study on this subject, Jiménez Vela and Pérez-Lorente (2005) apply a drastic reduction of the number calculated above. They assume the lowest index from the above three must be taken and reduced by half for sectors that may be destroyed in Barranco de Valdegutiérrez, mainly by plant action. They suggest the number of prints after cleaning the area would exceed 5000.

Cleaning this site would involve the destruction of vegetation covering 31,000 m² of virgin land. The impact on the natural environment would be so great that it is questionable whether the interest of such an

3.75. Barranco de Valdegutiérrez (BVG) tracksite in which direction and dip of strata with footprints are same as slope of hillside. (Left) Photograph of part of BVG (outcrops not shown include 1BVG, 25BVG, 26BVG, 27BVG). (Right) Topographic map with location of outcrops studied. To simplify presentation of data, not all symbols have been provided

objective merits such destruction. There is enough research work at other sites in La Rioja that needs to be done without having to begin any clearing or digging at Barranco de Valdegutiérrez.

During the initial exploration campaigns in La Rioja, it was difficult to think that occurrences of this size would be found. The sites have all been discovered since 1968, and even though there have been four research groups working on the footprints, discoveries continue to be made at a rate of several per year. Several known localities, with further cleaning of the layers, have revealed large outcrops with footprints. There are also tracksites that could be extended with a little extra digging. This means that although the chances of finding large sites are shrinking, discoveries of this nature may still be made.

Lithology

The Barranco de Valdegutiérrez and Corral del Totico footprints are in layers of mudstone (micritic limestone) of the Enciso Group, usually hard, and with the typical dark gray color. Weathering has transformed the gray on the outside of some strata (1TT) to yellow. The 2TT and Barranco de Valdegutiérrez stratification surfaces with footprints are in groups of calcareous layers interspersed with sandy and shale packets. According to Doublet (2004), the age of the sites, based on charophyta, is Aptian, and the sedimentary environment was emergent (or shallow) lacustrine carbonate in a shoreline area less influenced by waves from storms. It is likely that the tracksite areas were protected littoral areas, with sufficient standing water to maintain wet conditions, perhaps with mostly continuous water cover. The limestone layers extend over a large area (several kilometers), and so it is likely that the lakes were also extensive. The proportion of limestone compared with shale and sandstone is less in 1TT than 2TT and Barranco de Valdegutiérrez. The sand levels represent alluvial fans on the side of the lake and also contain footprints, but in this environment there are no extensive outcrops. There are natural casts filled with sand in some shale rocks.

Footprint-Related Features in Mud

At 1TT the tracks are shallow and record features of the soles of the feet and claw marks. Only the top part of the substrate (a few centimeters) stepped on by the dinosaurs was soft. At 1TT, the thickness of low viscosity mud was not much because the maximum depth of the tracks is about 10 cm or less.

At 2TT the mud layer was thicker and the viscosity lower; here there are footprints more than 25 cm deep as well as slide flood marks produced in the trampled mud. The footprints are generally deep and have a rounded outline. This site has structures showing mud collapsing into the shafts from the walls of the prints. These physical conditions of the mud (low viscosity and coherence) are most likely responsible for the

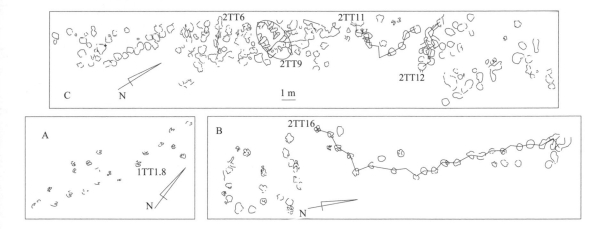

3.76. Cartography of (A) footprints of 1TT outcrop, (B) northern 2TT sector, (C) southern 2TT sector. Footprints and trackways cited in text are indicated.

rounded contours and the lack of structures in the soles of the feet at the bottom of the shafts.

At Barranco de Valdegutiérrez, the conditions differ greatly, depending on the outcrop. At 1BVG the footprints are clear, with correctly proportioned anatomic structures. However, at 2BVG the feet of some dinosaurs sank fully into the mud, and during the foot's kickoff (K phase), the print walls completely or almost completely collapsed. Consequently, observations cannot be generalized across the outcrop because the rock outcrops are small and have different characteristics.

Footprint Structure Distribution

The distribution of the footprints seems to be related to the physical properties of the mud. At 1TT there are no ornithopod or sauropod footprints (Fig. 3.76A). The theropod prints are shallow and show clear digital pads and claw marks. At 2TT (Fig. 3.76B, C) there are few theropod footprints (digitigrade and semiplatigrade) and no sauropod prints. Most of the tracks are ornithopod or are unidentifiable. The shafts are deeper that 1TT and are dominated by rounded shapes with poorly defined contours, which makes it difficult to establish the outlines of footprints when tracks overlap or are close, and have slightly sloping walls and sole surfaces without pads. This does not mean they have to be interpreted as undertracks; in some trackways there are footprints that are easily recognizable by their structures and others that are not. The footprints at 1TT show clear ichnotaxonomic features attributable to theropods, while the opposite occurs in 2TT because the prints here are mainly rounded footprints as a result of the characteristics of the mud and can predominantly be assigned to ornithopods.

In Barranco de Valdegutiérrez there are footprints from digitigrade theropods, ornithopods, and sauropods, as well as unidentified tridactyl tracks. Site 1BVG (Fig. 3.77) has a group of ornithopod tracks whose footprint outline is quite sharp and where the digit pads are clearly seen. However, at 2BVG there are theropod and ornithopod footprints with

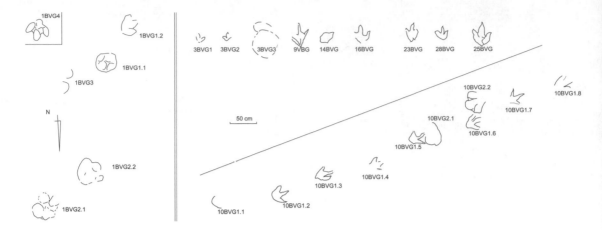

3.77. Representative examples of footprints found at Barranco de Valdegutiérrez (BVG). (Left) 1BVG outcrop. (Right) 10VBG outcrop and other selected footprints from various outcrops.

collapsed walls because the feet sank into mud of low viscosity and low coherence. Across the vast area of Barranco de Valdegutiérrez, it is likely that there are different sediment conditions depending on the sector.

Footprint Features (Ichnotypes)

ORNITHOPOD FOOTPRINTS

There are large ornithopod footprints at 2TT and Barranco de Valdegutiérrez. Most of them have three large, rounded front pads and one in the heel (*Brachyiguanodonipus* Moratalla, 1993, according to Díaz-Martínez et al., 2009). Between prints 11BVG1.5 and 11BVG1.7 there are two prints that probably belong to the trackway of an ornithopod with an interdigital web (10BVG2.2, Fig. 3.77), attributed to *Hadrosaurichnoides igeensis* Casanovas et al., 1993b.

In addition to the digitigrade ornithopod trackways there are two trackways 2TT6 and 2TT11, which are semiplantigrade or at least show metatarsal marks (Fig. 3.78). The metatarsus in 2TT6 has parallel medial and lateral margins with a rounded rear end, which suggests that the metatarsus supported the dinosaur's weight while walking. The mark

3.78. Ornithopod footprints with metatarsal marks, outcrop 2TT. (Left) 2TT6. (Right) Trackway 2TT11, ornithopod footprints, some with metatarsal impression. Acute rear portion of prints that is shallower than rest of track indicates metatarsal position was not horizontal.

from the metatarsus is significantly narrower than the width of the digitigrade foot, eliminating assignment to the ichnogenus *Deltapodus*. The 2TT11 footprints are larger than those of 2TT6 and also differ in the metatarsus area, which is pointed at the rear, suggesting that the metatarsus registered only because the foot sank deeply into the mud (Fig. 3.79B).

Different kinds of interaction between the trackmaker's foot and the substrate are evinced mainly in the back part of the footprint. The usual digitigrade tracks (Fig. 3.79A), in which the metatarsus is held off the ground by the trackmaker, have a short rear portion. If the foot sinks deeply into the substrate (Fig. 3.79B), the metatarsus will leave an elongated impression that will terminate in an angle. However, if the animal deliberately places its metatarsus against the substrate (Fig. 3.79C), the elongated extension of the print will have a rounded rear margin.

At 2TT, two ornithopod tracks with tail markings were observed, 2TT12 (Fig. 3.80) and the final part of 2TT16. The footprints are rounded, with broad, sinusoidal grooves superimposed on the tracks. No slippage striations were detected, and where there are expulsion edges, they are poorly marked. According to Torcida et al. (2003), the flat base and the width of the groove is typical of ornithopod dinosaurs. These two ornithopod trackways with tail markings would be two of fewer than 10 cases of this kind described so far (Jiménez Vela and Pérez-Lorente, 2006–2007). In La Rioja there are only two other trackways with tail markings, one in Era del Peladillo and the other in Peñaportillo.

OTHER FOOTPRINTS

The theropod footprints are both large and small. The largest (1TT1.8) measures 49.4 cm and the smallest (2BVG3.2) measures 24 cm. They have no special ichnotaxonomic features recognized to date. Sauropod footprints were also found, of which an example is the forefoot–hind foot pair in 3BVG3 (Fig. 3.77).

There are many prints whose rounded footprint outline and lack of structure on the base of the shaft made them unidentifiable. Many of them may be undertracks, or the behavior of the mud (low coherence) may give them this form. There is a set of rounded prints with grooves (Fig. 3.81) that has not yet been identified. They occur in a group of seven or eight rectangular footprints in two parallel lines connected by two channels. There are no criteria to suggest direction of movement or to suggest the type or types of appendages that left the marks. They must be related to some vital activity that is due to their regular distribution and type, although the nature of the trackmaker and the dynamics of the tracks' formation are a mystery.

Ichnotype Footprint Distribution

These sites show the general features typical of La Rioja tracksites. Small exposures (1TT, 2BVG outcrops) show footprints of a single type, or

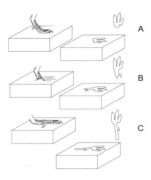

3.79. Different footprint morphologies reflecting different ways same foot can interact with substrate. (A) Typical digitigrade footprint. (B) Footprint with elongated "heel" made as digitigrade foot sinks unusually deeply into mud. (C) Semiplantigrade footprint with elongated "heel" with rounded rear margin. Dinosaur placed its entire metatarsus against substrate.

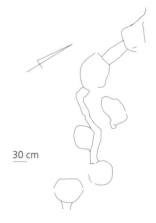

30 cm

3.80. Corral del Totico tracksite. Trackway 2TT12, ornithopod footprints with tail drag mark.

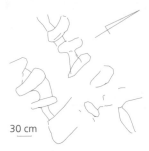

3.81. Corral del Totico
tracksite. 2TT9 trackway.
Subrectangular footprints?
accompanied by grooves.
Unidentified trackmaker.

30 cm

La Cuesta de Andorra

predominantly of a single type. Where the outcrops are larger in La Rioja (2TT, Barranco de Valdegutiérrez), a greater diversity of footprint types occurs. Site 1TT has shallow ichnites, and all are theropod footprints. Site 2TT has deep footprints, and many of them were made by ornithopods. This distribution of types of footprints according to the depth of penetration of the foot in the ground is also repeated.

One unknown factor is the existence of a strange track (Fig. 3.81) that has not been attributed to any animal so far. In interpreting the track, we assume that in 2TT there was always a layer of water covering the ground. There are no emersion marks (droplet impacts, desiccation cracks, plant root marks). There is no reason to suppose that all the trackmakers in 2TT traveled together at the same time, or that the water depth was always the same. If the water depth varied several meters, the number of animals that could leave marks and the water depth conditions are more extensive than those that are usually involved. Allowance should be made for other trackmakers.

History

We do not know who the first person to see the fossil tracks of La Rioja was. Presumably the dinosaur footprints in La Rioja have been known since the first settlers arrived. There are dolmens and many pre-Roman archaeological sites in the area. According to La Rioja University professor of philology Gonzalez Bachiller (unpubl. obs.), many place-names in the region end in -un (e.g., Navajún, Gatún, Bretún, Carnazún). These names are probably inherited from the Celtic word *dunum*, "village," and prefixes such as *nava-* are Celtic or Indo-European. The region therefore has been inhabited for well over 2500 years.

The site of La Cuesta de Andorra (AND) has been alongside a relatively busy road since ancient times, and so its footprints must have been seen. The first interpretations are not known, but oral evidence indicates that the people of Navalsaz and Ambasaguas thought the footprints were left by a giant lion (Pérez-Lorente, 2005). The footprints are similar to those of felids because they have three small, rounded forepads (without claw marks) and a larger one behind. The fact that cats have four, and not three, digital pads was apparently not considered important. Brancas, Martínez, and Blaschke (1979), who collected interpretations from Cameros residents regarding possible vertebrates who left the footprints, did not include lions among the candidates.

La Cuesta de Andorra was one of the first sites discovered and described in scientific publications in La Rioja. Casanovas and Santafé (1971) came to study it after hearing about the discovery of the footprints from a schoolteacher, Blas Ochoa, in 1968. At that time, it was common to assign to dinosaur footprints the name of the original animal assumed to have made them, in this case *Iguanodon*.

3.82. Part of La Cuesta de Andorra tracksite, with footprints AND1.15 through AND1.24 outlined in white. Bottom of Chusco ravine is between tree and trackway. Water flows from right to left.

Brancas, Martínez, and Blaschke (1979) recorded the site under the designation Lastra 6, mentioning three outcrops. However, from the number of footprints said to occur on the three outcrops (21 at 6A, 6 at 6B, 21 at 6C), Lastra 6 cannot be correlated with La Cuesta de Andorra because 6A and 6C also have the same number of footprints. Referring La Cuesta de Andorra to Lastra 6 was probably a typographical error. I have visited the site with two of the authors (Blaschke and Martínez) on several occasions, and they told me there were no more footprints on the site other than those shown in the study of Casanovas and Santafé (1971). The study of Brancas, Martínez, and Blaschke (1979) was the third published work on dinosaur footprints in La Rioja. Viera and Torres (1979) called it point 5, saying the site had already been studied by Casanovas and Santafé.

Site Description

The road (Fig. 3.72), a short distance from Navalsaz, has a steep slope called La Cuesta de Andorra. This runs parallel to a ravine (Barranco del Chusco), at the bottom of which is an elongated exposure with tracks. The position of the strata and the Barranco del Chusco is similar to the Barranco de Valdegutiérrez. On the east slope, there is a large expanse of gray limestone parallel to the topography of the hillside, similar to the site at Barranco de Valdegutiérrez. The slope is largely clear of brush and weathered detritus, but there are no footprints on the outcrops. In 2003,

3.83. Footprint AND1.22, showing typical morphology of prints in trackway AND1. Texture of study surface is partly due to presence of ripple marks, and partly to weathering and erosion of limestone. Seams of calcite fill joints in rock. Possible extrusion rim surrounds footprint. Print is marked by double outline. Outer line is drawing by Brancas, Martínez, and Blaschke (1979). Inner outline was drawn more recently.

the two sides of the Barranco del Chusco were explored and only two new, very small outcrops were found, one of which had sauropod footprints. The Cuesta de Andorra tracksite is on one side of the bottom of the ravine, and the trackway is parallel to the river course (Fig. 3.82). The distance between La Cuesta de Andorra and neighboring sites (Fig. 3.72) is as follows: 340 m from Barranco de Valdegutiérrez (2BVG); 350 m from Lastra 10; 460 m from Barranco de la Sierra del Palo, with the first semi-plantigrade footprints described in La Rioja; and 1400 m from Navalsaz.

Lithology

The rock with tracks is compacted limestone, which is black when freshly cut and yellow-gray on the outside. Apparently the layers are continuous and have preserved their thickness well. Its grouping (Enciso Group) is the same as Corral del Totico 2 (2TT) and Barranco de Valdegutiérrez. A notable difference is that there are desiccation cracks in the top of the layers at La Cuesta de Andorra, while there are none in 2TT and Barranco de Valdegutiérrez. Joints filled with white calcium carbonate (Figs. 3.82, 3.83) from the breakdown of the limestone are clearly visible.

The sedimentary environment is littoral–lacustrine in a setting not influenced by the waves produced by storms. The irregularities in the top surface of the rock (Fig. 3.83) are probably due to ripples. The rocks are emergent lacustrine carbonates (Doublet, 2004).

Track Features in Mud

The footprints were made by a large, stoutly built herbivorous dinosaur. The configuration of the trackway (Figs. 3.82, 3.84) indicates a heavy and slow animal with large footprints, relatively short steps, and a marked pigeon-toed walk. The shafts are shallow, contrary to expectation for large

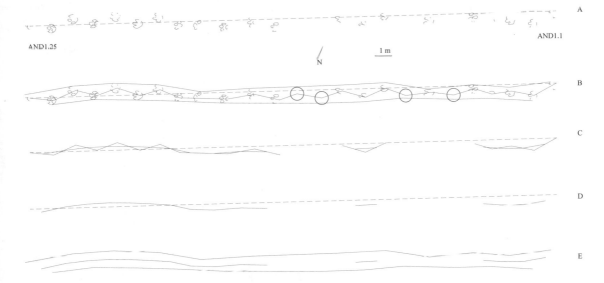

herbivorous dinosaurs. Theoretically, footprints made by heavy dinosaurs should be deep. This, however, is not always so: there are trackways with deep dinosaur footprints, whether the animal was large or small, and there are other trackways with similar types of tracks that are shallow. Trackways where the depth varies from one place to another are also known in La Rioja. All of this indicates that the depth of the tracks does not depend solely on the weight or size of the dinosaur. The simplest interpretations of this variation are the mud changing its physical properties according to the location (e.g., more or less dry), the surface mud properties changing over time; and the water depth varying from one place to another and from one moment to another.

3.84. Characterization of trackway AND1. (A) Sequence of footprints. As drawn here, dinosaur moved from right to left. Dashed line connects beginning and end points of trackway. (B) Trackway with line segments marking paces, and also outer width of trackway. Inferred position of missing footprints marked by circles. (C) Paces connecting preserved footprints in trackway, along with position of trackmaker's midline (determined by connecting midpoints of paces). Dashed line as (A). (D) Relationship between dinosaur's midline (solid curves) and overall orientation of trackway (dashed line), simplified from (C). (E) Relationship between outer width of trackway and trackway midline.

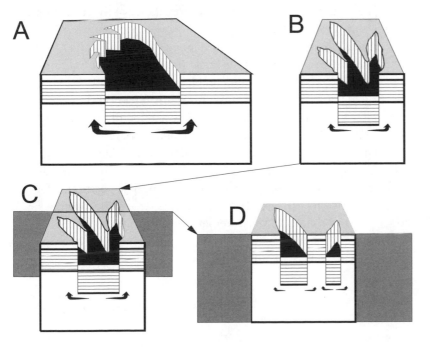

3.85. Resistance to flow of mud under feet of dinosaurs. Black arrows indicate resistance. (A) Wide footprint (sauropod). (B, C) Theropod footprint. (D) View of section that cuts theropod footprints under two toes. Resistance to mud flow is much higher under massive dinosaur pes.

It also seems logical that heavy dinosaurs are large, with big feet. The pressure (weight/area) exerted by all dinosaurs on the soles is a problem. The pressure under the soles of the feet of larger dinosaurs calculated by Alexander (1989) should be recalculated. It is logical that dinosaurs with wide toes or soles penetrate less in the mud than those with long, separate toes. This is because the flow of mud toward the sides under wide feet has longer to travel, and its mobility is therefore more limited. For example, the arrows in Fig. 3.85 represent the resistance to mud movement as a particle moves from beneath the foot toward the surrounding extrusion rims. The depth of the footprints depends on the physical properties of the ground and on the size and roundness of the soles of the foot.

Footprint Features (Ichnotypes)

The morphological features of the footprints are typical of ornithopod dinosaurs: the digits are marked by three broad pads with a rounded contour around the front margin of another larger pad, which contains the heel. Casanovas and Santafé (1971) classified them as *Iguanodon* footprints. They are currently considered ornithopod in the ichnogenus *Brachyguanodonipus* (Díaz-Martínez et al., 2009; Moratalla, 1993)

The AND1 footprints vary from 50 to 62 cm in length and 50 to 60 cm in width (Table.3.4). Deviation from average values in length and width are not correlated – that, is longer footprints are not always wider. This relationship can be seen in Fig. 3.86A, which shows the variation (above and below the mean value) of the length and width of the footprints. At the end of the trackway, the variation is similar, but the same cannot be said of the beginning. The footprints in the final part of the trackway are in better condition, and the boundaries between the digital pads are better defined. It is postulated that the irregularity of the first footprints has nothing to do with the pes morphology. The extrusion rims around all the footsteps are generally poorly defined.

Morphological features of the tracks (footprint outline, pad and claw marks) are irregular and are not complete in all of them. Three rhomboid digital pads with a broad and rounded heel pad can be seen in the footprints. There are rounded nail marks. The length and divarication of digits are difficult to measure as a result of the imprecision of the location of hypexes and the rounded shape of the distal digital ends. Digital pads II, III, and IV apparently have the same length, are slightly deeper at the front, and are accompanied by a small ungual constriction (visible in AND1.3, AND1.16, AND1.22, AND1.23, AND1.24, AND1.25). Applying Thulborn's (1990) formulas to measurements taken from footprints of large ornithopods, the height of the limb (H) at the acetabulum was estimated, with an average of approximately 390 cm.

Table 3.4. Measurements (cm, degrees, km/h) of trackway AND1 at La Cuesta de Andorra.

Footprint	l	a	P	Ap	z	O	Ar	Lr	II^III	III^IV	II^IV	H	z/H	v1	v2	z/l	(l – a)/a	Ar/a
AND1.25																		
AND1.24	52	55	136						37	30	67	347					-0.05	
AND1.23	59	59	129	147	253	0	19	96	28	27	53	397	0.6	2.7	3.6	4	0.00	0.32
AND1.22	60	50	131	150	252	-11	15	91	24	30	54	404	0.7	2.7	3.6	4	0.20	0.33
AND1.21	59	59	144	140	259	-8	22	100	24	24	48	397	0.7	2.8	3.7	4	0.00	0.38
AND1.20	59	56	127	145	257	-6	20	98	27	25	52	397	0.7	2.8	3.7	4	0.05	0.36
AND1.19	62	59	126	143	241	-10	20	99	23	30	53	419	0.6	2.5	3.4	4	0.05	0.34
AND1.18	50	50	119	157	240	0	11	89	22	30	52	333	0.6	2.5	3.4	4	0.00	0.23
AND1.17			126	170	280		6	76					0.7	3.2	4	5		
AND1.16	56	50	167	164	325	-5	11	85	25	29	54	376	0.8	4.1	4.6	6	0.12	0.23
AND1.15			145	160	307	0	12	90	30	27	57		0.8	3.8	4.4	5		
AND1.14		55				0												
—																		
AND1.11			150															
AND1.10			140	138	269	-15	29	100						3.0	3.8	5		
AND1.9					270				32				0.7	3.0	3.8	5		
AND1.7					260				24				0.7	2.8	3.7	4		
—																		
AND1.5	61	60	109									412					0.02	
AND1.4			125	155	227	-14	14	81	25	25	50	422	0.6	2.3	3.2	4		
AND1.3	62	55	130	142	200	-14	11	72	29	28	57	350	0.5	1.8	2.8	3	0.14	0.20
AND1.2	53	59			243		21	94					0.6	2.5	3.5	4	-0.11	0.36
AND1.1																		
Average	58	56	134	151	259	-7	16	90	27	28	54	387	0.7	2.8	3.7	4	0.04	0.30

Note: Abbreviations as in Table 2.1.

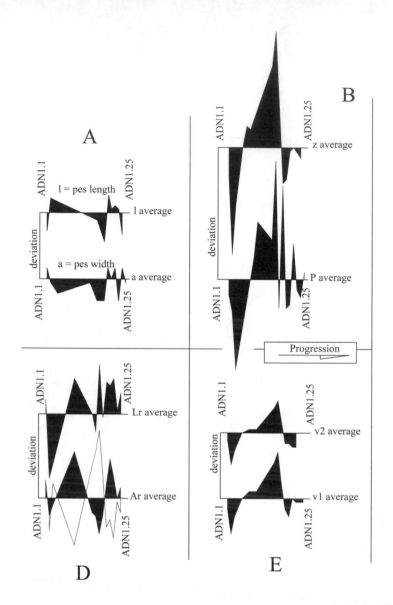

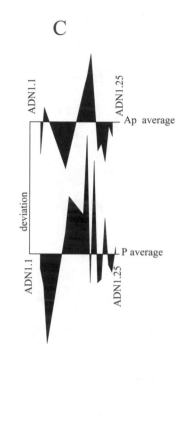

Trackway Features

The trackway measures 32 m and would have 25 prints if four intermediate ones were not missing (numbers 6, 8, 12, 13). It is one of the longest bipedal trackways in La Rioja. Trackway measurements are provided in Table 3.4. The pace (P) ranges from 109 to 167 cm (average 134 cm) and the stride (z) from 200 to 325 cm (average 264 cm). The pace angle (Ap) varies between 131° and 170° (average value 151°). The trackway deviation (Ar) is between 11 and 29 cm (mean 16 cm) and the trackway width (Lr) between 72 and 100 cm (mean 90 cm).

Variations in trackway parameters (Fig. 3.86) are quite substantial, typical of an animal with an irregular gait. The pace and stride lengths (Fig. 3.86B) are, unsurprisingly, directly and proportionally related: the higher the P, the higher the z. However, Ap behaves differently than expected at the beginning of the trackway (a higher value of P should lead to higher Ap). At the end of the trackway, the variations of Ap and P match (Table 3.4).

The variation in the values of Ar and Lr are represented in Fig. 3.86D (solid black fields). The variation must be consistent with and match Ap (Fig. 3.86D). The greater the Ap, the less Ar and Lr should be. The overlap in the broken line of Ap on the graph of Ar shows the expected inverse correlation. The variation of the three values indicated (Ap, Ar, Lr), which are out of synch with the pace and the stride lengths, implies irregularity in the layout of the trackway. This means that in this trackway, the longer (or shorter) paces/strides (P/z) do not consistently covary with larger or smaller values of the pace angle (Ap), or consistently vary with the width of the trackway (Lr) or separation of the feet (Ar) from the midline.

Footprint and Trackway Data

According to data from the works of Sternberg (1932) and Haubold (1971), dinosaurs with thin limbs have a high z/l value (cf. Pérez-Lorente, 2001a). The average value of the index for this trackway is 4, which suggests a dinosaur with thick limbs. The ratio z/l is between 3 and 6 at the beginning of the track and stays constant at 4 at the end of it.

According to the formulas of Alexander (1976) and Demathieu (1984, 1986, 1987a), velocity is a function of the stride, and consequently the comparison of v_1, v_2, P, and z are in complete agreement. Thulborn's (1990) ratio of relative stride (z/H) is used for calculating speed. The average value of this in AND1 is 0.7, which is consistent with an animal walking slowly. The walking speed of this dinosaur is quite low and is in accordance with the data from Casanovas et al. (1995a), which suggests that large bipedal dinosaurs were not those going the fastest in La Rioja (Fig. 3.170).

All data from La Rioja sites published by our team have calculations for the theoretical trackmaker speed according to the formulas of Alexander (1976), Demathieu (1984, 1986), and Thulborn (1990). A study of the variation of speed according to the size of the ornithopod footprints was carried out (Casanovas et al., 1995a; Extremiana and Lanchares, 1995). This study (55 ornithopod trackways; 10 trackways have pes length of more than 50 cm; seven of these trackways are at La Rioja sites) concluded that the minimum speed (1.1 km/h) is found in dinosaurs whose pes length is 50 cm (ranging between 45 and 55cm). The average speed in AND1 is between 2.8 and 3.7 km/h, and fits with the data.

Deviation of Midline

All long biped dinosaur trackways are sigmoid and do not follow a straight line. The oscillation of the midline is sometimes so clear that it seems to

3.86. Analysis of trackway AND1, La Cuesta de Andorra. (A–E) Deviations from mean value along trackway of various trackway parameters. Mean value is shown as horizontal line; deviations above this line indicate values greater than mean. Trackway begins with footprint AND1.1 and ends with AND1.25. (A) Footprint length (l) and width (a). (B) Trackway pace (P) and stride (z). (C) Trackway pace (P) and pace angulation (Ap). (D) Trackway width (Lr) and trackway deviation (Ar); pace angulation is plotted along with Ar as line, rather than solid black field of Ar. (E) Estimated walking speed of trackway according to methods of Alexander (v_1) and Demathieu (v_2). Average values of trackway parameters: $l = 58$ cm; $a = 56$ cm; $P = 134$ cm; $z = 264$ cm; $Ap = 151°$; $Ar = 16$ cm; $Lr = 90$ cm; $v_1 = 2.8$ km/h; $v_2 = 3.7$ km/h. See Table 3.4 for details.

be caused by factors other than the dinosaur itself (obstacles, trajectory adjustments in herds or in individuals traveling together, interfere with each other's progress). In all cases, the change is highlighted by drawing a straight line linking the center of the first and last tracks (Fig. 3.84). The deviation of the midline with respect to a straight line between AND1.1 and AND1.25 is normal and even lower than that found in other long dinosaur tracks (cf. Pérez-Lorente, 2003c). It is only pronounced in the first part of the trackway. The faltering progress in the trackway may be due to the trackmaker being old or sick or having a malformation.

The following conclusions can be drawn from the measurements (Table 3.12): all footprint and trackway values in AND1 are variable; the speed of travel is low and variable; some data show divergence from a regular trackway (see the relationships between l – a, or between P, z, Ap, Ar, Lr); and finally, although the variations and irregularities are clear in the initial part of the trackway, the final part is regular. This leads to the hypothesis that the irregularities were caused by the dinosaur being old or sick, or unable to defend itself.

Main Features of the Site

La Cuesta de Andorra was one of the first sites described in La Rioja. It may also have been one of the first with an oral tradition. The pathway has the slope called Cuesta de Andorra, which is the natural route between two large areas of different economic activity. The first is to the east, where esparto, cereals, and fruits are grown. However, this is an arid region, and rivers flowing through the area dry up quickly. The second area is to the west, where sheep farming and its more important associated woolen industry are located; along the Cidacos River, milling and fulling were carried out. The footprints and trackway run parallel to the pathway and are well marked. The path was well used in the area, and the tracks stand out clearly in the rock. The tracks were consequently seen, sparking speculation about their origin.

The second important feature is the length of the trackway. Information obtained from long trackways is all the more important the greater their length. Morphometric variations visible in the footprints of the same dinosaur and the variations in measurements of trackway (pace, stride, other data) allow us to better study the trackmakers. These outcrops also demonstrate the degree of imprecision that can be found when working with few footprints. The number of minutes of life of the dinosaur and conclusions about its behavior increase in proportion to the length of the trackway.

La Cuesta de Andorra has only one trackway, that of an ornithopod dinosaur. It does not follow that this dinosaur was alone and not part of a larger group because the bed outcrop shows only a thin strip limited to the bottom of a ravine to the west, and by the erosion that destroyed the strata area to the east. Indeed, the strip of rock that remains

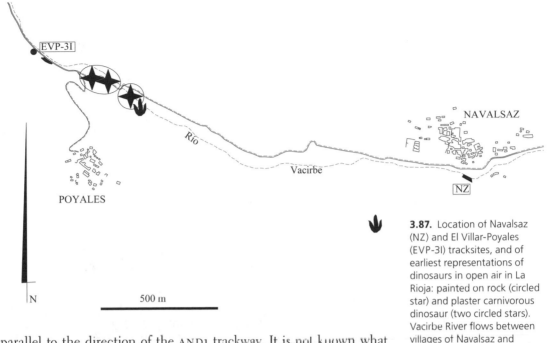

3.87. Location of Navalsaz (NZ) and El Villar-Poyales (EVP-3I) tracksites, and of earliest representations of dinosaurs in open air in La Rioja: painted on rock (circled star) and plaster carnivorous dinosaur (two circled stars). Vacirbe River flows between villages of Navalsaz and Enciso, near several important tracksites.

is parallel to the direction of the AND1 trackway. It is not known what was in the part that eroded, or what may remain in the part still buried underground.

History

The Navalsaz site (NZ) (Fig. 3.87) was first reported by Brancas, Martínez, and Blaschke (1979), who cited 79 tracks spread across three units that we have failed to identify. Within them are two important trackways, 2-A(I) and 2-A(II). This site was not mentioned in other studies made in the area (Casanovas and Santafé, 1971, 1974; Viera and Torres, 1979). Casanovas et al. (1993c) identified and studied 138 tracks (Fig. 3.88). The site was given the initials NZ and mapped. Correlation with the trackways defined by Brancas, Martínez, and Blaschke (1979) was established, such that NZ4 includes 2-A(I) and the last footprints in NZ8 are 2-A(II). More recently, there have been two publications, one with photographs and diagrams (Pérez-Lorente et al., 2001), and another with a compilation of sites that added a few more tracks (Pérez-Lorente, 2003b). The tracks are not as evident in this smooth rock as in Cuesta de Andorra, so although they are closer to town, they were less well noticed. The legend of the giant lions did not apply to this site because deeper and older tracks (NZ9) were hidden until 1992.

Description of Location

Navalsaz is in a stratum tilted 30° to the north, at the base of which a stream flows (Fig. 3.89), which becomes the Vacirbe River. Between

Navalsaz

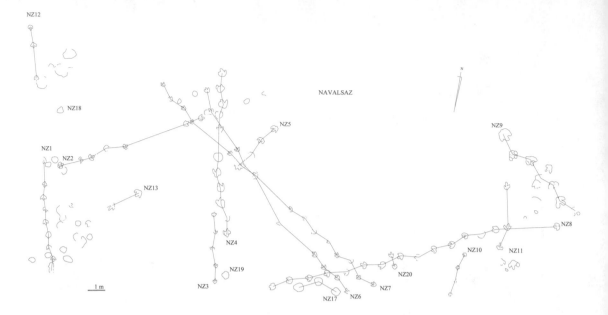

3.88. Map of Navalsaz (NZ) tracksite (Casanovas et al., 1993c). Symbols that identify trackways and isolated footprints are indicated.

Navalsaz and Enciso along this river are several important sites, including Barranco de Valdecevillo and El Villar-Poyales. Access to Navalsaz is easy because it is next to a paved road from Enciso, which goes through the listed sites and continues to the road leading to Los Cayos. To the east, nearby sites are Cuesta de Andorra, 1.4 km away, and Barranco de la Sierra del Palo, 1.2 km away. To the west, the first footprints described are 1.9 km away at El Villar-Poyales.

The tracks in Navalsaz are arranged in 14 trackways (NZ1 to NZ14), two isolated footprints (NZ18, NZ19), seven highly trampled patches, and an unidentified sequence. There are 12 ornithopod trackways and two theropod trackways (Fig. 3.88). The surrounding area contains a number of unidentified and apparently distorted footprints with rounded outlines. There is a sequence of irregular holes, parallel to NZ3 and NZ4, which have not been interpreted because no one knows whether they are tracks or sedimentary structures. The marks are elongated and partially destroyed.

Lithology

The footprints are in the top of a layer of dark micritic limestone, of the same type as at La Cuesta de Andorra, Barranco de Valdegutiérrez, and 2TT. As in previous cases, the thickness is greater than 25 cm and is cut by parallel joints filled with white calcium carbonate, which stands out after weathering. There are no sedimentary structures such as mud cracks visible to the naked eye. The roughness of the study surface may, however, be due to interference from ripples.

The lithology of the limestone is similar to that of La Cuesta de Andorra, Barranco de Valdegutiérrez, and 2TT, and likely was deposited under the same conditions, i.e., a lacustrine environment protected from

3.89. Overview of tracksite, before NZ9 trackway was discovered. Vacirbe River runs along base of Navalsaz tracksite.

Track Features in Mud

The depth of the tracks differs among the trackways. Some trackways have footprints deeper than others, even when they are the same size. In general, the footprints are shallow, with only the largest of them impressed more than 5 cm.

The upper shaft footprint contour is not sharp but rounded, so it is difficult to draw print outlines. Brancas, Martínez, and Blaschke (1979) wrote that the tracks were eroded, and for this reason they were now larger than they were originally. Despite the rounded contour of the footprint wall there are many prints with clear digital pads (one per digit and one more in the heel). Notches in the rear can be distinguished, usually between digit II and the heel pad.

The difference in depth of the footprints has nothing to do with their position on the site, as there are trackways with shallow and deep footprints that cross each other. No distribution of the tracks according to their structure has been deduced. Footprints can be distinguished only as being large or small, with or without digital pads, and with rounded or elongated digits. There is no preferred orientation of the trackways (Fig. 3.88).

The footprints may not be stamps; despite this, however, the digital pads are clear in many footprints. There may have been a thin layer above

3.90. Uninterpreted structures parallel to trackways NZ3 and NZ4. These structures are not mapped in Fig. 3.88. Note also peeling of rocks by weathering and colonization by lichens. Drawn squares are 30 cm on each side.

the study surface on which dinosaurs trod, i.e., the tracking surface may have been separated by a thin layer of sediment from the study surface. This thin layer – perhaps a laminated algal layer – must have had a consistency that enabled the pad structures to be transmitted. The extrusion rims around all the footprints are not well pronounced.

Track Features (Ichnotypes)

Only two types of footprints have been identified: ornithopod and possibly theropod. There is a sequence of prints with an elongated boundary (Fig. 3.90), probably with extrusion rims and mud sliding structures, whose origin (tracks or sedimentary structures) has not yet been established. Most of the ichnites are ornithopod footprints: they have digits with only one rounded pad, a rounded heel, and footprints that are as wide as they are long. Their ichnogenus is not clear owing to their different diagnostic features (*Brachyguanodonipus*, NZ9.1, NZ4.3; *Iguanodonipus*, NZ4.7), even in the same trackway. The ratios z/l and Ar/a (less than 0.3) are low for all of them, with the latter indicating a relatively narrow trackway with the feet treading the midline. The length of the tracks ranges between 25 cm (NZ3.4) and 75 cm (NZ9.1, Fig. 3.91). The average length of the footprints in trackway NZ9 is 63 cm, and the average width is 69 cm (Fig. 3.92, Table 3.5).

A theropod identification been assigned to those prints with digits long enough to not constitute a subcircular pad. The theropod footprints

NZ9

1 m

(NZ6, NZ7) are located in the center of the site (Fig. 3.88). The trackways are almost parallel but cross halfway along their lengths. NZ7 has theropod features (elongated, tapered digits) that are more pronounced than NZ6. Two trackways (NZ3, NZ10) with small footprints have some features similar to theropod trackways (e.g., the ratio z/l is high in both), but the footprint outline is ornithopod-like (Table 3.5). They may be shallow undertracks of theropod footprints.

Track Features and Size of NZ9.1

NZ9.1 is 75 cm long, which makes it one of the largest ornithopod footprints in both La Rioja and the world (Fig. 3.91). Ornithopod footprints in La Rioja range from as much as 80 cm in length (Sol de la Pita; Moratalla, Sanz, and Jiménez, 1997b) to as little as 9 cm (5PL2.1). Large ornithopod footprints are not usually stamps, and they generally have significant variations with respect to the true size of the foot of the dinosaur

3.92. Footprint NZ9.6 from Navalsaz site. (A) Overhead view. (B) Oblique view. This ichnite has well-defined ornithopod characters. One of the largest ornithopod footprints of La Rioja (l = 64 cm, a = 74 cm).

Table 3.5. Measurements of Navalsaz site trackways.

Print	l	a	II	III	IV	II^III	III^IV	P	Ap	z	O	Ar	Lr	H	z/H	v2	v1	(l − a)/a	z/l	Ar/a
NZ1	33	33	13	17	14	28	32	71	166	141	4	4	43	215	0.7	2.7	2.0	0.01	4.2	0.15
NZ2	36	34	19	17	18	8	27	96	165	184	−4	6	61	235	0.8	3.4	2.9	0.01	5.1	0.22
NZ3	27	26	12	16	11	25	28	96	166	192	0	7	41	174	1.1	4.1	4.4	0.11	7.0	0.31
NZ4	53	50	22	24	22	3	29	85	157	165	−8	8	62	355	0.5	2.5	2.3	0.07	3.1	0.16
NZ5	40	42	19	23	24	18	23	103	159	205	−10	9	58	262	0.8	3.5	3.0	−0.06	5.1	0.21
NZ6	34	31	188	19	14	19	35	99	155	185	−13	11	44	22	0.8	3.5	3.1	0.07	5.4	0.34
NZ7	30	27	15	18	11	21	29	96	150	182	−6	12	53	147	1.2	4.2	4.9	0.17	6.0	0.47
NZ8	39	46	17	17	15	19	29	106	159	209	−19	9	56	254	0.8	3.7	3.3	−0.08	5.4	0.19
NZ9	63	69	26	25	31	13	32	105	143	179	−10	16	115	424	0.4	2.4	1.4	−0.08	2.9	0.25
NZ10	25	22	12	13	8	6	32	83	164	162	−3	5	31	120	1.3	4.1	5.1	0.11	6.5	0.32
NZ11	44	39	15	24	16	41	16	123		232				288	0.8	3.8	3.3	0.12	5.3	
NZ12	36	32	15	17	13	13	17	101		194				234	0.8	3.5	3.1	0.05	5.4	
NZ13	50	54	15	24	15	19	34	167						333				−0.10		
NZ17	54	50						122						360	0.6	3.4	2.6	0.05	4.3	0.05

Note: Abbreviations as in Table 2.1.

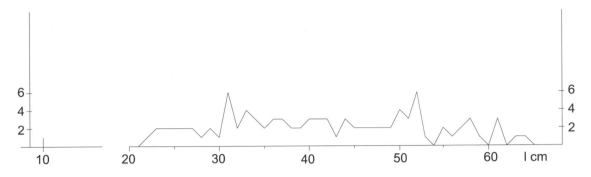

(the length of NZ9.6 is 64 cm; Fig. 3.92). The mean values from the trackways are more likely to be correct. Thus, the average length in NZ9 is 63 cm (Tables 3.5, 3.6). In total, 97 ornithopod trackways consisting of 220 footprints have been reported from La Rioja. In some trackways, only one footprint could be measured, as occurs with the one whose length is 64 cm. This length is not the average length of the three footprints in 2HR12 but the length of 2HR12.2, which is the only print that could be measured.

Large theropod dinosaur footprints have been cited in many places (Boutakiout et al., 2008; Boutakiout, Hadri et al., 2009), but there are few works that cite giant ornithopod footprints. The earliest such is *Tyrannosauropus petersoni* Haubold, 1971 (81.28 cm), first described by Peterson (1924), who attributed them to theropod dinosaurs. Lockley, Young, and Carpenter (1983) assigned them to ornithopods, and more accurately to hadrosaurid dinosaurs. Farlow and Hawthorne (1989:29) also found an ornithopod trackway "consisting of footprints some 70–80 cm long"; however, as at the Navalsaz site, "these are gosh marks." Following this was the finding of a cast in the area of south Préjano that measured 80 cm in length (Sol de la Pita; Moratalla, Sanz, and Jiménez, 1997b). In Regumiel, in the province of Burgos, large ornithopod footprints are also abundants. In this locality, trackways of quadrupedal ornithopods occur, with one of them showing the mark of the tail. Finally, Boutakiout, Ladel et al. (2009) cite an ornithopod footprint 80 cm long in the Upper Jurassic of Iouaridene (Morocco).

Of all these tracks, *T. petersoni* and the one in Sol de la Pita are casts, and would be larger than NZ9.1. The Iouaridene footprint was visited during a workshop (Dinosaurs in Demnat, held in 2009), and many experts doubted the attribution of this footprint to an ornithopod. The alternative is that it could be a sauropod footprint. NZ9.1 therefore may be the world's largest ornithopod footprint. No systematic worldwide analysis of the dimensions of ornithopod footprints has been performed. A size–frequency distribution for La Rioja seems to show two peaks, in the vicinity of 32 and 51 cm (Fig. 3.93). Between them is a wide depression. On either side of the two peaks are drops in abundance separating them from two other areas with footprints, one between 22 and 28 cm and the other between 54 and 64 cm.

Table 3.6. Ornithopod trackways of La Rioja arranged in decreasing order of length of footprints.

Trackway	l	O	z	z/l	H	z/H	v1	v2	Prints
2HR12	64	−7	240	3.8	430	0.6	2.2	3.2	3
NZ9	63	−9	179	2.9	424	0.4	2.4	1.4	5
CT2	61				413				2
CT4	61	−11	227	2.7	413	0.5	2.1	3.1	3
CT5	61	−14	181	3.0	413	0.4	1.4	2.5	3
LMG	59		237	4.0	258	0.9	4.1	4.4	7
CBC1	58	−18	227	3.9	239	0.9	4.0	4.1	5
CT6	58	−11	176	3.0	391	0.4	1.4	1.0	4
AND1	57	−7	264	4.6	230	1.1	3.7	2.8	21
STS21	57	−22	244	4	428	0.6	2.1	3.0	3
CT1	56	−7	194	3.5	377	0.5	1.8	2.8	4
2TT14	55		206	3.7	243	0.8	3.3	3.8	6
DR	55								4
MLV151	53	−9	149	2.8	351	0.4	1.3	2.2	9
2TT7	52	−5	148	2.9	345	0.5	1.5	2.5	4
NZ4	52	−8	165	3.2	349	0.5	3.2	1.4	11
BLC1	52	−22	183	3.5	349	0.5	1.8	2.5	30
UP3	52				347				2
LCH1	52		181	3.0	347	10.5	1.7	2.7	3
2TT16	51	−13	162	3.2	341	0.5	1.6	2.4	29
VA30	51	0	248	4.8	340	0.7	3.1	3.8	16
UP5	51	−3	272	5.3	342	0.8	3.5	4.1	3
15SV	50								4
2TT2	50	9	145	3	332	0.4	1.2	2.1	4
CT3	50				332				2
LCH2	50		175	35	333	0.5	1.7	2.7	3
2TT11	49	−1	138	2.8	320	0.4	1.2	2.1	11
PP150	49		186	3.8	294	0.6	2.1	3.0	7
LVC41	48		176	5.3	318	0.5	1.9	2.8	5
UP1	48	−10	218	4.5	320	0.7	2.7	3.4	14
2TT3	47	−6	116	2	311	0.3	0.7	1.5	12
1PL12	46		139	3.0	308	0.5	1.5	2.2	8
BLC5	46	−10	126	2.7	305	0.4	1.1	2.0	3
UP6	46	−10	229	5.0	302	0.8	3.1	3.7	4
CIV14	45	−5	229	5.1	311	0.8	3.7	4.2	3
2TT1	45		277	4		0.9	3.9	4.1	2
BLC6	44	2	183	4.2	293	0.6	2.2	3.0	3
VI	44								2
CVF1	44	−10	181	4	294	0.6	2.1	3.0	3
BLC3	43		204	4.7	286	0.7	2.7	3.4	5
TJG-R1	42								6
VAE3	42	−22	270	6	273	1.0	4.6	4.6	5
LCB4	42	0	187	4.4	280	0.7	2.4	3.1	6
1TT3	41	2	177	4.3	300	0.7	2.0	2.9	5
2TT4	41		96	2	266	0.4	0.8	1.6	2
CVF2	41	1	199	5	268	0.7	2.8	3.4	8
5PL13	40	−1	187	4.7	262	0.7	2.6	3.2	6
UP2	40	−3	239	6.0	262	0.9	2.9	4.1	4
14SV	40								5
2TT6	39	−3	117	3.0	183	0.6			5

Trackway	l	O	z	z/l	H	z/H	v1	v2	Prints
NZ8	39	−19	209	5.0	254	0.8	3.2	3.7	15
1PL11	38	−16	176	4.6	250	0.7	2.3	3.0	4
2PL173	38	−4	183	4.6	246	0.7	2.7	3.3	4
3PL21	37	−19	160	4.3	243	0.7	2.2	2.8	6
BLC2	37	−10	197	5.3	244	1.2	3.1	3.5	6
UP4	37	−12	194	5.2	244	0.9	3.0	3.5	3
NZ2	36	−4	184	5.0	235	0.8	3.9	3.4	7
BLC7	36	−7	219	6.1	221	1.0	4.1	4.1	5
LBG2	36	2	258	4.7	170	0.7	2.3	3.0	19
CIV5	35	−4	220	6.3	186	1.3	5.0	4.5	5
2SM4	35		206	5.7	229	0.9	3.6	3.8	3
3PL19	34	−12	97	2.8	220	0.4	1.1	1.8	12
UP7	34				218				2
LBG1	34	−19	243	5	177	0.7	2.6	3.2	5
CIV1	33	0	250	7.6	212	1.3	6.4	5.0	3
NZ1	33	4	142	4.0	215	0.7	2.0	2.7	9
VA9	33	−5	201	6.0	213	0.9	3.7	3.9	6
VA25	33	−8	176	5.3	215	0.8	2.9	3.4	5
4PL49	32		158	4.8	211	0.7	2.5	3.1	7
5PL12	32								2
3PL8	31	−7	148	4.1	234	0.6	2.0	2.7	4
3PL23	31	−12	84	2.6	204	0.4	0.9	1.6	4
4PL45	31		157	5.0	202	0.8	2.6	3.1	7
4PL51	31		151	4.9	198	0.8	2.5	3.0	4
VA11	31	3	144	4.6	204	0.7	2.2	2.8	5
2SM5	31				201				2
3ST6	30	−2	167	5.6	193	0.9	3.1	3.4	3
5PL7	29								2
SVR6	29	0	196	6.8	185	1.1	4.2	4.0	3
3PL22	28	0	141	5.0	180	0.8	2.5	2.9	9
NZ3	27	0	191	7.0	173	0.9	4.4	4.5	5
1ST11	27	9	141	5.0	176	0.8	2.6		3
3PL5	26		106	4.1	163	0.6	1.8	2.5	3
BLC4	26		142	5.5	168	1.0	2.8	3.1	3
NZ10	25	−3	162	6.9	136	1.2	4.4	3.9	4
3ST2	25	−15	173	6.9	128	1.3	5.3	4.3	5
1ST16	24				150				2
1ST17	24	16	153	6.4	150	1.0	3.6	3.5	3
3ST1	23	−3	170	7.4	150	1.1	4.3	3.9	4
3ST3	23		145	6.3	118	1.2	4.3	3.8	5
VA10	22		100	4.0	138	0.7	1.9	2.4	3
5PL2	10	−6	52	5.3	45	1.2	2.4	2.2	4
2TT8		−6	158						6
LCB2									2
LCB3		−14	137						3
2HR9									3
									220

Sources: Data compiled from the literature.
Note: Abbreviations as in Table 2.1. Evident are large negative values of orientation; low values of the z/l ratio (indicating thick to very thick limbs), and very low estimated speed.

Trackmaker Size and Speed

Applying Thulborn's (1990) calculations for the limb length of large ornithopod dinosaurs, the length of dinosaur NZ9 is in the range of 372 to 426 cm, which would mean it was a giant – perhaps 13 m in length. The speed range has also been calculated for this dinosaur (Fig. 3.170), in accordance with the data available. Increasing the size of the footprint decreases the calculated velocity with respect to shorter footprint lengths.

Footprint Distribution by Ichnotype

Ornithopod footprints dominate in this tracksite. Twelve trackways and most of the isolated ichnites inside the boundaries of the site are ornithopod footprints. There are only two theropod trackways and no sauropod footprints, with one possible trackway of unidentified footprints to be determined (Fig. 3.90).

Before assuming that the abundance of ornithopod footprints is due to herd behavior, the following questions should be considered. The concentration in this case does not necessarily imply any kind of common purpose: the footprints are of variable size, and the trackway paths are neither parallel nor trending in the same direction. However, if it is a group of dinosaurs in a herd, it can be added to similar ornithopod concentrations, such as that in 5PL, where the size of the footprints and the direction of trackways are also variable. Moreover, some tracks are deeper than others. There are trackways with deep footprints and others all of whose prints are shallow. There are several different factors to consider: variation in physical properties of the soil as the dinosaurs passed (passing at different times); variation in size of the dinosaurs; variation of paths; and possible variation of the trackmaker taxon. However, it can be considered that the physical properties of the sediment were not the same throughout the site and that the dinosaurs were part of a heterogeneous group regarding the size (and presumably age) of individuals, and at that time they were dispersed, without a preferred direction of movement. The hypothesis that the medium is suitable for bipedal ornithopods, and that therefore a number of socially unrelated individuals came to this place (not all at once), cannot be ruled out.

Site Environment

The study surface descends to the north under the bed of the Vacirbe River. Digging could be continued to the east and west in the search for new footprints. In surveys made within 200 m of the site, footprints have been seen in small areas of stratification surfaces. These new locations are not good for digging because the strata dip steeply into the subsoil.

Although the rock is relatively protected from the wind and sunlight, the layer is being destroyed (deplating and scaling) as the rock slowly disintegrates (Fig. 3.90). This is not due to joints or cracks that can be identified with the naked eye, so treating it is problematic. An attempt will be made to channel rainwater from the study surface by diverting it toward the sides in the upper stratum. We assume that this will prevent some of the rock erosion, although the role of lichens and similar organisms that can live inside the layer will have to be investigated. If these organisms are the major cause of the erosion, the outcrop conditions must be made as dry as possible.

Main Features of Site

Navalsaz is probably a site of undertracks, which clearly degrades many diagnostic structures to be used for ichnotaxonomic classification. The problem is the assignment of *Brachyguanodontipus* or *Iguanodonipus* as a result of the change in the outline of the footprints in the same trackway. In my opinion, it is not appropriate to define a new ichnogenus or new ichnospecies for these prints because one of them is of exceptional length. NZ9.1 is one of the world's largest ornithopod tracks.

This is one of the areas with the highest concentration of ornithopod footprints in La Rioja. There are places where the features (size and shape) of the ornithopod footprints are similar (1PL, El Contadero, La Barguilla, Los Chopos, Valdemayor). In others (5PL, Navalsaz), the ornithopod footprint concentrations are quite different, at least in size. In both cases, the trajectories have different orientations. The possibility of heterogeneous groups (of variable size) of ornithopods should be considered in these sites, perhaps also conditioned by the favorable habitat for the trackmakers.

History

The El Villar-Poyales (EVP) site does not consist of a single layer with tracks but rather several that have been given different names. The authors who have worked here have all described one to three outcrops, but they are not always the same outcrops (Table 3.7). In addition to the four layers there are at least three more, of which one is recognized for leaving loose fragments of casts on the debris. The first study, that of Casanovas and Santafé (1974), described an outcrop (Fig. 3.94) that has been successively called Lastra 1-A, 1a, and Icnitas 3 (3I). The second study, that of Brancas, Martínez, and Blaschke (1979), again described Lastra 1-A, but also Lastra 1-B (Fig. 3.95) and Lastra 1-C (Fig. 3.96). This study made a new contribution by finding six new footprints in outcrop 3I, five which are notable for their small size (Figs. 3.94, 3.97). In La Rioja, only at Barranco de Valdebrajés are there smaller theropod footprints. The third study was that of Viera and Torres (1979), who worked on the same outcrops and found more footprints in Lastra A (or 3I), which they called

El Villar-Poyales and Icnitas 3 Footprints

Table 3.7. Correlation between outcrops and trackways cited by authors who have worked El Villar–Poyales track site.

	Casanovas and Santafé (1971)	Brancas, Martínez, and Blaschke (1979)	Viera and Torres (1979)	Casanovas et al. (1992a)	Casanovas et al. (1993d, 1998)	Current name	Figure
East		Lastra 1-C (21 footprints)	Icnitas 4	1c (3 trackways)		1EVP	3.96
						2EVP	3.96
					EVP (131 footprints, 15 trackways)	EVP	3.96
		Lastra 1-B (7 footprints)	Icnitas 3	1b (2 trackways)		3EVP	3.95
						4EVP	
West	19 footprints	Lastra 1-A (25 footprints), footprints 3		1a (28 footprints, 5 trackways)	3I (22 footprints)	3I	3.94, 3.97

3.94. El Villar-Poyales. Outcrop 3I. Theropod stamps with marks of pads and claws. On right is trackway 3I2, which contains five small theropod prints. Ripples that cross study surface can readily be seen. Scale bar marked off in 10 cm intervals.

outcrop 1a. Casanovas et al. (1992a) reviewed 3I (Lastra 1-A), and they later found a new outcrop, Icnitas 4 (El Villar-Poyales, located between Lastra 1-B and Lastra 1-C) (Casanovas et al., 1993d). Later still, they broadened El Villar-Poyales by digging at the suggestion of Haubold to confirm the lack of forefoot prints in one of the trackways (EVP1) (Casanovas et al., 1998). Meanwhile, Pérez-Lorente (1993a) examined the semiplantigrade trackways in El Villar-Poyales.

After taking an inventory of the site in La Rioja in 1996, the nomenclature was standardized. However, the process has been difficult. It was proposed that all the outcrops be grouped under the name of El Villar-Poyales and the letters EVP used for all the samples, which was done in some studies. However, in the first official inventory, 3I was used for outcrop Lastra 1-A, 1a, or Icnitas 3, and EVP for the rest. It was decided in the subsequent inventory to continue with this nomenclature, i.e., 3I for one and EVP for the rest (Table 3.7). El Villar-Poyales itself is still being studied. Lastra 1-C, of Brancas, Martínez, and Blaschke (1979), contains more footprints and trackways than have yet been described (Fig. 3.96).

Description of Site

The site is alongside the road (Fig. 3.73, 3.98) from Enciso that goes through the sites of Barranco de Valdecevillo and Navalsaz, passes near Cuesta de Andorra and Barranco de Valdegutiérrez, and ends near Los Cayos. It consists of footprints over seven rock surfaces, four of which

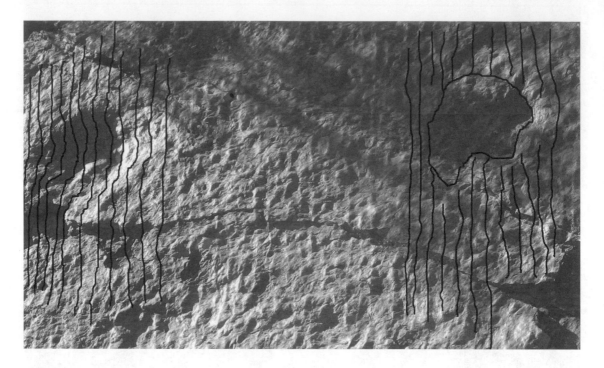

3.95. El Villar-Poyales track-site. Outcrop 3EVP. Ripples conform to depressions represented by footprints, which are therefore undertracks. At right is large depression filled by sediment, which therefore is a subtrack.

have been named in published papers (Table 3.7). From east to west, these four are as follows: Lastra 1-C or 1c (currently 1EVP), EVP (currently with the same name), Lastra 1-B or 1b (currently 3EVP), and Lastra 1-A or 1a (currently 3I).

Additionally, there is a layer between 1EVP and 2EVP with desiccation cracks and poorly preserved footprints. In addition, natural casts exist between EVP and Lastra 1-B or 1b (3EVP), of which fragments have been found. Finally, between 3EVP and 3I, there is another layer (upper intermediate or 4EVP) with one footprint. In relation to other sites, El Villar-Poyales is about 2 km west of Navalsaz, and almost 1.5 km east of Lastra 3-A, B described by Brancas, Martínez, and Blaschke (1979). Some 500 and 1500 m away there are two unnamed sites with poorly preserved footprints. Although the site is easily accessible, viewing some tracks is difficult because of the steep dip (40°) of the strata.

Lithology

3.96. El Villar-Poyales track-site. Outcrops 1EVP (labeled 1C, as named by Brancas, Martínez, and Blaschke, 1979) and EVP are labeled; outcrop 2EVP is not labeled but is situated between the other two. On 1EVP (1C) surface ripples can be seen, along with three trackways and an isolated footprint (highlighted). On 2EVP are desiccation cracks and two footprints. Trackway EVP1 (*Theroplantigrada encisensis*) stands out in surface EVP.

The composition of the layers with the tracks is variable. The bottom three (1EVP, 2EVP, EVP) are limestones, while the top three (3EVP, 4EVP, 3I) are lutitic sandstones. The sedimentary structures imprinted on the layers are also different. In 1EVP the top of the bed is undulated, probably as a result of interference from ripples; in 2EVP there are desiccation cracks (Fig. 3.96); in EVP there are no surficial sedimentary structures. On the top of the three upper beds (3EVP, 4EVP, 3I) are ripples (Figs. 3.94, 3.95). The lower layers were probably deposited in a lake protected from waves, while the upper layers are indicative of a more energetic, fluvial influence.

Track Features in Mud

All the footprints were made after the structures on the tops of the layers on the study surface. They are superimposed on the ripples and desiccation cracks. All of the prints except some from 3EVP and 4EVP are shafts.

There are various morphotypes of footprints at El Villar-Poyales. There are two different types with metatarsus marks (EVP1, EVP4); there are two with possible signs of interdigital web (EVP1, EVP18); tridactyl

3.97. El Villar-Poyales track-site. Outcrop 3I, left footprint (3I2.3), with digital pads outlined. Divisions of bar are 10 cm.

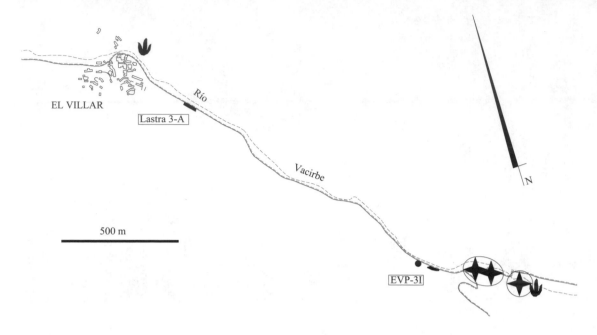

3.98. Location of tracksites of El Villar-Poyales (EVP-3I) and also of earliest attempts to depict dinosaurs in outdoor exhibits in La Rioja: rock painting of some dinosaurs (*Megalosaurus, Triceratops, Iguanodon, Ornithomimus,* one star), and plaster statue (two stars) of *Tyrannosaurus.* Site Lastra 3-A of Brancas, Martínez, and Blaschke (1979) and course of Vacirbe River are also shown.

3.99. Map of El Villar-Poyales (EVP) outcrop. Trackways and isolated footprints are labeled.

digitigrade footprints (such as EVP9) exist; and isolated (?) claw marks are observed (EVP12). It may be that the same kind of dinosaur left more than one type of these footprints.

Trackway EVP1 has 22 footprints (Figs. 3.99, 3.100) in the form of a delta. The wide front of the print shows the marks of three digits, with a sign of a lateral hallux in almost all, while the rear end is elongated and rounded at the end. These tracks have been interpreted as semiplanti-grade with an interdigital web that does not penetrate far into the mud (Casanovas et al., 1993d). Both the metatarsal and the digital part are the same depth. It is considered that the trackmaker was a digitigrade ani-mal adopting the semiplantigrade form to support its foot while walking (Farlow, 1987; Kuban, 1989; Pérez-Lorente, 2001a), so as not to fall over into the soft mud.

Trackway EVP4 (Fig. 3.101) also shows the metatarsus and hallux. In this case, the rear end of the tracks is sharp and not rounded, as in EVP1. These footprints are not interpreted as semiplantigrade because they

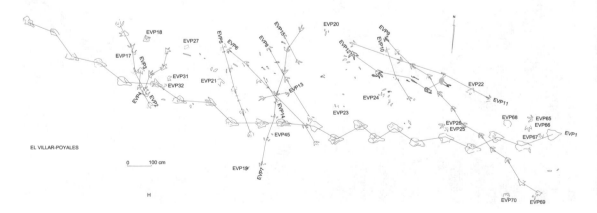

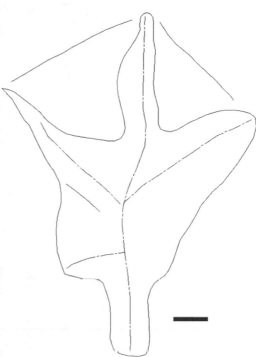

3.100. Photograph (left) and drawing (right) of footprint EVP1.17. Right print with depressed portion that connects digits (including hallux), but not metatarsal mark. Rear of metatarsal impression is rounded. Depression is interpreted as having been produced by an interdigital web. This footprint is holotype of *Theroplantigrada encisensis*. Scale bar = 5 cm.

show a long metatarsal impression. Although the foot penetrated deeply enough into the ground for the hallux to touch the mud and leave a mark, only the distal part, along with part of the metatarsus, also sank in (Fig 3.79). The metatarsus did not register horizontally in the mud but rather remained tilted at an angle to the substrate. In these footprints, the marks of the digits are narrow as a result of collapse into the interior of the shaft.

EVP18 is an isolated footprint (Figs. 3.99, 3.102) showing the marks of the three digits and two webs linking them. It can be interpreted as the track of a webbed foot and is located near two trackways (EVP2, EVP3) that do not have this feature. It is not far from another trackway (EVP4), which is deep, has the mark of the metatarsus and hallux, and does not have indications of a web. It is likely that the web structure is a deformation induced by the digits; the variability in forms due to changes in the physical properties of mud in this layer is quite wide.

EVP9 is a theropod trackway with noticeably asymmetrical footprints. The heel of the print is formed by joining together the imprints of digits III and IV at the rear, and digit II is separate from digit III. Interdigital angle II^III is less than interdigital angle III^IV (Fig. 3.103). Trackway EVP9, with all of its footprints of comparable depth, contrasts markedly with the isolated claw marks in its vicinity and with the disappearance of trackway EVP12 (Fig. 3.104). The claw marks are presumed to have been made by swimming dinosaurs, a hypothesis that has also been used to interpret the EVP12 tracksite. The supposed cause for co-occurrence in this outcrop of typical footprints and claw drag marks made by swimming

3.101. Right footprint EVP4.3 (EVP outcrop). Note pointed metatarsal mark and narrow digit impressions. These features are presumed to have resulted from collapse of footprint wall as dinosaur's foot sank into mud, or after foot was lifted out of mud. (Joint intersects proximal area of footprint.)

dinosaurs is fluctuation of water depth over the time the ichnites were made.

EVP12 begins (Fig. 3.104) with a relatively complete footprint (EVP12.1), but print shape changes along the trackway, and the prints finally disappear. EVP12.4 and EVP12.5 are simply grooves with mud accumulated behind them (Fig. 3.105). There is no EVP12.6 or any further footprints. This sequence is in an environment dominated by claw marks (Fig. 3.104), so it is assumed that when the dinosaurs left these marks, the water level was higher than when they left the complete tridactyl marks (such as EVP9, which also goes across the entire outcrop without interruption).

3.102. Isolated digitigrade, tridactyl footprint EVP18, showing clear interdigital depressions. Because this print does not occur in a trackway, it cannot be determined whether interdigital depressions were made by dinosaur foot (i.e., webbing between toes) or are indirect featuring caused by trampling and physical features of mud.

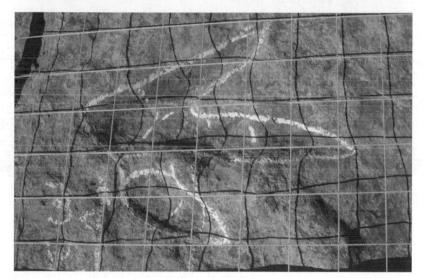

3.103. Asymmetrical left footprint EVP9.6 showing how marks of digits III and IV, but not II, join to form "heel" of print. Size and shape of print suggest possible ornithomimid trackmaker. Wire mesh squares are 5 cm on side.

3.104. Detail of Fig. 3.99, showing trackway EVP12, isolated footprints, and numerous other isolated features interpreted as claw marks. Trackway EVP12 is interpreted as footprint sequence made by dinosaur that begins to swim (see Fig. 3.105).

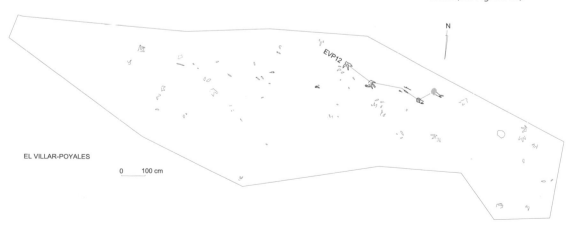

EL VILLAR-POYALES

0   100 cm

3.105. Left footprint EVP12.5 of trackway EVP12 (Figs. 3.99, 3.104). Toe tips dragged and accumulated mud at rear of ichnite. Dinosaur is thought to have begun to swim at this point. Chalk line grids are 30 cm on side.

At outcrop 3EVP, the ripples follow the deformation of the sedimentary surface, such that they are seen on the walls and continue without interruption along the base of the prints (Fig. 3.95). This means that the footprints are undertracks. On the study surface at 3EVP there are also tracks with no shafts, which are filled with material deformed during the track-making process, i.e., they are subtracks. This means that the tracking surface of the dinosaurs was above that studied. Undertracks (on the left) and subtracks (on the right) are preserved (Fig. 3.95). The footprint at 4EVP is filled in without any apparent structure, so it is probably a cast.

Of all the El Villar-Poyales tracksites, the footprints at 3I (Fig. 3.106) have the best-preserved anatomic structures. Digital pads show the lines of separation perfectly, with distinct lateral constrictions of the digital margins, indicating the separation between pads. Clear claw marks occur at the tips. Almost all the tracks are considered to be stamps. It is therefore not easy to explain why one of the trackways (3I2) appears abruptly near the middle of the site, because its prints were made after the ripples, and therefore after the last stage of erosion detected in the top of the bed. One can assume that the ground was entirely dry at the initial part of the trackway, because the first of the tracks (3I2.1) is the shallowest of all.

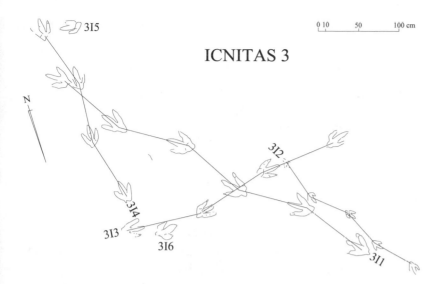

ICNITAS 3

3.106. El Villar-Poyales track site, 3I outcrop (Casanovas et al., 1992a). Small theropod trackway 3I2 is in bottom right-hand corner.

Footprint Distribution According to Their Structures

Exposures 1EVP, 2EVP, 3EVP, and 3I show the same style of preservation in all their footprints. This means that during the interval of time in which all at the footprints at these sites were made, the physical properties of the mud trampled upon were the same.

This is not so at El Villar-Poyales, where, in the same environment, some dinosaurs left only claw marks, while others left complete, well-defined footprints. One might suppose that some dinosaurs were small, could not touch the bottom, and therefore swam, while larger ones could touch the bottom and left their footprints. However, the opposite is true: a trackway that ends with claw marks only (EVP12) also left larger and complete foot marks about 37 cm long.

Ichnite Features (Ichnotypes)

The trackways at these exposures show different styles of preservation: typical digitigrade tracks, elongated prints with metatarsal impressions, and claw marks only.

On the basis of morphology, only EVP3 (Fig. 3.99), a trackway with three footprints, is identified as ornithopod (Casanovas et al., 1993d). The rest of the dinosaur ichnites are theropod footprints, both large (39–40 cm) and small (16 cm), with a majority of trackways having an average footprint length of more than 25 cm.

The 35 cm long footprint EVP18 (Fig. 3.102) and footprints in the EVP1 trackway have their digits linked by depression areas. Until recently, these footprints were considered to contain an interdigital web. However, some studies have shown how such structures could be formed without webs, depending on the physical properties of the ground (Falkingham et al., 2009; Manning, 2004). In EVP18, this influence and dependence

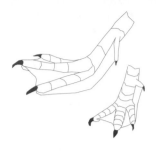

3.107. Interpretation of webbed feet (EVP1 footprints), according to Pérez-Lorente et al. (2001). Web is structure connecting digits I–IV.

are arguable because it is a single impression, but this is not the case in EVP1.

EVP1 INTERDIGITAL FOOTPRINT DEPRESSIONS

All the tracks in EVP1 have depressed interdigital regions. In 10 footprints, the depression extends not only across the front digits (II, III, IV) but also joins to the hallux (I). The same is thought to occur in the remaining 12 prints, but the shape of the footprint outline is not clear enough. The depression stops at the hallux (Figs. 3.99, 3.100) and does not extend onto the shaft of the metatarsus. It has been postulated that an interdigital web linked the four digits of the dinosaur (Fig. 3.107, cf. Pérez-Lorente et al., 2001).

Several scenarios were considered in which a depression might form in EVP1, similar to that produced by an interdigital web (Casanovas et al., 1993d; Manning, 2004). Perhaps during the step the surface was slightly elastic as a result of the presence of an algal or other biofilm capable of acting as a flexible mesh. In this case, the pressure of the toes would have extended into the space between them. The interdigital areas therefore sank, driven in by the toes. Even without a biofilm, if the physical behavior of the deformed surface was flexible, the effect would be similar to the previous situation. Examples of this phenomenon have been described before in the tracks of birds (Garcia Raguel et al., 2009). Finally, mud can also flow to the base of the track during or after the K phase. As the dinosaur pushes down and moves its foot, an induced thixotropic movement of mud can flow into the interior, especially in the interdigital parts.

These interpretations, which are viable alternative explanations for the existence of depressed regions in print EVP18, cannot apply to EVP1 because the location of the depressed area is selective. The depression is formed only between the four digits, not between the marks of the hallux and metatarsus. All these possibilities predict that depression would be favored when the distance between the segments connected by it are closer together (distance between the digits) and more pronounced the closer together they are (hypex area). Consequently, none of these alternatives explains why there is a depression between digits I and II, but not between digit I and the metatarsus (Fig. 3.100).

Because it is impossible to explain the depression as an extramorphologic structure, the prints became the basis of a named ichnotaxon, *Theroplantigrada encisensis* Casanovas et al., 1993d, with the trackmaker's foot reconstructed with webbing between all its toes (Fig. 3.107), as drawn by Pérez-Lorente et al. (2001).

The footprints in EVP1 have the marks of four digits and the metatarsus. Pérez-Lorente (1993a) compared these tracks with other semiplantigrade footprints known at the time. There had been enough studies published to distinguish such prints with separate digits from those with a depression linking them. Trackway EVP1 provided the only example of semiplantigrade prints with a depression linking the three large digits' impression (digits II–IV).

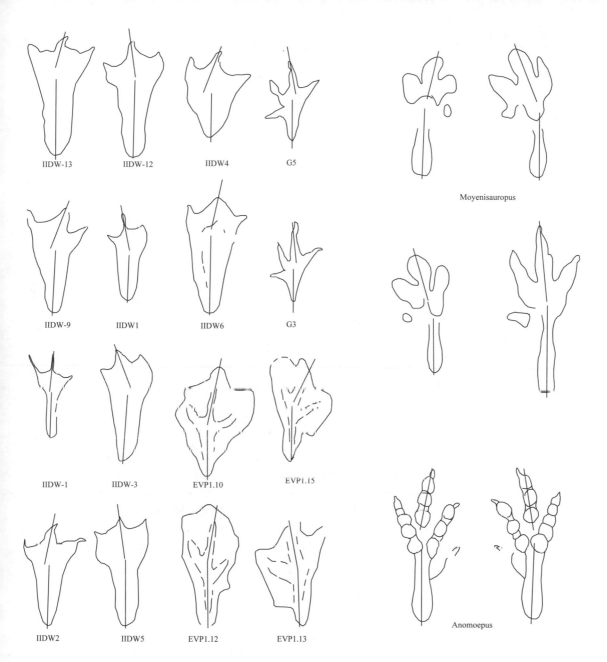

IIDW-13   IIDW-12   IIDW4   G5

Moyenisauropus

IIDW-9   IIDW1   IIDW6   G3

IIDW-1   IIDW-3   EVP1.10   EVP1.15

IIDW2   IIDW5   EVP1.12   EVP1.13

Anomoepus

SEMIPLANTIGRADE CHARACTER OF EVP1

The tracks in EVP1 have marks of digits II to IV at the front and the mark of the hallux (digit I) at the medial position. The rear end of the print is rounded (indicating a metatarsus resting on the ground) and not angular (typical of an inclined metatarsus). The trackway shows a rather high estimated moving speed (7 km/h) when compared with other biped dinosaur trackways, consistent with the general features of a semiplantigrade trackway (Kuban, 1989; Pérez-Lorente, 1993a). The trackway has a variable width (Lr), with the speed decreasing as the Lr increases (Thulborn, 1990).

3.108. Angle between axes of metatarsal and digitigrade footprints. Examples of semiplantigrade and deep footprints with metatarsal mark. Examples of Paluxy River (IIDW; Farlow, 1987), *Anomoepus* (Hitchcock, 1848), *Moyenisauropus* (Ellenberger, 1974), El Frontal (G; Aguirrezabala and Viera, 1980), and El Villar–Poyales (EVP).

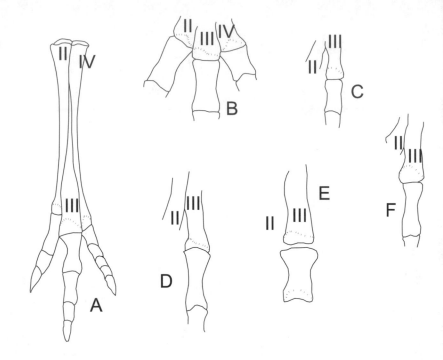

3.109. Pedial skeletons showing asymmetry of distal articular surface of metatarsal III (cf. Pérez-Lorente, 1993a) of theropoda (Weishampel et al., 1990). (A) *Struthiomimus altus*. (B) *Harpymimus oklad-nikovi*. (C) *Ingenia yanshini*. (D) *Deinonychus antirrophus*. (E) *Allosaurus fragilis*. (F) *Borogovia gracilicrus*.

The axis of the acropodium or digital part of the foot in EVP1 forms an angle with the axis of the metapodium or metatarsal part. Pérez-Lorente (1993a:figs. 5, 7) noted that the same was true of nearly all other bipedal dinosaur trackways with metatarsal impressions published. This included semiplantigrade prints both with and without depression in the interdigital regions—that is, not just those footprints with depressions at the front but also those with separate digits (Fig. 3.108). The angle occurs both in semiplantigrade tracks with the metatarsus uniformly impressed into the substrate over its entire length and in prints in which the metatarsal impression slopes downward anteriorly into the mud (footprints G3 and G5; Fig. 3.108).

Theropod trackways usually have footprints with a negative rotation, i.e., the axis of the digital foot points inward, toward the trackway midline. The position of the axis of the digital part of the foot is thus not parasagittal with the metatarsus or the dinosaur's body. In footprints with an inclined metatarsus, as well as in those with the metatarsus flat on the ground, the metatarsal and phalangeal areas (metapodium/acropodium) are at an angle. The anteroposterior (craniocaudal) plane of motion of the metatarsus does not align with that of the phalangeal area in any of the three metapodial orientations.

This means that rotation of the digits and metatarsus was not co-planar. Pérez-Lorente (1993a) deduced this from comparing trackways but also found that the rotation (via digitigrade and metatarsus marks) is reflected in the metatarsophalangeal joints of theropod foot skeletons. The condylar surface of the joint of the first phalanx of digit III and the distal end of metatarsus III should correspond with the motion described.

Pérez-Lorente (1993a) reported examples that show an asymmetry of metatarsus III, which has a greater anteromedial extension on the condylar surface (Fig. 3.109).

Angulation of the digital part at the foot with respect to the metatarsus means that the narrowness commonly seen in theropod trackways reflects more than the configuration of the hip joint. The foot would be positioned under the body while walking simply as a result of the structural arrangement of the distal components of the limb. This implies that while walking, the center of gravity shifts little toward the sides, and that balancing amplitude should not be wide. In theory, this means the dinosaur may be standing on one leg at any time while walking, without needing to balance vertically.

FOOTPRINTS WITH IMPRESSIONS OF INCLINED METATARSUS

Footprints in EVP4 trackway, like those of EVP1, have marks for the three front digits, the hallux, and the metatarsus located in an interdigital depression. Even so, they are quite different from EVP1.

The metatarsus print is not rounded at its proximal end but pointed (Fig. 3.101). The outline of footprints forms an acute angle at its rear end (Fig. 3.79B). The prints in EVP4 are not stamps; they are distorted because the mud walls of the track fell into the shaft. However, if the walls of the metatarsal area had been parallel, the collapse would have left parallel or linear metatarsal tracks (Figs. 3.3, 3.29). Pérez-Lorente (2001a) concluded that the angular rear end of the print is a reflection of carriage of the metatarsal section, including soft tissue (tendons). The suggestion is that the foot penetrated deeply into the ground with the metatarsus inclined – the normal position of the metatarsus in this dinosaur. It is not clear whether the foot slid forward as it sank into the mud.

The hallux is medial and directed backward. The position of the hallux mark is not necessarily a reflection of its true position. If one assumes that the foot slid forward during the T phase, the hallux would makes a mark during this phase. It would touch the ground with its distal part. As the foot penetrated the substrate, the hallux sank and moved forward (Gatesy et al., 1999).

The digital part of the print is marked more or less uniformly, with clear impressions of the three digits. Mud collapsed around the toes, probably immediately after the foot sank into the substrate, resulting in narrow digit marks and an interdigital depression. There are no extrusion rims as a result of the low viscosity of mud, which allowed the flow and horizontal movement of the mud to penetrate under the foot.

The problem of interpreting footprints with metatarsal impressions in terms of the depth reached by the foot was treated by Aguirrezabala and Viera (1980). They studied a similar trackway with the mark of the hallux, with an interdigital depression, and with a prominent and angular rear end, from a tracksite in in Bretún (in the nearby province of Soria).

They concluded that the foot penetrated so deeply into the ground that the hallux was able to leave an impression.

TRACKWAY EVP12

After all of the footprints at this set of complex tracksites have been assigned to unambiguous trackways, a more complicated sequence of footprints (EVP12) remains, along with a seemingly unrelated series of claw marks and isolated footprints (Fig. 3.104).

EVP12 is a sequence of tracks with the following features: the first footprint (EVP12.1) is of normal shape; the second (EVP12.2) has inordinately long toe marks, probably resulting from the foot slipping on the bottom; the third (EVP12.3) consists of three long, roughly parallel toe marks that do not meet at the proximal end of the footprint; and the last two (EVP12.4 and EVP12.5) are simply drag marks of claws and the tips of toes that slid along the ground, accumulating mud dragging behind them. This sequence (Fig. 3.104) shows that the strongest pressure of the sole of the foot occurred in the first print; the sliding tips of the toes are evident in the second print and the claws in the third. In the last two footprints, only two of the three claws, and perhaps the most distal part of the ungual pads, made contact with the ground. The claws in the last footprints in the trackway (EVP12.4, EVP12.5) scratched the mud, accumulating it behind the marks left by them (Figs. 3.104, 3.105). The study surface is the tracking surface, as can be seen in this action. The claw marks are the last tracks of the trackmaker.

Casanovas et al. (1993d) interpreted this sequence in the same way as Pienkowski and Gierlinski (1987). It was made by a dinosaur walking on a submerged slope that started to swim. The alternative interpretation of Lockley (1991), that these prints are undertracks, does not apply because the footprints of EVP12 are real impressions with direct structures.

This is the first location in La Rioja with claw marks attributable to swimming dinosaurs. The second is the site of La Virgen del Campo (Pérez-Lorente et al., 2001), where apparently disordered claw marks were found, as well as marks indicating a trackway left by a dinosaur that swam (Ezquerra et al., 2004, 2007).

EVP9 FOOTPRINTS

Trackway EVP9 comprises nicely preserved tridactyl footprints with an average length of 28 cm. Evident are clear digital pads delimited by creases in the toe marks and by constrictions in the margins of the toes. The ends are tapered, with clear claw marks. The claw marks of the three digits are comparable in size. The outlines of the proximal ends of digits III and IV join together. There is a shallowing of the proximal part of digit III, such that the continuation of digit IV into the heel is indisputable. The outline of the proximal end of digit II is separated from the bases of digits III and IV or the joint between II and III to IV is at least less clearly defined (Figs. 3.99, 3.103). The digits are thin, the heel is projected backward, and the proximal pad of digit II is well ahead of digit IV and separated from the digit III to IV joint. Farlow (1987) and Thulborn (1990) give three examples of footprints attributed to *Ornithomimosaurus* with the same features. The shape and size of the *Ornithomimipus angustus* Sternberg, 1926, footprint are similar to EVP9.

ICNITAS 3 FOOTPRINTS (FIG. 3.106)

All the footprints in this study area have the same morphological features. The digits are long and separate despite having low interdigital angles (II^III 14–17°, III^IV 19–21°). Interdigital angle II^III is significantly less than interdigital angle III^IV. The digital pads are well separated by horizontal creases and sometimes by side constrictions of the footprint outline. There are clear signs of claws. Digit II is further from digit III (the heel) than digit IV – that is, its proximal end is more distally positioned than the proximal end of digit IV. Finally, digit IV is joined to digit III, and its proximal part makes the heel of the footprint. Two track groups can be distinguished by the lengths of their tracks. One trackway (3I2) has small footprints (16 cm long), and the lengths of the rest of the tracks range 26 to 35 cm.

The 3I footprints are assigned to the same ichnotaxon. Their morphological similarity led Brancas, Martínez, and Blaschke (1979) to propose that trackway 3I2 is of a youngster from a group of others. Casanovas et al. (1992a) found no reason to dispute this conclusion but could not find sufficient reasons to confirm it because the trackways are not parallel, nor do they go toward the same side of the site; trackmakers at the same site, and even of the same taxon, need not be gregarious; and there may have been species in the same habitat with similar foot shapes but with different adult sizes.

General Information

It does not seem appropriate to report an overall track density for EVP/3I because the result would not be informative. According to available data there are six layers with tracks (shafts, casts, subtracks, undertracks) and at least one that is not visible providing cast fragments. All these layers are superimposed on each other such that the separation between bedding surfaces within footprints is of the order of decimeters in the interior of the sets 1EVP/2EVP/EVP and 3EVP/4EVP/3I. In other words, a hypothetical count would have to consider the number of outcrop tracks and covered tracks containing the seven layers throughout the site. A calculation with outcrop footprints would not show the proper proportion.

The repetition of layers with tracks is normal in many parts of the basin. However, what is interesting at this point is that the number of layers with footprints is greater than can be deduced from the average of sites studied. Also, there are many footprints, even in locations with different types of sediment (limestone in 1EVP to EVP, and siliceous sediments in 3EVP to 3I).

Main Features of Site

El Villar-Poyales shows the importance of the number of observations regarding footprints, i.e., the recommendation of Sarjeant (1989) to define a new ichnotaxon. Thus, if one trackway has more footprints than another, or if the number of features is greater (semiplantigrade versus digitigrade), the conclusions and deductions multiply. If the depressions between the digits are due to an interdigital web, the maker of *Theroplantigrada encisensis* was a dinosaur with an interdigital web that extended not only between the three front digits but also reached the hallux. This same trackway, being semiplantigrade, provides information regarding the anatomy of the limb as well as the posture and form of movement of the lower parts of the hind limbs.

EVP18 is a footprint with the digits and depressed interdigital area well defined. However, it is isolated because it forms no part of any trackway, and there are no similar ones on the site. This situation means that EVP18 cannot be ideally considered in order to define the presence of an interdigital web.

The presence of two trackways with a metatarsus mark at this site leads to the conclusion that the best explanation for the way a semiplantigrade animal walked is with its metatarsus flat against the ground. Such prints were not mere products of unusually deep penetration of the foot into the mud (Farlow, 1987).

In some places, the structures of the footprints indicate that the physical conditions of the sediment and/or the depth of water during the track-making process were unchanged, while in EVP the opposite occurred. There are probably three different stages that cannot yet be put in chronological order: a stage with a water depth of around 2 m (claw marks and trackway EVP12); a phase of extremely soft mud (trackways with the metatarsus mark); and a more viscous mud stage (digitigrade trackways and footprints).

First Dinosaur Model of La Rioja

El Villar de Enciso and Poyales are almost deserted villages where only four families live—and even then, not all at the same time. Some of their descendants decided to form an association (Asociación de Amigos de Poyales) and construct a dinosaur to show to visitors, using mainly plaster, logs, dry branches, and paint. The dinosaur was placed on the crossroads between Poyales and Navalsaz (Fig. 3.98, double star). It is difficult to pin down exactly what type of beast it was. However, it is probably a relatively large and fierce carnivorous theropod dinosaur, perhaps a *Tyrannosaurus* (Fig. 3.110A). Plaster and dry wood are not the most durable of materials,

3.110. (Left) *Tyrannosaurus* model done by Asociación de Amigos de Poyales in 1981. (Right) *Megalosaurus* painting on side of Vacirbe River, made by the team of Brancas, Martínez, and Blaschke (1979) on the basis of a classic painting by paleoartist Neave Parker.

however, and after the rain, snow, and frost of the first winter, the sculpture fell apart.

This was not the only attempt to get people interested in dinosaurs. In 1979, a wall was decorated with four dinosaurs (Fig. 3.98, single star). The trackmakers and footprints they left were painted by the Brancas team while collecting data (Brancas, Martínez, and Blaschke, 1979). One of the dinosaurs was a *Megalosaurus* (Fig. 3.110B). The paintings have since been removed.

History

Barranco de Valdecevillo

According to local sources, at least one footprint at the Barranco de Valdecevillo site (va4.3) was discovered when removing stone on a hillside of the Valdecevillo ravine to make repairs to the Enciso–Navalsaz road. There is no record of the work, but it was presumably done in the first half of the twentieth century. Among the local population, the ichnites were said to be the footprints of giant chickens. They must have known of the footprints: they are located beside the old road linking the two major river valleys, along which manufactured products for the local population are transported.

Casanovas and Santafé (1974), who described the site for the first time, called it Barranco de Valdecevillo. They reported 19 footprints, including a set of three large theropod footprints (va4.2–va4.5) thought to be *Megalosauripus* and three small prints (probably a part of the va9 trackway) that they attributed to a medium-size ornithopod. Brancas, Martínez, and Blaschke (1979) called the site Lastra 5-A. They wrote that one of them (Brancas) had made molds of footprints from this site in 1960, and later, in 1975, had unearthed a new track (va4.1) with a jackhammer. They found a total of 44 footprints at Barranco de Valdecevillo.

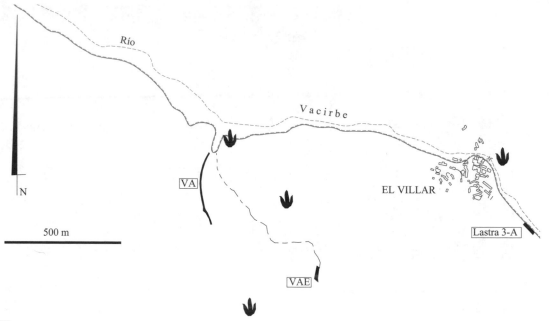

3.111. Location of tracksites of Barranco de Valdecevillo (VA), Barranco de Valdecevillo Este (VAE), Lastra 3-A, and La Senoba (LS). Black prints show outcrops that have not yet been studied. Map also indicates Vacirbe River, town of El Villar, road from Enciso to Los Cayos, and path of access to VAE.

Viera and Torres (1979) found two new outcrops, one at the bottom of the Valdecevillo ravine and the other with two small theropod footprints on the other slope. In 1980, a new team from Colegio Universitario de La Rioja (Fernández Ortega, Pérez-Lorente, Sáinz Ruiz de Zuazo) formed a group with some of the aforementioned authors (Casanovas, Blaschke, Martínez Flores, Santafé) and began digging here. They found previously undescribed footprints, almost all of them part of trackways. The published study cited 168 footprints (Casanovas et al., 1989).

Description of Site

The site of Barranco de Valdecevillo is on an outcrop (Fig. 3.111) exposing the top of a bed that runs along part of the west side of this ravine. The width of the bed varies but averages around 4 m with a length of 330 m, and so forms an elongated strip. The dip of the bed is a bit steeper than the slope of the ravine; therefore, you climb up the slope in order to reach the site. The trackways on the lower part of the exposure are oblique to the direction of the strip and contain few footprints, but above the level of trackway VA13 there are nearly parallel trackways containing more footprints.

The other tracksites closest to these footprints (Fig. 3.111) are less than 1500 m away (Lastra 3-A is at a distance of 1400 m; East Valdecevillo at 500 m; and La Senoba at 1000 m). In the same ravine there are other points with footprints (Viera and Torres, 1979) identified as unstudied sites. Alongside the road there are two other areas with tracks and casts that also have not been studied. There are also fossils of fish (scales,

3.112. Dinosaurs in Barranco de Valdecevillo area. (Top) *Iguanodon* model from an exhibition, donated by Fundación Cajarioja. (Middle left) Model of *Tarbosaurus*. (Middle right) Detail of head and neck of *Brachiosaurus*. (Bottom) *Iguanodon* family.

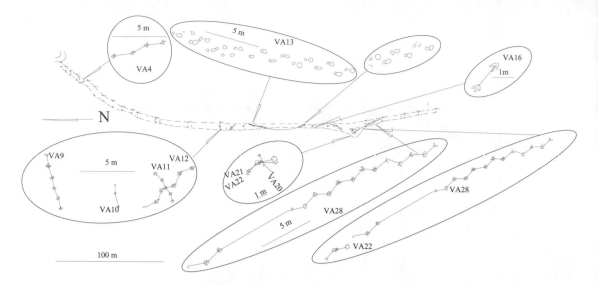

3.113. VA tracksite. Tracksite is elongated and curving strip in center of drawing. Certain areas have been expanded to show most important trackways.

vertebrae), invertebrates (bivalves, gastropods), and plants (stems, soil with roots).

The area is accessed by a path with a wooden railing that runs along the site. The path turns, returning to the starting point, going past six life-size dinosaur models (Fig. 3.112). The family of three ornithopods and the sauropod were scaled according to measurements obtained from footprints at Barranco de Valdecevillo; the theropod (*Tarbosaurus*) was the first dinosaur installed and was made by an amateur sculptor using old diagrams. Finally, there is an ornithopod model that came from an exhibition, which was protected with resin to withstand the outdoor conditions.

There are four series of drawings of footprints, published between 1974 and 1989, of Barranco de Valdecevillo, each of which improved and supplemented earlier references (Fig. 3.113). More footprints on the site have been found recently. The mapping used as a basis for this discussion dates from 1989 (Fig. 3.113).

Lithology

The study surface is on dark argilaceous sandstone that runs along both sides of the ravine. The rock is resistant to erosion and overlies a shale layer that weathers and erodes more easily. The tracks are on the west slope of the ravine because here erosion has easily removed debris, revealing a strip of 4 m. This has not happened on the eastern slope, where only a cross section of the footprint bed appears (Fig. 3.111, dashed line). If on the east slope the inclination of the terrain varies, this may favor the appearance of footprint outcrops (East Valdecevillo) in the same strata. There are no sedimentary structures (ripples, mud cracks, or any other) on the top of the bed. There are many signs of bioturbation by invertebrates around trackways VA4 to VA9 and in the upper tracksite area. Vertical burrows passing through the top layer give off radial galleries at their bases.

3.114. VA tracksite. Sauropod trackway VA13. Footprints visible in 1989 in black. Since 1989 part of talus on left has eroded enough for several additional right footprints (gray) of same trackway to appear.

Footprints Features in the Mud

The footprints are not filled with hard material (casts or subtracks), so it cannot be established whether they are real or undertracks. Four pads can readily be distinguished in many of the ornithopod footprints, but few details can be discerned in the theropod and sauropod tracks. There are no pronounced mud extrusion rims. On one side of the site, erosion of the surface continues to expose new footprints in sauropod trackway VA13 (Fig. 3.114). The shafts of the new footprints are filled with shale that could be interpreted as nondeformed sediment, as there are no internal structures. If so, the VA13 fillings would be casts, and the holes would be direct structures. Sediment viscosity and consistency of the site varied for all tracks. An example of this substrate behavior is described below in terms of variation of consistency (trackways VA4–VA28, with walls of different stability) and changes in viscosity (trackways VA28–VA29, tracks of different depth).

COLLAPSED OR MALFORMED DINOSAUR FOOTPRINTS

The VA4 footprints have abnormalities that have been interpreted to be products of erosion (Casanovas and Santafé, 1974), or alternatively as malformations caused by injuries experienced by the dinosaur (Brancas, Martínez, and Blaschke, 1979). Portal Paleontológico Argentino (http://www.mesozoico.com.ar/Forum/viewtopic.php?f=41&t=369) reports that a malformation can be seen in the two left footprints of the track: "The internal digit was damaged and therefore had swollen and became distorted, as well as having one of the phalanges with claw missing. . . . The footprints seem to correspond to a megalosaur of 120 million years ago." According to Casanovas and Santafé (1974), the anomaly was at the back of footprint VA4.3. This was also observed by Brancas, Martínez, and Blaschke (1979) in VA4.3 and VA4.1, although they do not specify the type of malformation.

There are several features seen in the trackway footprints. The first is the asymmetry (Fig. 3.115) of the interdigital angles (angle II^III < angle III^IV), which are also seen in many other theropod footprints (Pérez-Lorente, Fernández, and Uruñuela, 1986). This asymmetry is evident (Fig. 3.116) in the topographic reproduction of Valle (1993). There is a deformation in the tracks (Figs. 3.115, 3.116) produced by the mud falling into the shaft interior (Farlow et al., 2012; Pérez-Lorente, 2001a). The collapse of the mud significantly thins the medial and distal parts of digit III, so that it seems the digit had a wide and very short proximal part and the claw was greatly developed. Finally, the collapse of the mud also thins all the digits of the left side of the trackway (digits IV and II). In conclusion, there is probably no malformation of the foot. It is a deformation of the footprints in the K phase or after the exit of the foot. Casanovas et al. (1989) have reviewed the site and the published works and did not mention any malformation.

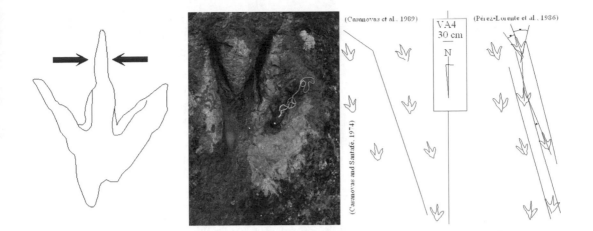

CAVITIES THAT HAVE NOT COLLAPSED

Trackway VA28 (level 100, section 2, of Viera and Torres, 1979) consists of 16 or 19 footprints (depending on whether trackway VA22 forms part of the same sequence). The first 13 footprints are shown in detail in Figs. 3.113 and 3.117. The tips of the digits penetrate into the soil, forming cavities like small holes with roofs that did not collapse (Fig. 3.118). Here the coherence of the mud was high because the holes in the walls of the shaft at the tips of the three digits are maintained. At least part of the claw of the front digits penetrated into the mud, which stayed in the same position without collapsing. The depth of the footprints is about 5 cm.

SHALLOW FOOTPRINTS

Next to VA28 are some quite shallow trackways (less than 2 cm deep), such as VA29, which crosses it. Other shallow rocks occur repeatedly at the lower part of the site. There are also shallow footprints to the north and south of VA4, between VA4 and VA9. It does not seem possible to establish temporal distribution criteria for the footprints on the basis of their mud structures. This means that the ground conditions probably varied across the study surface and at different times, when the dinosaurs passed.

Footprint Features (Ichnotypes)

ORNITHOPOD FOOTPRINTS

There are several trackways attributed to ornithopods. Their tracks are distinguished by the shape and number of pads (one pad per toe) and by having a blunt distal end. Trackways VA9, VA10, VA11, VA20, and VA25 were originally labeled as *Iguanodon* (Casanovas et al., 1989) but today (Figs. 3.113, 3.119) are classified as *Brachyiguanodonipus* (Díaz-Martínez et al., 2009) despite showing a substantial difference in size. Different species or subichnotypes should be established within this ichnogenus. The medial and lateral toes are more nearly parallel to the central one in VA25 than in the group VA9-VA10-VA11.

3.115. Trackway VA4. (Left) Drawing and (middle) photograph of footprint VA4.3 (left foot). Digit IV and distal end of III are narrow because walls of shaft have fallen inward. Digit IV contacts "heel." (Right) Two successive interpretations of trackway (which is right trackway of two illustrated). Casanovas et al. (1989) illustrated footprint (VA4.1) discovered by Brancas, Martínez, and Blaschke (1979) as first track in this sequence. Later interpretation of trackway by Pérez-Lorente Fernández, and Uruñuela. (1986) clearly shows negative rotation of footprints (sensu Leonardi, 1987), narrowness of trackway, and asymmetry of pedal digits (in terms of interdigital angle and relative positions of proximal ends at impressions of digits II and IV).

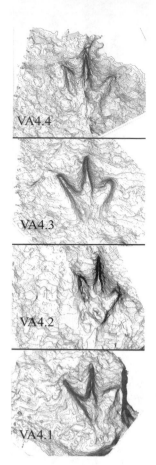

3.116. Photogrammetric reproduction of VA4 (Valle, 1993). Contour interval 2 mm. Prints 1 and 3 are lefts, and 2 and 4 are rights. Hallux (clearly visible in silicone mold at Enciso Museum) is gently suggested by some.

3.117. *Facing, above.* Different visions of VA28 trackway. (A) Casanovas et al. (1989) only drew footprint outlines; trackway progression is from left to right. (B) Work camps in 1980, anonymous report with depressions in right part of prints, from right to left. (C) Superposition of two previous drawings with depressions interpreted as claw marks; from left to right. (D) Synthesis of footprint outline of Casanovas et al. (1989) and claw marks of 1980 with elimination of erroneous lines; trackway progression from left to right.

Trackway VA28 (Figs. 3.113, 3.117, 3.118, 3.119) was known to the authors who worked on the footprints in the area. Before 1989, a map of part of VA28 was included in an anonymous report of fieldwork from 1980. Twelve footprints were recognized (Fig. 3.117B), the outlines of which were consistent with the dinosaur moving south. Blaschke told me before the dig in 1980 that from the first footprints, the impression was that the dinosaur was going in the opposite direction to that considered now. Erosion has transformed much of the outlines of the first four footprints (VA28.1–VA28.4), and this interpretation was not out of the question. Casanovas et al. (1989) drew the outline of the footprints (Fig. 3.117A) superimposed to the 1980 interpretation (Fig. 3.117B), thus providing a more complex picture (Fig. 3.117C). The simplest and clearest interpretation is that of Casanovas et al. (1989) with the front-end claw marks (Fig. 3.117D).

The VA28 footprints show the four round pads without claw marks common to ornithopod footprints. However, they differ from typical ornithopod footprints from La Rioja in the following features: three separate digits (the interdigital area is well defined–more defined that in other ornithopod footprints); the pads of digits II and IV are as wide as that of digit III, but shorter; the central digit (III) protrudes farther in front of a line connecting the distal ends of the lateral toes (II and IV); the axes of the three digits are parallel (Fig. 3.119); and the claw marks are generally wide and flat. They have not yet been allocated to any ichnotaxon.

The remaining ornithopod footprints show no clear characters, either because they are components of trackways with few footprints or because their footprint outlines are irregular or not continuous.

THEROPOD FOOTPRINTS

Apart from those in trackway VA29, all theropod footprints would be classified as large (Thulborn, 1990). The average length for most is over 30 cm, with the largest being 60 cm (VA4.4, VA16.1). Casanovas and Santafé (1974), who only studied VA4, assumed that they were footprints from a *Megalosaurus*, and therefore, they included them in the ichnotaxon *Megalosauripus*. Currently, and according to the study of Lockley, Meyer, and Santos (1998), this classification is maintained (Figs. 3.113, 3.115, 3.116).

SAUROPOD FOOTPRINTS

Casanovas et al. (1989) identified three sauropod trackways. They labeled two trackways H and I (Figs 3.113, 3.114), currently named VA13. Their trackway H has 35 footprints, and their trackway I contains 13. Both include forefoot and hind foot. VA13 is clearly a sauropod trackway, and Casanovas et al. (1989) thought it was made by a type of brachiosaur. They then attributed both trackways to the same dinosaur, assuming that trackway H was the continuation of trackway I. Erosion has removed part of the sediments on the western boundary of the site (Fig. 3.114), with the result that in 2007, 10 more footprints were uncovered, five forefeet and five hind feet, all of them right feet. If erosion continues at the same rate,

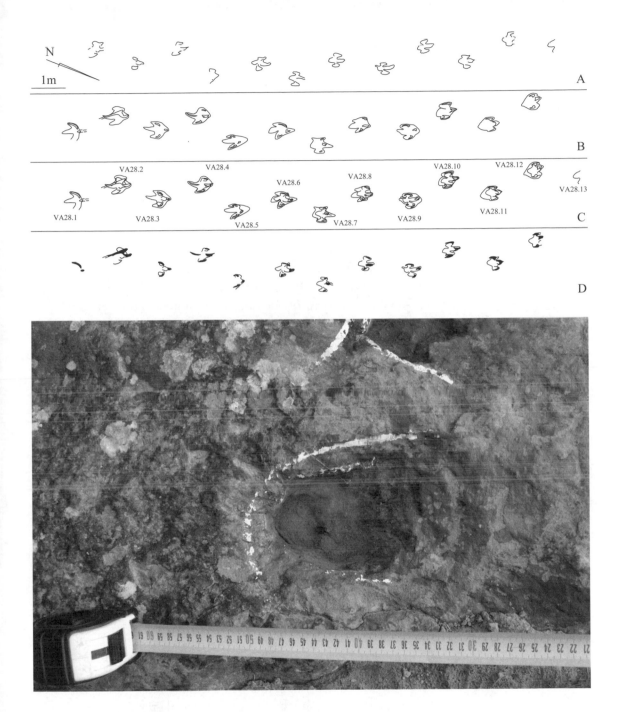

A

B

VA28.2　　　VA28.4

VA28.6　　　　　　VA28.8　　　　VA28.10　　VA28.12

VA28.13

VA28.1　　　VA28.3　　　　　　VA28.5　　　VA28.7　　VA28.9　　VA28.11

C

D

perhaps within 10 years, the two sides of VA13 (trackways H and I) will be joined together.

A dinosaur was constructed using the dimensions deduced from the trackway. The assessment made in 1979 was that the glenoacetabular distance for a walking dinosaur with an ambling gait would be 241 cm, and therefore the maximum length would be 17 m. The sculptors scaled it to 19 m and then to 22 m (Figs. 3.112, 3.114) because, according to the press at that time, other sculptors had made a sauropod sculpture of 21 m,

3.118. Distal part of digit III in VA28.11. Depression made by claw penetrating more deeply into substrate than fleshy pad of toe. There was no collapse of toe mark walls after claw was withdrawn from sediment.

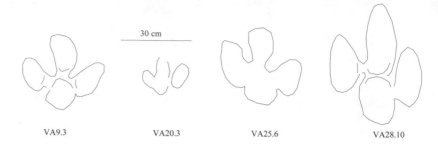

3.119. Footprints from Barranco de Valdecevillo tracksite attributed to ornithopods. VA20.3 is a right, and other prints are lefts. These ichnites are assigned to ornithopod footprints on basis of sole pad configuration. There are four such relatively large and rounded pads, three of which correspond to digits II–IV and the fourth of which is a "heel" mark. This rear "heel" pad differs generally in size from toe pad impressions.

30 cm

VA9.3   VA20.3   VA25.6   VA28.10

and they had no intention of theirs being smaller. In the opinion of the time, this dinosaur walked with an ambling gait (Casanovas et al., 1997a) and was represented thus. In subsequent references, VA13 was classified as *Brontopodus* Farlow, Pittman, and Hawthorne, 1989, or a wide-gauge sauropod (Farlow, 1992).

Casanovas et al. (1989) described trackway K_6 (VA22) as that of a biped whose footprints had three toes directed forward and two backward (Fig. 3.113). They assigned it to a sauropod walking bipedally. The trackway is not clear, and my most recent personal review noted that it could be a continuation of VA28 (Fig. 3.113).

CASTS

In 1999, as part of a footprint protection campaign, a side road was built alongside the whole western part of the site. During this work, two natural casts of a bipedal dinosaur track surfaced (Fig. 3.120). The left cast is the front part of a right footprint with the tip marks from three rounded digits visible. The right cast had only the rear part preserved. This has been interpreted as marks from a dinosaur whose claws were not sharp but blunt. There are no criteria for determining whether the trackmaker was a four-legged dinosaur. However, if compared with the set of footprints from Cameros, both footprints would be allocated to a bipedal ornithopod dinosaur walking toward the photographer.

Ichnotype Footprint Distribution

It could be argued that the large theropod footprints (VA4, VA11, VA16) (Fig. 3.113) have the same features (except for the collapse in VA4) and belong to three solitary dinosaurs. The smaller theropods of Barranco de Valdecevillo do not seem to belong to any social group.

VA13 is a sauropod trackway located parallel to the edge of the site (Figs. 3.113, 3.114). It cannot be concluded whether the dinosaur was accompanying a herd (that was moving to its left along the part eroded by the ravine). Also, no conclusions can be drawn about its social behavior.

VA9, VA10, and VA11 (Fig. 3.113) are three parallel ornithopod trackways that go together across the middle of the site (at 140 m from the start in the north and 180 m from the south end). It has been interpreted

(Pérez-Lorente et al., 2001) that the three dinosaurs were moving together, with the two large ones on either side and the small one in the middle. The height of the dinosaurs would have been approximately 2.2 to 2.1 m and 1.15 m, which was taken as the reference for the three ornithopod dinosaurs shown in Fig. 3.112.

3.120. Barranco de Valdecevillo tracksite. Natural casts in wall of lateral tracksite path. Front of right footprint (left) and rear part of left footprint (right). Layers are seen folded and cut by action of print, and front mark of two digits. This dinosaur probably had rounded nails.

Behavior

In addition to the hypothesized family behavior of the ornithopod markers of VA9, VA10, and VA11, some other aspects are highlighted. VA12 is a trackway of a large theropod. The footprint length is 57 cm. It seems paradoxical that VA12 crosses VA9, VA10, and VA11 (ornithopods) with no indication that the herbivores were alarmed by the carnivore; the small ornithopod (VA10) would seem particularly vulnerable to the theropod. Perhaps dinosaurs passed by at different times, thereby not interfering with each other's paths. Alternatively, perhaps the theropod dinosaur would not attack an ornithopod above a certain size because it was primarily a piscivore. There are three locations with skeletal remains of a *Baryonyx* or a similar form in the Cameros Basin: Igea (the location of La Era del Peladillo), Enciso (at the same site as La Virgen del Campo;

3.121. Segment of sauropod trackway VA13. (A) Line segment indicates successive positions of trackmaker's midline, as defined by manus (thick line) and pes (thin line) footprints. (B) Lines connect prints of manus and pes on each side of trackway, showing how the dinosaur would have supported itself if it walked with an amble gait.

A

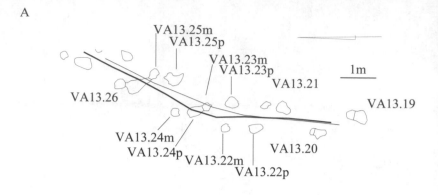

B

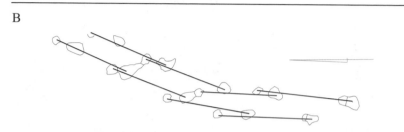

Le Loeuff, unpubl. obs.), and Salas de los Infantes (Burgos province, near the western boundary of La Rioja).

Another type of behavior is recorded in sauropod trackway VA13. Looking at the placement of the forefoot and hind foot prints, the following may be observed (Fig. 3.121). First, beginning at VA13.20, the marks of the forefeet are further away from the hind feet. This happens even though the estimated walking speed of the dinosaur decreases (from about 6 to 3 km/h) in VA13.22–VA13.24. Second, the hind foot print of the pair VA13.24 (VA13.24p) is in front of the mark of the forefoot of the other side (VA13.23 m). It would have to be placed under the body to adopt this position.

The trackway shows the phenomenon of off tracking described by Ishigaki and Matsumoto (2009) for *Breviparopus* in Morocco. The comparison is taken no further because the width of the trackway with forefoot and hind foot is different in the two cases and because the overlap of prints is also different. The foot in VA13.24p is placed under the dinosaur's body—something that should not occur in the Moroccan track. Off-tracking, the gap of the front and rear wheels of the trace, is observed in turning four-wheeled vehicles: "When a right turn is made, the trace of the front wheel is at the left of the rear wheel and opposite" (Ishigaki and Matsumoto, 2009:8). The large sauropods have this type of midline gap, and he "estimate[s] that the walking direction of large sauropods is controlled mainly by forelimbs, not by hindlimbs." However, in quadrupedal vertebrates that have a short glenoacetabular distance, the opposite is true. In my opinion, the walking direction is always controlled by the forelimbs, and the midline gap is controlled by hind limb superposition or the relation between glenoacetabular and limb lengths.

It can be deduced in VA13 that the dinosaur decreased speed and quickly changed its direction of travel. To visualize the change and the phenomenon of off-tracking, the traces of the midline of the hind feet (thick) and forefeet (thin) have been indicated in Fig. 3.121. In the lower part of the same figure, the lines connecting the left and right forefoot–hind foot pairs supported at every step have been drawn. According to this model, if the dinosaur walking style is an ambling gait, the location of the footprints is justified. However, this type of gait is more consistent with narrow trackways than with those of the *Brontopodus* type. The center of gravity in ambling must be above the line joining the forefoot and hind foot on the same side, or the animal's body position must be exaggerated for it to keep its balance. Because the problem is the body's balance, perhaps movements of the neck and tail could accommodate the ambling gait and wide track.

Theropod trackway VA4 (Figs. 3.115, 3.116) also evokes a picture of the trackmaker and its behavior along its path. The average length of the footprint (l) is 55 cm, its rotation is negative (−13°), and the stride is long (306 cm). The pes length (l) data gives the height of the acetabulum as 242 cm, and the ratio relationships of the data give the following results: the trackway is quite narrow ($Ar/a = 0.47$); the limbs were not thick ($z/l = 5.6$); and the dinosaur was walking ($z/II = 1.3$).

The speed of travel according to Alexander (1976) was 6.5 km/h, and according to Demathieu (1986) it was 5.5 km/h. The dinosaur that left VA4 was a carnivorous dinosaur about 7 m long and 2.5 m high, with virtually no swaying of the body (inferred from the narrow trackway and negative orientation) while it was walking unhurriedly.

General Information

In the valley formed by the Barranco de Valdecevillo there are several places where isolated dinosaur footprints and other fossils can be found. They include remains of fish, gastropods, bivalves, and ostracods, as well as plant roots as much as 10 cm thick. The succession of rocks includes limestone, marl, shale, and sandstone, all dark or black in color as a result of organic content. According to Doublet (2004), these rocks are indicative of the environment of the Enciso Group: deep, open lake areas, lake bottoms showing the influence of storm surge waves, protected lacustrine areas, lake borders and shorelines, and areas of vegetation. The footprints were made in a strip formed along the lake borders and shorelines, between the deeper part of the lake and the vegetation zone (Doublet, 2004). A fluvial environment is also recognized in the strip.

Erosion is now revealing more of the site toward the west, as noted in the description of the sauropod footprints (Fig. 3.114). It is unlikely that other areas will suffer erosion because the rock is less weathered. Digging the top to bring out more footprints is not easy or recommended for three reasons. First, the thickness of the sediment above the footprint bed increases rapidly to more than 2 m. Second, weathering, and therefore the

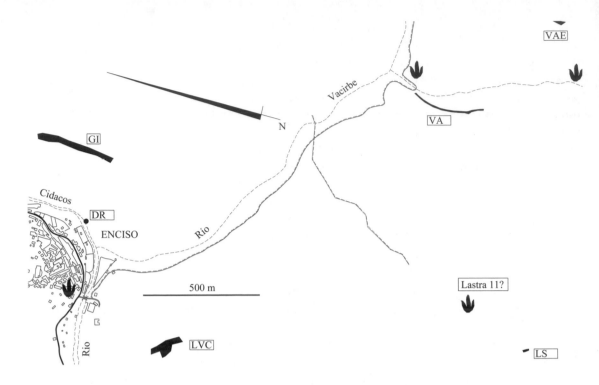

3.122. Location of La Senoba (LS) and nearby tracksites: Barranco de Valdecevillo (VA), Barranco de Valdecevillo Este (VAE), La Virgen del Campo (LVC), Gilerra (GI), and Del Rio (DR). Map also shows village of Enciso, Vacirbe River, and three tracksites not yet studied.

friability of the ground, decreases significantly with depth. At less than 30 cm deep, the rock is no longer decomposed, and it is very hard. Finally, as the hardness of the rock increases, differences in hardness between the layer containing footprints and the shale in the layer above are slight and not evident while digging. The filling in the recesses of the footprints is not released but resists digging.

This site is one of the three first tracksites studied in La Rioja (La Cuesta de Andorra, El Villar-Poyales, Barranco de Valdecevillo) and the place where giant chicken footprints were observed by inhabitants of the region.

La Senoba

History

The site was discovered in 1986 by a shepherd, Domingo Lafuente, the discoverer of many other Enciso sites (East Valdecevillo, Corral de Valdefuentes, Barranco de Valdeño). It was dug and worked on from 1986 and 1988; results were published in 1989. The site was beneath a clay covering remains of walls–and, according to tradition, tombs and treasure. Humans have lived in Enciso since at least pre-Roman times (Celtiberian ceramics have been found). The archaeological team carrying out the dig found neither tombs nor treasure. The walls were of ancient huts for shepherds, and under the clay cover were the footprints that Lafuente found. The name of the site seems to be Latin and therefore is relatively recent in the history of this area. The most likely interpretation (Gonzalez Bachiller, unpubl. obs.) is "white path," with *sen* from *semita* (Latin) or

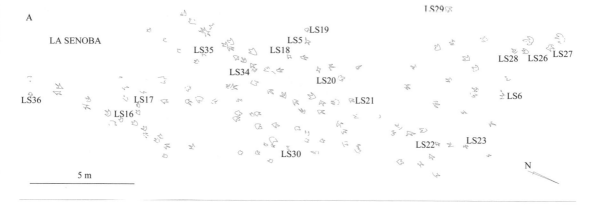

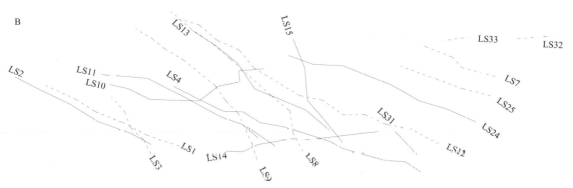

senda (Spanish), "path," and oba from alba (Latin) or blanca (Spanish), "white." The sedimentary layer overlying the track site rocks (black claystones with minor limestones) is white when weathered.

3.123. Dinosaur tracks of La Senoba. (A) Individual footprints and labels of isolated prints. (B) Direction of travel and labels of trackway sequences. Trackway labels are placed at beginning of trackways. Trackways shown as solid lines head north and dashed lines south.

Description of Site

La Senoba (LS) is at the top of a hill on a gently sloping layer (Fig. 3.122). It is 1000 m from the site of Barranco de Valdecevillo, 1420 m from East Valdecevillo, 1550 m from La Virgen del Campo, and 2000 m from Guilera and Del Río. To the south, the nearest are La Muga and Carreterra de Soria, both outside La Rioja. Access is relatively easy during the dry season. During the winter and after rains, the road from Enciso to La Senoba is impassable or difficult for some vehicles.

The site is not protected from the weather; as it is a relatively flat layer, it is vulnerable to winter snow and rainwater. Its location also means it is exposed to the sun, which is directly overhead at certain times of the day during the summer. La Senoba has at least two surfaces with footprints, but they are well preserved only on one surface (Fig. 3.123). This is the top of a layer that is just underneath overlying layers dipping westward. Despite this, further digging to expose more prints is not recommended because the strata are not sufficiently weathered, and the contact between them is so strong that breaking through the overlying rock would likely lead to fragmentation of the footprints beneath. A few meters from

La Senoba, there is a still higher layer and a site with an abundance of gastropods.

Nearly unweathered siliciclastic sediments predominate below the layer with the footprints. Erosion of the southern ravine is vertical or downward, and the slope in that direction is steep. The stratification surfaces are at a counterslope, so there are no large outcrops. It is not an ideal place for sites with footprints. To the north, black shales with interbedded minor limestones cover the track layer. Clay shales provide clayey soil. There are no footprints here either: the rock is crumbly, and they are deep-water sediments from an open lake.

Between this site and Barranco de Valdecevillo is a fault that makes correlation difficult. Nevertheless, by mapping, it can be deduced that La Senoba is stratigraphically lower than Barranco de Valdecevillo. Between La Senoba and Barranco de Valdecevillo are two types of sediments. The first is from an open and deeper lake (black claystones and limestones), and above this are siliciclastic areas with fossil roots of emergent zones separate from the lake.

Lithology

The rock containing footprints is a dark siltstone that in outcrop is almost black when freshly exposed. At first glance, no type of visible fossil can be distinguished, but needles of phyllosilicates are visible. Upon this are several layers of sandstone and a layer with abundant turritellid gastropods, but not in life position. The thickness of the layers and strata is not great (less than 50 cm), and no ripples or mud cracks can be seen. The sedimentary environment is probably the edge of a lake (Doublet, 2004) or siliciclastic shoreline. A certain degree of emergence must be recognized to account for the footprints, but the lack of aerial structures (mud cracks, rain drop impacts, etc.) means that it cannot have been totally exposed to the air. The underlying beds are predominantly siliciclastic, while those above are much more argilaceous with thin, interbedded limestone layers.

The shells of ostracods can be easily distinguished in some levels of these limestone layers. According to Doublet (2004), the sedimentary environment prevailing in the lake would have been deep, quiet water conductive to mud accumulation. Erosion gullies cut into these beds, which when freshly exposed are black, probably as a result of high organic content. However, after weathering, the black color changes to light gray or almost white. Perhaps this is the origin of the name of the site: white along the path from La Senoba to Enciso.

Footprint Features in Relation to Mud

At La Senoba are both large and small footprints, with deep or shallow shafts and with narrow or wide digits. All footprints have digits that are

3.124. Four footprints in trackway LS1; footprint labels are written in chalk on rock. Prints LS1.4 and LS1.6 are lefts, and LS1.3 and LS1.5 are rights. (A) Print LS1.4 is broad with wide toe marks. (B) Toe marks of LS1.6 are much narrower due to inward mud collapse but walls of toe marks have not completely sealed shut. Note elevated rear "heel" area. (C) Print LS1.3 shows shallow hypex, thinning of toe tips due to mud collapse, and rearward projection on "heel" area. (D) Print LS1.5 shows extreme mud collapse and near obliteration of toe marks, as well as shallow "heel" region.

separated and relatively long and tapered at the ends. They are identified as theropod footprints.

Several footprints at La Senoba have digits that are quite narrow, and some have a central rear projection, which is also long and narrow (Pérez-Lorente, 2001a; Pérez-Lorente et al., 2001). For example, LS2.2 has three long, narrow digits; digit III is 30 cm long with a width of 2 cm over more than half of its length. Two or more ichnotypes can be defined in many La Senoba trackways if the shape of the footprints is considered a parataxonomic feature. The last four footprints of trackway LS1 (Fig. 3.124) can be considered typical for the site.

The footprint in LS1.3 is 40 cm long with a triangular outline; the toe marks are short and sharp; the depth of the shaft is about 5 cm; the joining area between the toes is wide, with the hypexes far from the heel. LS1.4 is wider both across the hypexes and in the heel zone. The digits are almost parallel, and they seem to be longer because the hypexes are marked closer to the heel than in LS1.3. The depth is about 1 cm, and the footprint length is 36 cm. LS1.5 has digits that are narrow and long, with the mark for digit II missing. The depth of the footprints is about 1 cm and the length is 42 cm. LS1.6 has impressions of digits III and IV, which are long and narrow; digit III is 27 cm long; the depth of III and IV is about 5 cm, and the width is between 1 and 3 cm; digit II is wider and shallower; the footprint has a groove in the central rear position in the metatarsal position; and there are bulges (rounded areas in the hypex points) resulting from the flow of mud in the interdigital spaces. The footprint length is 45 cm.

NARROW AND CLOSED DIGITS

Many La Senoba footprints have narrow digit marks. The length of the marks varies between 15 and 30 cm and the width between 1 and 3 cm. Casanovas et al. (1989) categorized them as leptodactyl footprints. They can be deep (up to 5 cm in LS1.6) or superficial (about 1 cm in LS1.5). The deep and narrow digital marks of LS1.6 and similar footprints are due to partial collapse of the footprint. The walls of toe marks do not come into contact, or they do so only at the base of the shafts. LS1.5 and similar prints have narrow but shallow digit marks, which can also be explained by the total collapse, leading to the joining of the two walls. The depth of the footprints is therefore highly variable. The present depth of the narrow digital La Senoba footprints should be less than when originally impressed. The direct structure generated on the bottom by the sole of the digits must be buried.

Footprints with narrow digits can alternate in the same trackway with other tracks that have digits that are wide and deep (e.g., LS1.3, where only the distal part of the digits has fused together), or with digits that are wide and shallow. The appearance of the latter (LS1.4) are more similar to an undertrack than a real footprint.

Narrow digit marks are not due to the real shape of the digits but to the collapse of the mud (Fig. 3.124). The fall of the mud into the footprints is something that occurs at many sites. In La Rioja, these include Barranco de Valdecevillo, Era del Peladillo 2 (2PL), and La Virgen del Prado, as well as many other sites. The degree of collapse is variable in these sites. In VA4 (Fig. 3.115), the collapse mainly affects the front (distal) part of digit III, giving the footprint the appearance of having a digit III that is wide and short, and that has a long and powerful claw. In 2PL169.1 (Fig. 3.29) and in *Filichnites* of La Virgen del Prado (Fig. 3.5), the collapse of the mud affects the entire interdigital space and fuses the walls of the digits. At 2PL (Fig. 3.30) and Santisol (Figs. 3.188. 3.189) there are footprints where the collapse is especially noticeable in the hypex area, which is fused.

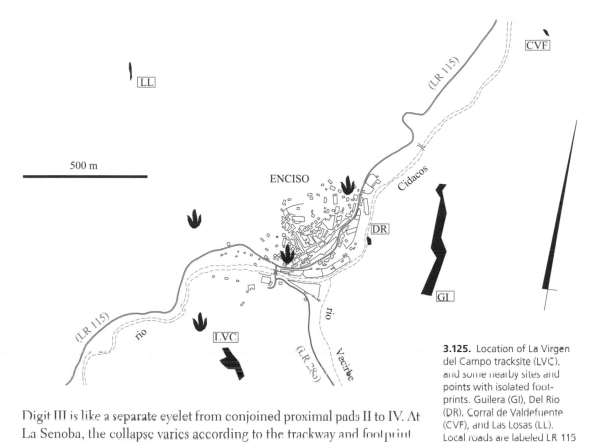

3.125. Location of La Virgen del Campo tracksite (LVC), and some nearby sites and points with isolated footprints. Guilera (GI), Del Rio (DR), Corral de Valdefuente (CVF), and Las Losas (LL). Local roads are labeled LR 115 and LR 286.

Digit III is like a separate eyelet from conjoined proximal pads II to IV. At La Senoba, the collapse varies according to the trackway and footprint

VERY LONG DIGITS AND REAR GROOVE

The length of the digits and overall footprint length vary greatly from one footprint to another, even in the same trackway. In LS1.3 and LS1.4, the digits and feet prints are shorter than in LS1.6. It cannot be said that the shorter marks are due to the collapse of mud because LS1.3 and LS1.4 actually have wider toes than LS1.6. The elongation therefore does not reflect the size of the foot but rather the slip of the foot between the T phase and the K phase.

Some tracks are long and narrow at the back. Casanovas et al. (1989) thought such marks to be impressions of digit I. This interpretation, however, is wrong. The narrow rear part of the track is rather due to collapse of mud into the shaft produced by the metatarsus. In footprints at La Senoba, the trackmakers' feet slipped in the substrate during the T phase before becoming fully supported. This slip between the "touch moment" and the "fully supported moment" created a groove.

If the foot slides forward in the mud during and after penetrating the sediment, it pushes mud forward that accumulates in the interdigital spaces. The first point where the heel touches the ground is behind its final position when the foot is fully supported (phase W). In the interdigital spaces, the displaced mud, as well as mud extruded by downward pressure exerted by the foot in the mud, accumulate. The foot is raised in the W phase, then leaves the ground; then the accumulated mud falls into the footprint and forms lobes (Gatesy et al., 1999; Pérez-Lorente, 2003c; Pérez-Lorente and Herrero, 2007).

COLLAPSE OF MUD

The amount of collapse is a function of the cohesiveness of the sediment and the depth of the walls of the footprint. The less the coherence and the deeper the depth of the wall, the more easily the mud collapses. One might expect that the depth of footprints in the same trackway would be nearly the same in most consecutive footprints. Consequently, the inward fall of mud would also be similar among them. However, the gap between the shaft walls of prints at La Senoba is variable in footprints in the same trackway. It is just a few millimeters in LS1.6, while in LS1.2 the mud in the walls nearly obliterates the gap. In the first case the footprints are narrow grooves (the width of which is in millimeters only) of up to 5 cm in depth, while in the latter (with widths of 1 to 3 cm) they are shallower, about 1 cm deep.

The inward fall of the mud may or may not be the same even in the same footprints of La Senoba. There are footprints of 5 cm in depth where the collapse is limited to the tip of the digits (LS1.3), while in other prints (LS12) can be 10 cm without collapsing. There are numerous footprints representative of each case at the site. Thus, the footprints in La Senoba trackways do not always have the same features. The variation depends on the behavior of the mud, which in turn depends on the physical variables, including consistency, viscosity, and thixotropic behavior. The major cause of such variation in physical properties of the substrate is the difference in water content of the sediment; it seems unlikely that changes in sediment composition in such a fine sediment could be responsible. Places in the tracking surface with more water would be less viscous, and the foot would penetrate more deeply in them. The depth of penetration of footprints in the sediment might also have been affected by the extent to which algal mats covered the surface, thereby affecting its physical properties.

All this means that the mud coherence conditions varied from one point to another (causing changes of structures in the same trackway) and over time (resulting in such variation as deep footprints without mud collapse in trackway LS12, next to collapsed footprints of other trackways LS8 and LS13). Variability in substrate conditions over time would likely involve changes in water content. Variability among footprints made at the same time might involve changes in the proportion of sand to mud in the sediment.

If substrate conditions changed from one point to another, it might be expected that in La Senoba there would be distinguishable boundaries between areas with collapsed footprints and those without such tracks. It could be checked from mixed trackways containing prints of variable features whether the footprint changes were due to differences in space or time. This procedure was followed in the Las Losas tracksite to separate potentially flooded areas from those without any water cover. In La Senoba there are insufficient data to consider this possibility; the site is too narrow and the trackways have a predominant direction (i.e., the trackways do not cross each other). Furthermore, the variability of footprints depends on both the conditions of the tracking surface and the physical conditions up to several centimeters below this surface. In other words, this variability could be due to changes located several centimeters below the tracking surface.

Footprint Features (Ichnotypes)

All the footprints in La Senoba are attributed to large theropods except for one isolated footprint and the trackways LS9 and LS14. The ichnogenus *Fillichnites* Moratalla, 1993, has been described from prints at various points in the Cameros Basin (La Virgen del Prado and a tracksite in Soria province) but has not yet been formally published. The ichnogenus is said to have a Tithonian–Barremian occurrence. Examples to which this name has been applied in La Rioja (La Virgen del Prado) are structures completely changed by the inward collapse of the mud, especially during the K phase of the footprint. Because they are not stamps, such footprints should not be used to define any ichnotaxon.

The relative depth to which the footprints in La Senoba penetrate is lower than at La Virgen del Prado and lower in absolute terms than at VA4. At La Virgen del Prado there are tracks with collapsed marks of the metatarsus and hallux, which are not seen at La Senoba. The VA4 tracks are deeper than those of La Senoba, although the foot in VA4 is much wider than in La Senoba (foot length/width/depth in VA4 = 60/55/10, and in LS1 – 47/35/5).

Footprint Distribution by Ichnotype

The orientation of the trackways (Fig. 3.123) seems preferential, with most trackmaker routes parallel to the north–south direction. Eight of the dinosaurs moved toward the north and seven toward the south. Some of the animals going in the same direction intersect (LS15 with LS5 heading north, and LS1 with LS3 heading south). Of those going north there are two trackways with wide footprints, which are deep and without collapse structures (LS4 and LS11) and which intersect with others that do show signs of collapse. They are all large theropod footprints. Of those going south, there is one with wide and deep footprints (LS12) and a small theropod footprint (LS9). The rest of the trackways are in the north–northwest direction and running both ways. There are also large and small theropod footprints, and footprints with and without mud collapse.

The estimated speed of the trackmakers heading in the same direction varies between 4 and 11 km/h for those going north, and between

3.126. Overview of La Virgen del Campo and its outcrops. Limit of outcrops indicated by continuous white line. There is also a gateway for visitors and two dinosaur models, which represent an attack by a carnivore on an herbivore.

3 and 9 km/h for those going south. The velocity does not depend on the size of footprint. LS7, whose average footprint length is 35 cm, goes at 3 km/h; LS9, whose average footprint length is 22 cm, runs at a speed of between 5 and 7 km/h.

It is difficult to justify the hypothesis of herd behavior in view of the differences between the groups. They are probably different dinosaurs passing at different times, with individual animals that vary in size, speed, and time of passing that go in the same orientation but the opposite way. It can be assumed that the preferred direction was determined by something natural.

Conservation

La Senoba is at an elevation of 1000 m, and the attitude of the strata is almost horizontal, neither of which is favorable for the preservation of the rock. Snow and rainwater collect easily, and sunlight is often directly above the track surface. The climate is also extreme: frost in winter is common, and there is a big difference in temperature between the rock heated by the sun during the summer (ambient temperature of 40°C) and cooled at night (about 17°C).

After the archaeological digging excavation, an oak was seen to have roots at the junction of the two breaks that divide the rock into four sites. The tree's roots were splitting the four blocks, so the tree was cut down when this was discovered in 1986. Later, fractures were cleaned and filled with mortar. The frost split the rock into pieces of varying size, and the daily changes in temperature are the probable cause of the flaking. The rock is black; the maximum temperature reached must thus be at least 5°C above the ambient temperature, so that the diurnal temperature range in summer is about 30°C. There have been four conservation works

3.127. LVC outcrop with situation of significant areas. Lb mud tongue (see Fig. 3.131); Fr, fracture (Fig. 3.132); Sp, slump (see Fig. 3.132); Fb, fracture (see Fig. 3.133); Rp, grooves and striae area (see Fig. 3.135).

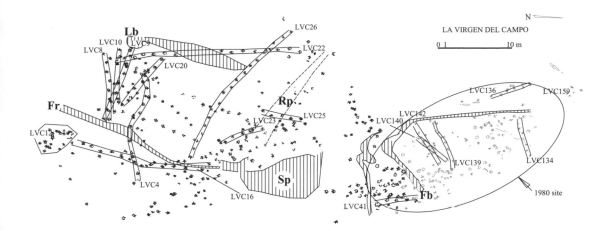

with mortar, silicone, and epoxy resin over several years, all in an attempt to delay the destruction of the site. There is also a fence to keep livestock and people away from the footprints.

History

The first published reference to the La Virgen del Campo (LVC) site is Brancas, Martínez, and Blaschke (1979), although the location was not clearly stated (Fig. 3.125). It was said that Lastra 11-A corresponds to La Virgen del Campo on the map, but this identification must be wrong because Lastra 11-A is at 1000 m of elevation if the outcrop is assumed to be in the right place on the La Senoba (Fig. 3.121). Lastra 12-A would be more likely because it is said to have some 60 prints, with plenty of ripples and tail tracks. Currently the site (Fig. 3.126) is composed of four outcrops (LVC, 2LVC, 3LVC, 4LVC) named according to the order of their being described. The first is LVC, the letters of which stand for the initials of La Virgen del Campo, which is the name of a nearby chapel. The following outcrops were named 2LVC, 3LVC, and 4LVC where subsequently recognized lateral extensions were discovered. Pérez-Lorente, Fernández, and Uruñuela (1986) observed probable tail drag marks (La Virgen del Campo), and in unstudied parts of 4LVC there are grooves that have yet to be described as a result of slump movements of the sediment.

In 1980 (after the study of Brancas, Martínez, and Blaschke, 1979), more work was done at La Virgen del Campo, where a small and unpublished part was exposed. Casanovas et al. (1985) reported the precise situation, and some trackways at La Virgen del Campo as well as at other sites in La Rioja were officially made known to Spanish paleontologists studying dinosaur tracks in La Rioja. In 1980, La Virgen del Campo occupied 350 m², and according to this study, after the summer of 1984, the cleared surface reached 2000 m². Three trackways were analyzed in La Virgen del Campo, tracks 1, 2, and 3, which correspond to LVC134, LVC16, and LVC41 (Fig. 3.127).

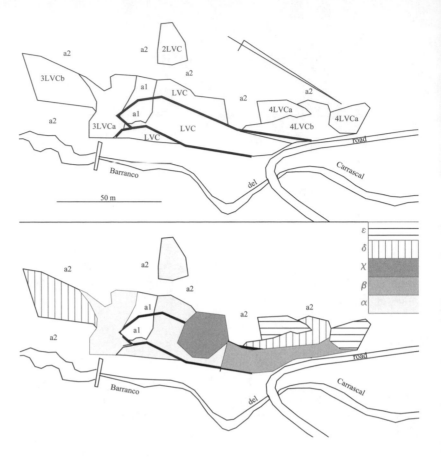

3.128. Outcrops (top) and subsites (bottom) of tracksite of La Virgen del Campo. Description of study surfaces and sectors is provided in Table 3.8 and in text. Subsites are indicated by Greek letters.

Casanovas et al. (1989) studied the footprints of vertebrates at La Virgen del Campo (Fig. 3.127) and increased the number of footprints and trackways described, providing maximum, minimum, and mean data of these footprints and trackways. Casanovas et al. (1991c) extended the study to the south of the site by including an adjacent part (2LVC) of 80 m². Seven theropod trackways were added to the site.

The clearing work that followed uncovered new outcrops (Figs. 3.126, 3.128) to the east (3LVC) and west (4LVC). 3LVC has not yet been studied; it consists of three surfaces with tracks. One of these is the same as the previous outcrops, LVC and 2LVC. Another is more recent and contains a lot of mud deformation and flow structures, while the third is an upper layer with three recognized trackways with shafts and casts preserved.

Work began at 4LVC in 1989, but the marks that came out were strange, some of which are still not explained. Some are small shafts without a fixed shape (between 5 and 20 cm in size); there is a group of nearly parallel and probably concentric striations; and claw marks are evident. The claw marks at LVC and 4LVCa (Figs. 3.129, 3.130) were attributed by Pérez-Lorente et al. (2001) to swimming dinosaurs. In later fieldwork, Ezquerra et al. (2004, 2007) found the continuation of two series of striations in the eastern part of 4LVC (Fig. 3.130B). After clearing the site, other tracks also emerged (Fig. 130C), thought to have been

3.129 Slide marks of two claws with accumulation of mud in posterior part (LVC, χ area; see Fig. 128). Scale provided by 60 mm camera lens cap

produced by crocodiles (Ezquerra and Pérez-Lorente, 2003). The claw marks left by one of the swimming dinosaurs and the trackway of one of the crocodiles converged. The point of convergence was under sediment that could be removed. The crocodile seems to have been moving toward the dinosaur. In 2011, the area where the trackways cross was excavated, revealing that the dinosaur passed by after the crocodile did.

Description of Site

To better locate what is to be described at this complex site, La Virgen del Campo has been divided into sectors (Fig. 3.128), which are briefly described here (Table 3.8). There is a stratification surface that extends across LVC, 2LVC, and part of 3LVC, which will be called the main surface and will serve as a reference. The main area is on the top of layers of sandstone called the sandstone layer (Fig. 3.128, α, β, and χ). Below are various layers of sandstone without tracks called the lower sandstone layer (a1 of Fig. 3.128). Below this layer is a black shale level with gastropods. On the main surface at La Virgen del Campo, there is fine-grained shale with an abundance of bivalve fossils (apparently internal molds filled with sand) and strata with sandy casts (Fig. 3.128, δ). Finally, there is a top layer of sandstone with ripples and swimming tracks (Fig. 3.128, ε). The most extensive surface with the largest number of footprints is the top of the sandstone layer—the main surface. The intermediate shale and sandstone layers have three trackways preserved as sandstone casts in 3LVCb and isolated casts at 4LVCb. In the upper sandstone layer there are several types of footprints in 4LVCa.

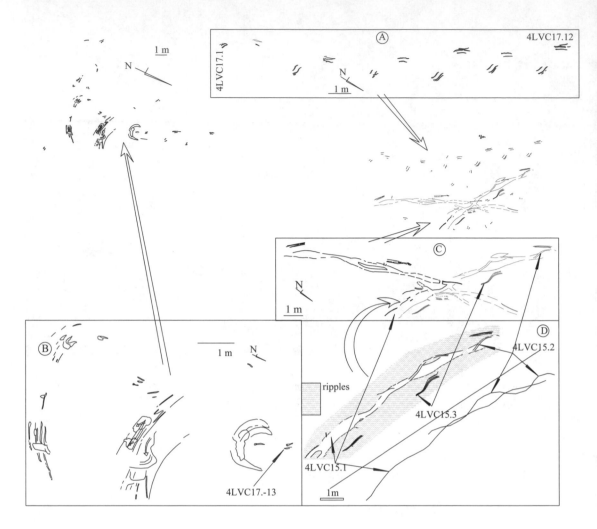

3.130. 4LVCa outcrop with details of sectors. (A) Detail of dinosaur swimming trackway (movement from left to right). (B) Concentric striae and grooves, and first ichnite of swimming trackway (4LVC17.–13). (C) Two semi-flotation trackways attributed to semifloating crocodiles (direction of progression from left to right). (D) Detail of trackway 4LVC15 from (C). Upper portion of (D) (above diagonal line) shows crocodile trackway 4LVC15. Ripples are preserved in shaded region but are absent in sector shown in white. Lower part of (D) shows axes of ventral sliding surface made by crocodile's body (4LVC15.1) and by its tail drag (4LVC15.2).

La Virgen del Campo (which includes LVC, 2LVC, 3LVC, and 4LVC) is spread over about 5000 m² and contains sectors still not dug and footprints not yet studied. The following sites (Fig. 3.125) are located at less than 1500 m from La Virgen del Campo: Del Río (750 m), Guilera (820 m), Corral de Valdefuentes (1015 m), and Las Losas (1250 m). A little further away are La Senoba (1550 m) and Barranco de Valdecevillo (1500 m).

Above La Virgen del Campo is a thick unit of shale and limestone strata that is the same as at La Senoba. The units of La Virgen del Campo and La Senoba correlate laterally. Although they are in the same stratigraphic position, a perfect match for the La Senoba level cannot be established with any of those in La Virgen del Campo because they belong to an active sedimentological area, where there are a lot of flat lenticular bodies and large-scale cross-stratification. Except for La Senoba, all neighboring sites previously listed are stratigraphically higher. At about 2.5 km, on the border with the province of Soria, is the site of La Muga, stratigraphically underneath. La Muga is on the highest part of the Urbión Group or in its transition with the Enciso Group.

Table 3.8. Categorization of tracksite LVC outcrops, sedimentary layers, and notable features.

Lithology	Sector	Content: footprints and other structures
Debris	a2	
Upper sandstone	4LVCa (ε)	Vertebrate and invertebrate ichnites. Claw marks, slide marks, irregular shafts. Real tracks.
Intermediate shale and sandstone	4LVCb, 3LVCb (δ)	Vertebrate footprints and undertracks. Grooves of diagenetic origin.
Sandstone layer	LVC, 2LVC, 3LVCa, 4LVCb (α,β,γ)	Vertebrate and invertebrate ichnites. Real tracks and undertracks. Structures of synsedimentary mechanical origin (fractures and slump).
Lower sandstone	a1	No visible fossils.

Outcrop	Sector	Content: footprints and other structures
4LVC	4LVCa (ε)	Vertebrate and invertebrate ichnites. Claw marks, slide marks, irregular shafts. Real tracks.
	4LVCb (δ)	Vertebrate footprints and undertracks. Grooves of diagenetic origin.
	4LVCb (β)	Vertebrate and invertebrate ichnites. Real tracks and undertracks. Structures of synsedimentary mechanical origin (fractures and slump).
3LVC	3LVCb (δ)	Vertebrate footprints and undertracks. Grooves of diagenetic origin.
	3LVCa (α)	Vertebrate and invertebrate ichnites. Structures of synsedimentary mechanical origin (fractures and slump).
2LVC	2LVC (α)	Vertebrate and invertebrate ichnites. Structures of synsedimentary mechanical origin (fractures and slump).
LVC	LVC (α,β,γ)	Vertebrate and invertebrate ichnites. Plant ichnites? Real tracks and undertracks. Structures of synsedimentary mechanical origin (fractures and slump).
a1, a2		Without footprints.

Study surface	Outcrop	Lithology
ε	4LVCa	Upper sandstone
δ	4LVCb, 3LVCb	Intermediate shale and sandstone
γ	LVC	Sandstone layer
β	4LVCb, LVC	Sandstone layer
α	LVC	Sandstone layer

Lithology

The rocks of the stratigraphic unit with footprints are mixtures of sand and silt with one or the other predominating, depending on the layer considered. The strata with the highest sand content (sandstone layer and upper sandstone; Table 3.8) have real tracks and undertracks on the top. In the strata with the higher proportion of shale (intermediate shale and sandstone; Table 3.8), undertracks and casts are seen. According to Caro and Pavía (1998), the sandstone layers are composed mainly of quartz

3.131. Gravitational mud tongue Lb. Mud slid from top of photograph toward viewer ending (bottom) in lobule of accumulated mud. Area affected by slide is delimited by lift-off surface (top) and lateral sliding surface (right, near pen; see Fig. 3.134 for detail).

grains, with lesser amounts of feldspar and phyllosilicates. There is quartz recrystallization and possible neoformation of phyllosilicates as a result of metamorphism in the area. The mineralogical content in the shale layers is similar, but the proportion of quartz grains and feldspar is much less. According to Doublet (2004), on the main surface of La Virgen del Campo, storm wave crests are seen, which would be deposits from the top of a deltaic lobe in a beach area.

Tracks of vertebrates and invertebrates, as well as slide marks and grooves of known or supposed origin (root and branches of floating trees), are superimposed on the sedimentary structures as a result of water movement (ripples). There are also synsedimentary structures due to whole-mass movements of sediment after the passage of the dinosaurs.

Footprint and Mud Structures

DEFORMATION DUE TO ROCK FLOW AND RUPTURE

The Virgen del Campo study surface is the main surface or top of the sandstone layer (Fig. 3.128, a, β, and χ), and extends through 2LVC and part of 3LVCb and 4LVCa. There are sedimentary structures on top of the bedding (ripples) with the following superimposed: bioturbation resulting in small craters or holes that produce a uniform pockmarked; bioturbation by dinosaur footprints; and deformation due to sediment flow and

rupture (fractures, tongues of mud, slump), mainly in zone a (Figs. 3.127, 3.131, 3.132, 3.133).

Neither the intermediate shale and sandstone layer nor the upper sandstone layer are bioturbation structures of type 1 or type 2. The intermediate shale and sandstone preserve the breaks and flow of the rock (sandstone layer). Some literature (Pérez-Lorente, Fernández, and Uruñuela, 1986; Pérez-Lorente et al., 2001) suggests that the breaks, tongues of mud, and slumps may have been caused by a Cretaceous earthquake.

The history of the rock (sandstone layer) would have to be written from the time when the movement of sand grains produced ripples of various scales. There are ripples affected by the bioturbation of invertebrates (*Planolites, Scolithos*) with dinosaur footprints above. The sequence of events is as follows. First, sedimentary structures (ripples) are created; second, invertebrates colonize; third, dinosaur make tracks; and last, plastic mud movement and fractures occur. According to this account, the sediment moved after the first dinosaurs passed and before the sedimentation of the intermediate shale and sandstone. There are fracture lines in an approximately north–south direction, with others cutting across that are more irregular. Slump, lobes of mud, and mudflows (Figs. 3.131, 3.132, 3.133) are related to the fractures. These structures only affect the main surface (sandstone layer) and do not affect the upper strata.

One of the faults (Fr) starts (Figs. 3.127, 3.132) on the side of a slump (Sp), and as it opens, the deformed sediments form an increasingly large mound. Between Fr and the northern edge of the site there are raised

3.132. Fracture (Fr in Fig. 3.129) developing to left (south) part of LVC outcrop that started in Sp. Symbols as in Fig. 3.128: α, sandstone layer, main surface; δ, intermediate shale and sandstone; a1, lower sandstone.

3.133. Mud flow between edges of fracture with mud tongue in center of photo. Length of black line, 4 m. Structure is that labeled Fb in Fig. 3.127. There are no footprint marks where mud flows.

3.134. Edge of mud tongue (Lb in Fig. 3.131) cutting across footprint LVC9.1. Part of footprint in tongue (left half) has been removed by movement of mud. Black groove separating two parts of photo (left and right) is lateral surface of tongue that flowed Lb. Structure Lb is illustrated in Fig. 3.131 and situated in upper left part of Fig. 3.127 (LVC outcrop).

3.135. Irregular grooves and striae attributed to drag of roots or branches of floating tree trunk (Rp band in Fig. 3.127).

spaces between the sedimentary laminae due to the movement of mud, accompanied by recrystallization of carbonates in the gap it leaves. The wall of the fracture (Fr), from its inception and without breaks, is a structure of this type. The raised spaces and the wall of the fracture can be confused with plant stems and at first were interpreted as such (Pérez-Lorente, Fernández, and Uruñuela, 1986).

Some La Virgen del Campo footprints have been eliminated, cut off, elongated, or flattened according to their position. In areas of strong sediment flow, the tracks disappear. This is seen in stretched areas of the slump (Sp), and on the inside of the tongue (Lb) and location (Fb) of mudflow (Figs. 3.133, 3.134).

Footprint and sedimentary structures are obliterated within the limits of the flow, such as in trackway LVC9 (Figs. 3.127, 3.134). A stable sector preserves digit impressions of print LVC9.1 (Fig. 3.134, right) but the footprint has been erased in an area of mud that moved. Where the motion was less destructive, the tracks are narrower and longer or shorter and wider, depending on their orientation. In LVC12.7, the sediment slipped perpendicularly to the longitudinal axis of the foot. In LVC12.3 and LVC12.4, the footprints have their longitudinal axis subparallel to the direction of the mud movement. Estimates of the magnitude of shortening or stretching cannot be made because the exact direction of maximum compression is not known. Even if the footprints were from the same dinosaur, the length of the shafts and interdigital angles would probably be different.

3.136. Straight grooves at LVC presumed to have been made by the sliding tail of a vertebrate (Pérez-Lorente, Fernández, and Uruñuela, 1986).

ICHNITES OF INVERTEBRATES AND PLANTS

Numerous invertebrate traces occur at La Virgen del Campo. These include the ichnotaxa *Skolithos* and *Planolites*. The traces were made after the ripples but before the passage of the dinosaurs. There are also grooves and slide marks, which are parallel and discontinuous, straight or sinuous (Figs. 3.127 [Rp], 3.135). They occur in a band with a minimum width of 50 cm over a length of 15 m. Across this band, grooves appear and disappear, and it is not clear whether any of the discontinuous grooves were made by the same object or objects. These marks may have been made by the branch tips or roots of floating trees. Elsewhere in the region, dropstones in muddy sediments have been interpreted as fallen from clay trapped in the roots of floating logs (Doublet, 2004). Other parallel marks occur in bounded bands (Fig. 3.136). Pérez-Lorente, Fernández, and Uruñuela (1986) suggested that these might have been made by the sliding tails of dinosaur or other large reptiles.

FOOTPRINT STRUCTURES IN MUD

Footprints at La Virgen del Campo occur as both open and filled shafts. Among the open shafts are stamps, deformed footprints, and undertracks. Among the filled prints are casts, more or less destroyed by erosion, which are faithful copies of the real hollow left by the foot.

The stamps are characterized by preserved marks of the pads; they show no deformation. Some trackways contain stamps and other true

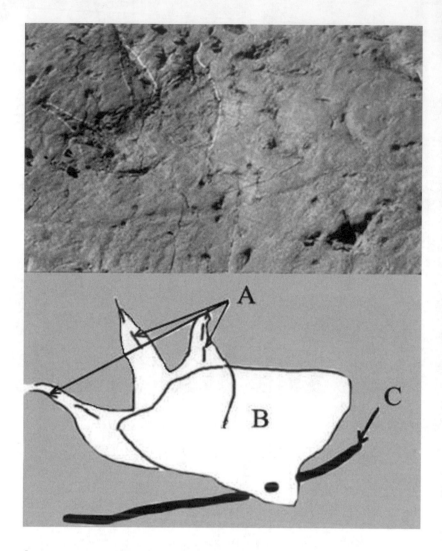

3.137. Footprint LVC22.3 showing incisions (A) left by claws while lifting foot (K phase) and striae (B) through sole of foot during T phase. An unrelated groove (C), probably made by an invertebrate, preceded formation of footprint.

footprint types. Changes in the properties of the mud or deformation of some parts of the site may lead to variations. Examples of stamps are the footprints of LVC8, some footprints of LVC10, LVC4, and LVC25, and probably the LVC124 and LVC14 trackways.

In addition to footprints consisting of direct structures but deformed by the movement of the sandstone layer (or main surface) there are other changes to the footprints. Associated with the T phase are the parallel grooves in print LVC22.3 (B zone of Fig. 3.137) that run from the back to the center of the track. This footprint was made after invertebrate traces (e.g., Fig. 3.137, mark C). In LVC159 (Fig. 3.138), at the back of the footprint is a smooth surface. In both LVC22.3 and LVC159, the sole of the foot slid across the ground before being supported, resulting in slide marks or smooth areas located at the rear of the footprints.

In theory, the foot reaches its greatest depth in the W phase, when the foot experiences maximum weight loading. However, the depth of the footprints in La Virgen del Campo varies in a more complicated manner.

3.138. Footprint LVC159. Smoothed area left by sole pad during T phase. Mud was pushed forward in W phase.

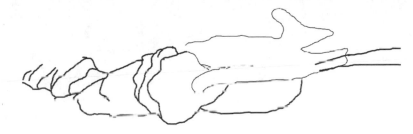

Footprints can vary considerably in depth, even over the same trackway. In LVC26, some intermediate tracks (LVC26.4, LVC26.5, LVC26.6) are so shallow that they are barely visible. However, the rest of the tracks are clear and much deeper (Requeta, 1999).

At LVC and 4LVCb (Fig. 3.128), the main surface is exposed with true tracks and undertracks. The undertracks are usually deeper. This is the case because the depth of the footprints does not depend on the weight of the dinosaur but rather on the viscosity of the ground beneath its feet. A dinosaur cannot change its weight as it walks. Yet the viscosity of the mud does not always increase over time; there are undertracks of dinosaur footprints that penetrate deeper into the ground than others made beforehand (larger and true LVC140 footprints are more shallow than the smaller undertracks at the same surface). The viscosity of the ground (sandstone layer) decreases after spending time buried under modern shale layers (intermediate shale and sandstone), perhaps because of the increased water content of the sediment.

At LVC, during the W phase, there was also sometimes mud movement induced by the treading of a dinosaur (Fig. 3.138). A slide may have begun before the foot was fully supported on the ground, toward the end of the T phase. The claws could scratch forward or backward on the ground when the dinosaur raised its foot (K phase). The latter is what is

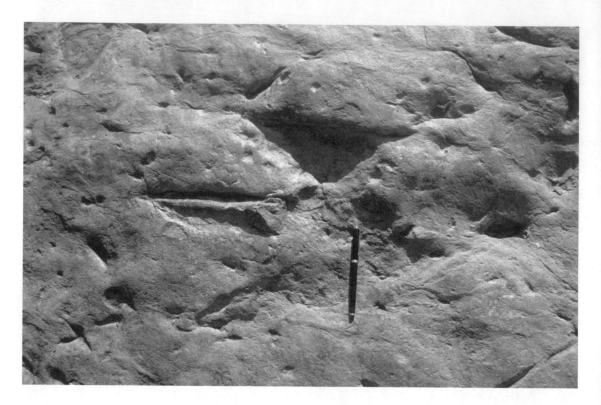

3.139. Left footprint LVC25.2. Mud is dragged by adhering to sole two digits of foot.

seen in footprint LVC22.3 (Fig. 3.137A), where three grooves that start at the tip of the mark of the nails scratch backward through the footprint.

In addition to these structures, in various trackways (LVC4, LVC8, LVC20) are footprints where mud stuck to the soles of the feet and was pulled from the bottom of the shaft during the K phase, leaving the shaft partially filled (Fig. 3.139). This filling ends at the top of a single acute line running lengthwise through the digit mark. In the intermediate shale and sandstone layer of 3LVC and 4LVC there are footprint shafts filled with sandy sediment after the K phase, leaving casts. In 3LVC there are three such trackways of small tridactyl theropod footprints (Fig. 3.140). No pads or other anatomic structures can be discerned in either the occasional casts or the shafts left.

At 4LVCb, the footprints are larger and deeper. There are both tetradactyl and tridactyl footprints (Fig. 3.141A, B), which bend the underlying sediments, with the shafts later filled by sand. The previously formed footprints produce undertracks (Figs. 3.142, 3.143) as hollows with a subcircular outline that is also expressed on the main surface. Undertracks are found in all sectors of the Virgen del Campo outcrop but are better developed in the western part of LVC (Fig. 3.128, β) and in 4LVCb (Fig. 3.128, β, δ).

SWIMMING MARKS

3.140. Part of unnamed trackway of sandy casts in 3LVC in shaly matrix.

Expanding on the categories of phases of footprint formation of Thulborn and Wade (1989), Pérez-Lorente (2001a) introduced an S phase to

3.141. 4LVCb. Tridactyl cast (A) and fragmentary cast (B) showing sandy matrix of fill. Underlying shale layers were deformed by creation of footprint.

3.142. Formation of true tracks, casts, and underprints on same surface, in sites composed of sandstone and shale. Column A: A layer of sand is deposited (panel 1). Discontinuous layers of mud (b) are subsequently deposited upon low-lying portions of sandy layer (panel 2). A dinosaur walks across this surface (panel 3), leaving true footprints both in mud and sand. In sandy layer, only true tracks are made (gray arrows; d). In track surface with patches of mud, true tracks are made in mud, but underlying sand is also deformed to make undertracks (white arrows; c). At some point in the future, erosion removes shale that formed from mud (cf. panels 4 and 5), destroying true tracks in surface of shale. Removing sandstone (panel 5) therefore preserves both undertracks (areas c) and true tracks (area d). Column B. A dinosaur walks across sand layer (panel 1), making a series of true tracks (gray arrows; d). Tracked layer is buried by layer of mud (panel 2), after which another dinosaur walks across this continuous muddy surface (panel 3), making true tracks and undertracks. True tracks in mud layer are then filled with sand (panel 4f). Modern erosion creates complex study surface (panel 5). In some places, only true tracks (d) originally made in sand layer are preserved (panel 5, left). In other places, both true tracks made in sandstone (d) and also undertracks (c) transmitted from now-destroyed overlying shale layer remain (panel 5, middle). In the most complex portion of study surface (panel 5, right), casts of prints made in mud layer (f) occur along with true tracks in sandstone (d) and undertracks (c).

encompass structures produced by the foot sliding on a mud surface under water when a vertebrate is swimming or semifloating. In areas χ (Fig. 3.128) and ε (Fig. 3.129) of La Virgen del Campo there are many structures left by two or three autopod claws that contacted the substrate. No relationships among the marks can be discerned for determining the direction of progression of any trackmaker.

At 4LVCa there are claw marks of chaotic and isolated distribution, as well as a sequence of claw marks (Fig. 3.130C) forming a trackway (4LVC17). The sequence is clear on 4LVCa and eroded in sector 4LVCb. The western part of 4LVCa contains a trackway consisting of repeated footprints in the form of groups of two or three claw marks (Fig. 3.144). These footprints have been interpreted as marks left by a dinosaur while swimming (Ezquerra, Costeur, and Pérez-Lorente, 2010; Ezquerra et al., 2007).

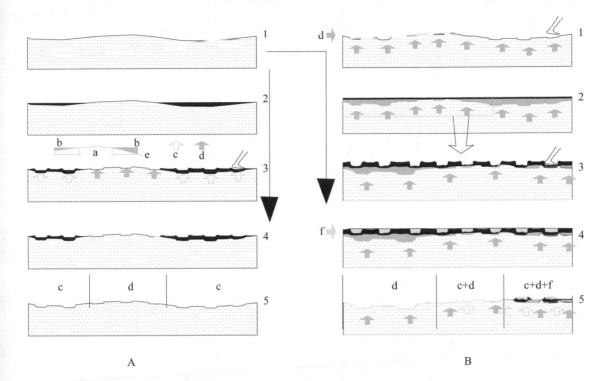

A   B

Trackway 4LVC17 begins further back than described by Ezquerra et al. (2007). If footprints are numbered backward with a negative sign (e.g., 4LVC17.–1 precedes 4LVC17.1), the first print would be 4LVC17.–13 (Fig. 3.130B). No such slip marks are seen further to the left. On the other hand there are two groups of striae and a contour line structure similar to the horns of a bull, which have not yet been interpreted (Fig. 3.130B).

Other drag marks could also be assigned to the S phase (Fig. 3.130C, D). They have been interpreted as marks left by two crocodiles partially floating in the water (Ezquerra and Pérez-Lorente, 2003). The trackways are composed of three different types of structures. In trackway 4LVC15 (Fig. 3.145), marks identified are those of the belly (4LVC15.1), the tail (4LVC15.2), and the hind limbs (4LVC15. 3). The first is a smooth, wide band that destroys the ripples, the second is a long groove between 5 and 20 cm wide, and the last are parallel sets of grooves (three distal ends of pes digits; Kumagai and Farlow, 2010) 1 cm deep and 100 to 140 cm long. None of the structures is straight. The midline of the smooth band is sinusoidal, as is the tail mark (Fig. 3.130C, D). In accordance with the movement of the parts of the body, the midline of the belly mark and the mark of the tail wavelength are the same, with the wave amplitude of the first being less than the second. Foot marks are discontinuous and sigmoid, and they show the tips of the digits, perhaps made by the claws of the hind feet. There are no forefoot marks. Unlike other tracks of crocodiles swimming (grooves and holes left by tips of the digits) and crocodile gait (traces of hand, pes, body, and tail) described by Farlow and Elsey (2010) and Kumagai and Farlow (2010), in the two tracks 4LVC15 and 4LVC16 there are all types of marks except manus marks.

3.143. LVC. Main surface with many undertracks (see Fig. 3.142, footprints labeled "c").

Track Distribution by Structure

The footprints of vertebrates and invertebrates, and the deformation structures of strata due to rock movements, occur in different parts of the site (Pérez-Lorente et al., 2001). The distribution can be examined in the light of the rock surface where the ichnites are found. Throughout the main surface (sandstone layer) there are small visible bioturbation craters (Figs. 3.133, 3.135–3.139, 3.143), distributed evenly. The distribution of the other ichnites is different. Real tracks (stamps and others) are more common than undertracks in the 2LVC outcrops and the eastern part of LVC (Fig. 3.128, α and χ). Real tracks are less common than undertracks in the western part of LVC and 4LVCb (Fig. 3.128, β). *Planolites* and claw marks are more abundant in area χ (Fig. 3.128) of LVC.

Casts are preserved in the intermediate shale and sandstone outcrop area. The sedimentary layers of shale have axial downfolds (Loope, 1986) under the casts. The complete profile of a cast has not yet been seen, so it is not known whether in the track shaft walls there are levels cut off or crushed above the study surface.

At the top of the upper sandstone (Fig. 3.128, ε) occupied by 4LVCa there are several intersecting trains of ripples of variable amplitude. Above them are irregular holes, probably made by bivalves; claw marks; and, on the bottom, drag marks of various parts of vertebrate (dinosaur and crocodile) bodies.

Track Features (Ichnotypes)

FOOTPRINT SHAFTS

Many footprints with rounded contour lines and lacking acuminate digit marks were drawn in the study by Casanovas et al. (1989:125). These authors assumed that the prints were not stamps "due to the unreliability of the shapes." Although they all have ornithopod features, only the footprints in trackway LVC41 were assigned to an ornithopod. They are very irregular and would now be classified as unidentified. The rest of the dinosaur footprints identified in La Virgen del Campo are of theropods. There are a large number of unidentified footprints.

No names have been assigned (either ichnogenus or ichnospecies), and ongoing studies need to be completed before applying a more logical classification than that used to date, if this is even possible. There are several ichnotypes that are quite different in size and shape.

The ichnogenus *Megalosauripus* (Lockley, Meyer, and Santos, 1998) or *Megalosauropus* (Colbert and Merrilees, 1967), may be applied to all large theropod footprints, whether deformed or not, at the La Virgen del Campo tracksite. All tridactyl footprints at the La Virgen del Campo track site are more than 25 cm in length. However, care must be taken with the use of these names because they are not fully supported (Romero-Molina, Pérez-Lorente, and Rivas, 2003; Thulborn, 2001), and because some have

3.144. Print 4LVC17.7. Dinosaur claw marks made while swimming. Trackway 4LVC17 is drawn in Fig. 3.130A.

3.145. Trackway 4LVC15, attributed to a crocodile. (A) Overview of trackways, with dark shading in its components (sliding mark of ventral surface of body, tail, and autopodial digits). (B) Close-up of initial part of trackway (without shading).

been defined using deformed footprints–for example, the *Megalosauripus* of Casanovas and Santafé (1974) on trackway VA4.

There are three morphotypes of large theropod footprints (Fig. 3.146) longer than 40 cm. One of these is trackway LVC8, which has the following distinctive theropod track features: relatively long, separate digits; more than one pad per digit; acuminate ends; proximal part of digit II forward with respect to the heel; and interdigital angle II^III less than III^IV. They also have the following features, common to this ichnogroup, which are different from typical theropod footprints: wide mark of the foot ([l − a]/a = −0.06), a consequently wide II^IV angle (65°); and short and wide digital pads.

The second of the three morphotypes is more common, as it is recognized in several LVC trackways (e.g., LVC5, LVC18, LVC25) and in all trackways of 2LVC. It differs from the first morphotype (Fig. 3.146) in that the foot is narrow ([l − a]/a = 0.33, 0.27), the digital pads are long, and the divarication angle of the digits is less (II^IV = 47–60°).

Finally, the third of the three morphotypes comprises a large theropod trackway (LVC140) with faint markings, which have theropod stamps from the next type (Figs. 3.127, 3.146) superimposed on them. The prints in this trackway are the largest stamps at the site (l = 50 cm) and have wide digits and digital pads.

The other type of undistorted footprints that can be seen in some trackways (LVC134, LVC136, LVC139) are assigned to smaller theropods whose footprints exceed 25 cm in length and whose digital pads can be distinguished (Fig. 3.146). As in the larger theropod prints, the forward position of digit II is clear. There is no previously named footprint taxon to accommodate such tracks, only the generic ichnotaxon *Megalosauripus*. Casanovas et al. (1989) assumed that the ratios ([l − a]/a; III/l, length/ digit width, etc.) were more important than the size of the footprint. They

3.146. Types of IVC left theropod footprints. Scale = 1 m.

LVC8   LVC18 LVC25   LVC5 LVC140

1 m

LVC136 LVC134   LVC139

classified these footprints within the theropod group in an ichnogroup called T_3. The following features are notable: a footprint as long as it is wide; tapered digits; relatively wide II^IV angle; and prominent and irregular heel (strong indentation in digit II). The projection of digit III of Demathieu (1990) and Weems (1992) is perhaps relevant, as it is higher in small theropods than large ones. The interdigital angle is greater the larger the projection of digit III.

Toward the end of one of the trackways (LVC22) there are metatarsal marks in some of the footprints (Fig. 3.127). The metatarsal mark is at a higher level than the digitigrade sole. This can probably be explained in the same way as for footprints at the Las Losas site. as it moves forward, the dinosaur first lifts the metatarsus, so the animal's weight falls on the digits, which occupy a smaller area than the semiplantigrade footprint, and they sink deeper.

CASTS

The casts with recognizable autopod shape are tridactyls with long, tapered digits. There are casts with rounded contours where the shape of the sole cannot be discerned. Only theropod casts have therefore been identified.

Casts are assigned to two groups depending on their size and depending on the outcrops where they are found: three trackways of small footprints in 3LVCb and several, apparently disordered, large casts in 4LVCb. All the casts are filled with fine to medium sand, and their hollow sits in dark shale of varying color shades depending on its state of weathering (gray, green, blue).

Some casts are shallow (Fig. 3.140). There are well-separated digit marks of varying thickness, which is probably due to diagenetic deformation. The length of the tracks is around 20 cm. Two of the trackways are parallel, and the third is perpendicular to these. They have not been studied or counted, but there are expected to be more than 30 footprints.

The large casts measure between 40 and 60 cm in length. The measured depth is about 30 cm. Sometimes digit marks can be identified

(Fig. 3.141A), including the hallux. The structures are similar to those described at the site of Costalomo (Torcida et al., 2006) in the sense that the digits stick out of the mud at the back, leaving gaps at the front. These footprints are responsible for the undertracks of sector b in LVC (Figs. 3.142, 3.143). Under the casts, the intermediate shale and sandstone are deformed, as is the top of the sandstone layer (i.e., the main surface), leaving lots of undertracks.

Natural Barriers and Water Depth Variation

The development of the site cannot be said to be simple – not even that of the main surface. By analyzing the organic and inorganic structures, the following comments can be made (Fig. 3.142).

The formation of the sandstone layer (layer a) occurred under conditions determined by the composition of the rock (shale/sandstone) and the sedimentary structures (ripples) it contains. Colonization by bivalves and other invertebrates occurred after the previous sedimentation and ripple formation phase. At some point, the invertebrates disappeared and dinosaurs crossed the surface, modifying the burrows.

The passage of the dinosaurs suggests that the water depth was not the same throughout the main surface because there are both deep and shallow stamps, depending on where they occur. There are marks indicating swimming, sliding of skin, and floating of bodies. The water depth must have varied so that the dinosaurs' feet, at least at some points, did not touch the bottom in sector χ in LVC (Fig. 3.128).

On the same main surface, the distribution of tracks and undertracks indicates that the amount of sediment (intermediate shale, sandstone) deposited above the basal sand layer was not uniform (Fig. 3.142), so that there are areas with and without undertracks, depending on their location.

A water depth of more than 120 to 150 cm in 4LVCa (Fig. 3.130) is well indicated by the marks of swimming dinosaurs and semiswimming crocodiles.

It is difficult to know from the evidence at the site whether all the dinosaurs passed at the same time, indicating gregarious behavior, even for those trackways with parallel paths. There are overlapping footprints and variations in depth of some trackways with respect to others, with some seemingly contradictory features. For example, LVC140 was produced by a large theropod dinosaur (50 cm footprint length), but it is shallower than LVC142, made by a smaller theropod (34 cm).

General Information

FOOTPRINT DENSITY

The number of vertebrate footprints on the site cannot be calculated. The number of claw marks, smooth bands due to dragging along the bottom, or crocodile footprints have not been counted. One reason is that it is unclear whether or not some of the striae marks and smooth bands are vertebrate ichnites.

The footprints recorded are an arbitrary number of stamps, true footprints, and undertracks. During the first dig, boundary lines for certain casts were drawn. The ability to draw these lines accurately was so limited in some cases that the casts were not included. Consequently, there are many shallow, rounded undertracks left uncounted. In addition, the count is to some extent atificial because some undertracks have been described as footprints and others have not. It is also difficult or impossible to know the limit for considering certain tracks as real footprints or undertracks.

If the recognized tracks in La Virgen del Campo tracksite are taken, the footprint density is 0.36 per square meter; an area of 10 m² is thus likely to include about four tracks.

CONSERVATION

There are two problems in conserving La Virgen del Campo. One is that, like all other sites in La Rioja that are not covered, it is subject to attack (e.g., water, frost, temperature changes in the rock, colonization by plants), leading to fragmented, flaked, and chipped surfaces. Another is landslip of the west slope to the northeast. From the junction with the road and to the north, the Barranco de Carrascal (right part of Fig. 3.128) is incised several meters, leaving the sandstone of La Virgen del Campo without support in its lowest part. There are open cracks in the rock caused by the movement of the slope toward the ravine.

La Virgen del Campo is one of the sites where attempts to maintain the area have been made in recent years. The edges of the strata have been reinforced (Fig. 3.145A), loose fragments have been stuck down, and open cracks have been cleaned and filled with mortar and silicone. The area is protected by a wooden railing and a walkway set into the rock by gravity, without any anchoring or attachment (Fig. 3.126).

Main Features of Site

WAS THERE REALLY A FIGHT?

Trackway LVC16 and the group of tracks in LVC12 (Fig. 3.147) have been interpreted as having been made during a fight between a carnivorous dinosaur (LVC16, part of LVC12) and an herbivorous dinosaur (part of LVC12). There is no possibility of further observations in the surrounding area because the sandstone layer or main surface is eroded around the meeting point. A theropod trackway advances in one direction, reaches a point where it is interrupted because the dinosaur steps, then returns to step over the area in such a way that some of its footprints overlap with each other. In the same area, shorter tracks with much wider digits do the same and overlap. The speed of the theropod trackway is that of a slow

3.147. LVC12 and LVC16. Attack. Trackway (LCV16) of large theropod footprints progressing from right to left (up to LVC12 group). In LVC12 is a chaotic mix of long and short footprints, which are superimposed.

walk at the beginning, moderate to rapid speed in the intermediate area, and finally a large step before entering the LVC12 group. The interpretation of a carnivorous dinosaur attacking an herbivore may be correct, but perhaps there is another possibility.

Looking at the deformation in the sandstone layer (Fig. 3.127), we see that fracture Fr is beside LVC12–LVC16 and that there is a rock slip toward the east in the part occupied by the footprints. It may be that, depending on their location, some footprints decrease in width (LVC16 and those parallel to it) and others increase (perpendicular to Fr). Might the same trackmaker therefore have made both the putative theropod and ornithopod footprints? Any hypothesis is possible. The reason for the concentration of footprints in LVC12 and their overlapping remains a mystery.

WAS THERE REALLY AN EARTHQUAKE?

The main surface is deformed by flow and rupture. The rock is brittle in the fractured surfaces, plastic in the slump, and fluid in the mud tongues and liquefaction points. Its deformation gives rise to folds like slumps or is clearly split. The flow may be because the sandstone layer thins, so that it moved like a liquid, or there may have been a mudflow.

The footprints either break or disappear depending on the elasticity and plasticity of the area in question. Almost all these structures (fractures, mudflows) are parallel or contiguous with each other (Figs. 3.127, 3.132). There is no upper sediment trapped in the tongues of mud. The deformation occurred when the rock was not covered by later sediments; it also occurred after the dinosaurs passed and before the sedimentation of the upper layer. It could have been submerged, but there was no erosion between the time the footprints appeared and when they were deformed.

The movement of the sandstone layer could be gravitational as a result of the loss of stability of the deltaic lobe in which it occurs. The cause of the loss of stability may be a small change in the slope of the deltaic lobe due to movement in the sedimentary basin or erosion at the base of the slope. Interestingly, the movement is toward the south in mud tongue Lb and eastward in Fr, Sp, and Fb. In other words, it is moved away from the area where the dinosaur claw marks are concentrated on the main surface, or centrifugally with respect to the deepest part of LVC (Fig. 3.128, χ), just the opposite of what would be expected. It is therefore likely that the rupture and flow of the sandstone layer is indeed due to movements in the sedimentary basin—that is, an earthquake.

A resident of Enciso found the site in 1996 while walking through the countryside. Only a few footprints were seen among the vegetation and soil (Fig. 3.148A). The man who found the tracks was in an area separating two plots of land that had been cultivated for some time. In July 1997, the slope between the plots was cleared and relevant data obtained. Las Losas (LL) covers an area of 400 m² and could be extended because the footprints go under one of the plots and laterally toward the west.

The initial study was performed by A. López for a thesis at the University of Granada in 1999, when news of it was first published (Fig. 3.149). After the visit of the late paleoichnologist W. A. S. Sarjeant to La Rioja, and in view of the difficulties in working at the site, a working party consisting of Sarjeant and part of the team at the Fundación Patrimonio Paleontológico de La Rioja was formed. Sarjeant participated actively in compiling the results, which were shown in two posthumous publications (Romero-Molina et al., 2003; Sarjeant et al., 2002). The use of the word "semiplantigrade" rather than "plantigrade" for footprints to indicate the support of the metatarsus was one of his contributions to the terminology of dinosaur ichnology.

Las Losas is the site with the largest number of semiplantigrade footprints in La Rioja. The first such semiplantigrade trackway discovered in La Rioja was at another site, the Barranco de la Sierra del Palo, found by Brancas, Martínez, and Blaschke (1979). These authors assumed that the elongated mark at the back part of the foot was made by the hallux of a dinosaur (similar to an *Ornitholestes* or to an *Ornithomimus*) placed directly backward. A map of the Barranco de la Sierra del Palo trackway made in 1979, with a revision made 15 years later, was published by Pérez-Lorente (1993a).

Description of Site

Las Losas is on a layer of sandstone with a N164°E strike direction and a 26°E dip. The study surface has been farmed and shows signs left by iron plows (Fig. 3.150B) that are not to be confused with natural features. Access is by foot, either from the road surrounding the village of Enciso or by taking a path from the Enciso castle and passing by the side of two defense pits (probably Celtic). There are no known track sites within a kilometer of Las Losas. However, there are some outcrops at the same rock layer known further west and another about 200 m to the northeast (Fig. 3.151). The nearest studied tracksites (Fig. 3.151) are La Virgen del Campo at 1250 m (off the figure), Del Río at 1100 m, and Guilera at 1400 m. To the north and northeast, at more than 1500 m, are Barranco de Valdeño and Corral de Valdefuentes, while at more than 4000 m to the north are Peñaportillo, La Canal, Malvaciervo, and others.

3.148. Las Losas tracksite. (Top) Site in 1996, the year of its discovery. (Middle) During cleaning as part of summer courses and work camps in 1997. (Bottom) At end of 1997 work camps.

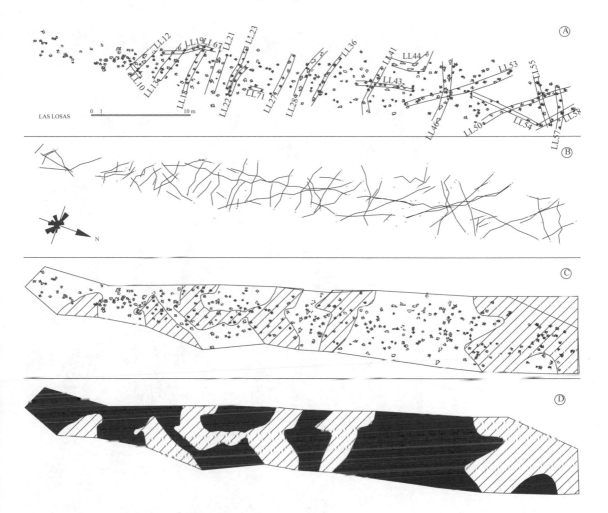

Lithology

The rocks at Las Losas are gray to dark brown sandstone. The thickness of the layer is greater than 20 cm (which is the depth of the deepest footprint). No sedimentary structures, such as ripples or mud cracks, occur, although there are many mud deformation structures syngenetic with the footprints. In addition to the tracks are numerous surface irregularities, such as small craters, possibly caused by bioturbation. Stratigraphically, Las Losas is in a siliciclastic unit above La Virgen del Campo, from which it is separated by a lacustrine interval indicated by a black claystone. The site is located a few meters higher than the level of the roots, which Doublet (2004) described as being above the black claystone near La Senoba track site.

Footprint Features in Mud

The footprints at Las Losas range from completely filled with sediment, partly filled, or completely free of sediment. The filling is argillaceous

3.149. Las Losas tracksite (LL). (A) Map of footprints and trackways; trackway labels are placed in the beginning of trackways. (B) Mapping of pace sequences of trackways. (C) Distribution of footprints in terms of their overall character; typical tridactyl prints occur in cross-hatched sectors, and atypical (irregular and semiplantigrade) ichnites occur in unmarked sectors. (D) Variability of substrate conditions of track surface at time dinosaurs crossed it. Inferred places with soft mud are black, and places with firm mud are cross-hatched.

3.150. (A) Footprint LL23.8, showing striae produced by claws (K phase). (B) Scratch marks caused by plow during farming.

sandstone, forming casts (depressions filled with sandstone without any internal structure) or subtracks (depressions filled with deformed sedimentary sandstone layers). In some footprints (unfilled depressions as well as casts) there are grooves caused by claws (Fig. 3.150A), which should not be confused with marks left by farmers who plowed the land (Fig. 3.150B). In the study area there are also irregular gaps in the rock, which may be attributed to the activity of invertebrates.

In the ichnites of the tracksite are structures related to the three footprint formation phases. These range from the movement of the foot through the ground before and during penetration (T phase), through total support and support variation (Fig. 3.154) (W phase), to lifting of the foot and subsequent collapse of the mud (K phase). The subtracks, with deformed fillings, have to be the result of the three phases of the tread, but identifying which phase is responsible for observed footprint features is complicated.

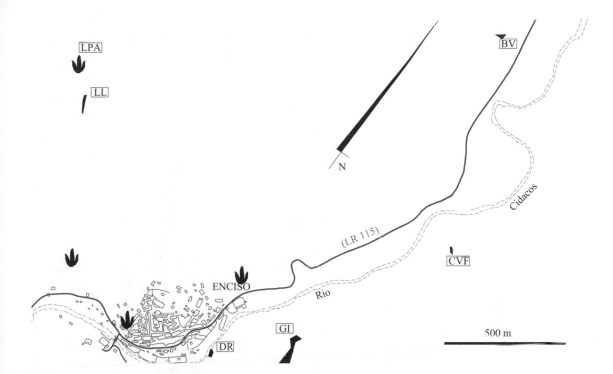

3.151. Location of Las Losas (LL) and neighboring tracksites (BV, Barranco de Valdeño; CVF, Corral de Valdefuente; GI, Guilera; DR, Del Rio; LPA, Las Peñas Amarillas) and other outcrops. Local road is labeled LR 115.

T PHASE STRUCTURES

Smooth structures made by the foot moving through the substrate, before being supported, can be seen at the rear of some footprints (Fig. 3.152) in both semiplantigrade (LL44.3) and digitigrade footprints. The rear sliding indicates that this surface was the tracking surface, at least for those tracks where it occurs. LL44 (Fig. 3.153) is a trackway where all footprints are semiplantigrade. There are signs of sliding forward even after the foot sinks into the ground. Semiplantigrade footprints in the trackway LL44 are highly variable in shape (Fig. 3.153); the total length of the print, the distance between the position of the hallux (T), and the proximal or distal metatarsus are all variable. This means that the foot also slid after it had begun its entry into the mud.

W PHASE STRUCTURES

The W stage is when the digital pads make a deeper mark. This phase also produces features in La Rioja that were first seen at this site. For example, there are semiplantigrade footprints where the depth of the digit marks is greater than that of the metatarsus (Fig. 3.154). This is because in the W phase, the weight is supported by the metatarsus and the toes before the metatarsus is removed. Elevation of the metatarsus puts the full weight of the dinosaur on the digitigrade part of the autopod. As the pressure increases, the digits penetrate further into the mud, increasing the depth of the front part of the footprint. Only some of the footprints at Las Losas show digital pad marks (Table 3.9); these are mostly footprints with lateral constrictions on the digital impressions that correspond to

3.152. Footprint LL44.3, semiplantigrade ichnite with slide zone at rear of print, made during T phase of foot's interaction with substrate. (A) Overall view of print; scale bar = 30 cm. (B) Detail of rear portion of print with drawing showing parallel slide grooves.

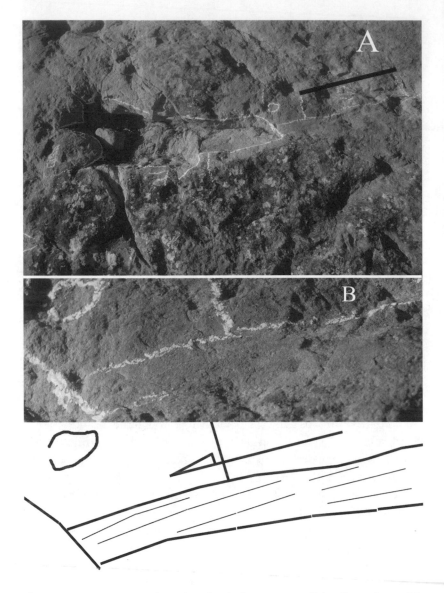

the creases separating digital pads. At least some of the footprints with such lateral constrictions can be considered as stamps, imprinting directly onto the study surface.

K PHASE STRUCTURES

K phase structures at Las Losas are many and varied. Some were produced directly by the foot, e.g., certain claw marks. The exit of the foot also deformed the subtracks. The exit in the true footprints could be followed by partial or total collapse of mud into the footprint holes; sometimes foot kickoff and mud collapse occurred simultaneously.

Several trackways (Table 3.9) contain footprints with a groove running along the axis of the digits (Figs. 3.150A, 3.155), like a slot cut through the mud. It is interesting to note that the claws do not leave parallel streak marks but rather follow the digits in converging toward the back of the

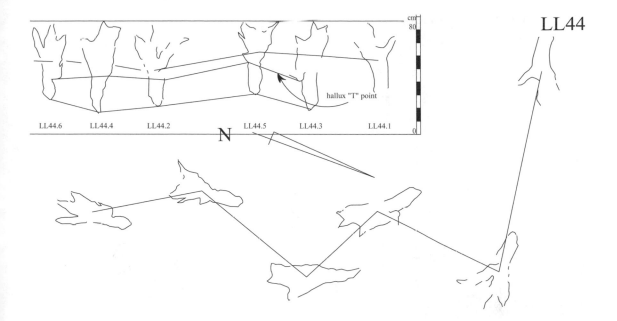

3.153. Trackway LL44. Upper left box shows variability of LL44 footprints. Distal tip of each footprint has been placed along same line; center of footprints, position of rear of hallux, and that of rear of metatarsal imprint, indicate foot movement as it entered mud. Rest of figure shows change in direction of LL44 trackway. Trackway is located at top of Fig. 3.149A.

foot. There are cases where the grooves cut through mud collapse structures in the shaft of the footprint. However, the opposite is the norm: the mud falls after the claws have left their marks, as in LL55.5 and LL36.2 (Fig. 3.155). Claws sometimes accumulate mud at the back, as is also seen in footprint LL55.5. Some tracks then fill with sand without deforming, which preserves the claw marks.

Footprint shafts more often show mud collapse structures when the viscosity and consistency of the mud being walked through was low, even during the K phase. In some cases, it is not easy to distinguish between claw marks or collapse of the walls of the footprints as being responsible for the grooves seen at the tips of the digits.

Ichnotypes

TRIDACTYL DIGITIGRADE FOOTPRINTS

The footprints at Las Losas all show theropod features. There are relatively long, tapered acuminate digits, many of which show lateral constrictions and creases delimiting digital pads on the sole. The heel mark

Table 3.9. Trackways of Las Losas showing particular features.

Feature	Trackway
Visible pads in the trackways	LL76, LL75, LL59, LL57, LL53, LL52, LL51, LL50, LL47, LL43, LL40,LL38, LL37, LL35, LL34, LL27, LL25, LL24, LL22, LL18, LL15, LL14, LL10, LL8
Claw stria	LL68, LL58, LL57, LL55, LL53, LL52, LL49, LL48, LL36, LL30, LL26, LL23, LL18, LL13
Semiplantigrade tracks	LL5, LL13, LL16, LL21, LL22, LL25, LL27, LL28. LL31, LL37. LL38. LL41. LL43, LL44, LL46, LL47, LL60, LL70, LL71, LL88, LL98

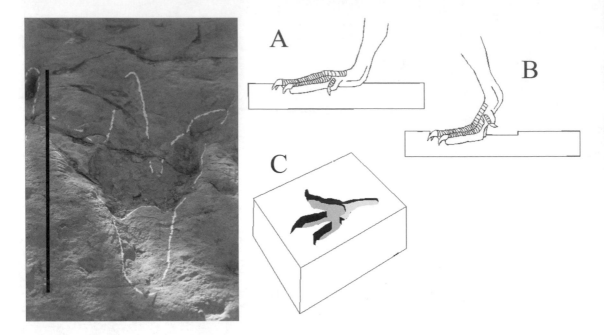

3.154. Left footprint LL46.6. Acropodium shaft is deeper that metapodium. (Right) Presumed mechanism of formation of LL46.6: initially there is semiplantigrade foot support (A), followed by elevation of metatarsus, resulting in deeper penetration of acropodium (B), creating depth distribution (C) observed in preserved print (left). Scale bar = 30 cm.

is prominent, sometimes showing a clear individual pad and sometimes a bilobular rear end (Fig. 3.156; LL19.1). The largest digitigrade footprint (Fig. 3.156; LL58.1) is about 47 cm long, and the smallest (LL10.2) is 20 cm long. There are tracks of all intermediate sizes, with no suggestion of distinct peaks in the print size–frequency distribution (Romero-Molina et al., 2003).

There are no ichnotypes distinguished by the difference in shape of the footprint outline. Tracks sometimes change shape along the same trackway. For example, trackway LL21 includes prints with both broad and narrow digit impressions. Trackways LL10, LL50, and LL53 include footprints that have a prominent heel pad as well as prints with no such pad. The large footprints would fall under the ichnogenus *Megalosauripus*, but the small ones can only be identified as theropod prints. There is no sharp break in either shape or size between large and small prints.

SEMIPLANTIGRADE FOOTPRINTS

The term "semiplantigrade" is used instead of "plantigrade" for certain footprints on the recommendation of the late W. A. S. Sarjeant. In March 2002, during the study of Las Losas, he wrote (unpubl. data), "Personally I think 'semiplantigrade' is preferable; otherwise, what does one call prints where the whole heel is impressed: 'ultraplantigrade'? This may not happen often or at all, with dinosaurs; but the term used should not, I feel, have a special definition for dinosaurs and a different one for all other animals!" "Semiplantigrade footprints with long, separate, acuminate digits, without pads, with or without a hallux mark" were recognized as a distinct morphotype by Pérez-Lorente (1993a:198). All the footprints are rounded at the rear, from which it can be deduced that the metatarsus was placed in a horizontal position and was fully supported on the ground. In

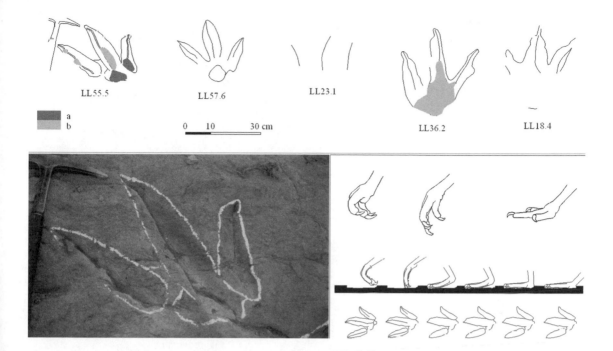

several such prints (Fig. 3.157), the divarication between the orientation of the metapodial axis and the digitigrade foot is seen clearly.

Las Losas is the site with most footprints and trackways (Table 3.10) of this kind in La Rioja. It is noteworthy that the semiplantigrade footprints and trackways at Las Losas are not a distinct category apart from other

3.155. Striations created by claws drawing along axes of toe marks. (Top) Examples of footprints with such grooves that converge toward rear of print. (a) Dark gray regions of prints indicate zones of accumulation of mud dragged backward by claws. (b) Light gray indicates zones of mud collapse after kickoff (K phase). (Bottom left) Photograph of footprint LL55.5. (Bottom right, right to left) Sequence of events and corresponding footprint features made during kickoff (K phase).

Table 3.10. Trackways, pairs, and isolated ichnites containing semiplantigrade footprints.

Trackways containing:

Three or more prints	LL5	LL5.3, LL5.4
	LL13	LL13.3, LL13.4
	LL16	LL16.1, LL16.2, LL16.3, LL16.5
	LL21	LL21.1
	LL22	LL22.5
	LL25	LL25.7, LL25.9
	LL27	LL27.4
	LL28	LL28.2, LL28.3, LL28.4, LL28.5, LL28.6
	LL31	LL31.1, LL31.2, LL31.5
	LL37	LL37.6
	LL38	LL38.3
	LL41	LL41.1, LL41.2
	LL43	LL43.2, LL43.3, LL43.4, LL43.5
	LL44	LL44.1, LL44.2, LL44.3, LL44.4, LL44.5, LL44.6
	LL46	LL46.1, LL46.2, LL46.3, LL46.4, LL46.5, LL46.6, LL46.7
	LL47	LL47.1
Two prints	LL60	LL60.1
	LL70	LL70.2
	LL71	LL71.2
Isolated prints	LL88, LL98	

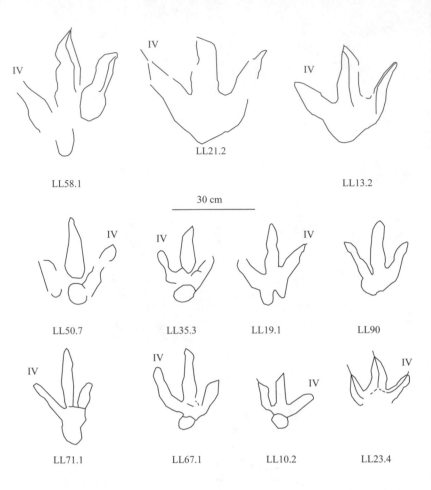

3.156. Las Losas digitigrade footprint types.

morphotypes. Although there are trackways at Las Losas where all the footprints are semiplantigrade, there are many cases where digitigrade and semiplantigrade prints alternate. At Las Losas, the semiplantigrade footprints are also placed alternating and interspersed with digitigrade ones in the same trackway. This shows that a single dinosaur can make both kinds of footprint (Fig. 3.158). Of the 16 trackways (sequences with more than three footprints) with footprints showing a metatarsus mark there are 43 semiplantigrade footprints; in the three pairs of related footprints (LL60, LL70, and LL71) there are three semiplantigrade footprints; and finally there are two semiplantigrade footprints not associated with any trackway. At Las Losas there are in total 48 footprints with a metatarsus mark. This is the first report in La Rioja of isolated semiplantigrade footprints, i.e., those that are not part of a trackway.

Footprint Structures

TRACKWAYS WITH SEMIPLANTIGRADE AND MIXED FOOTPRINTS

Different explanations have been offered for the creation of semiplantigrade footprints. Kuban (1989) thought such prints were made in the

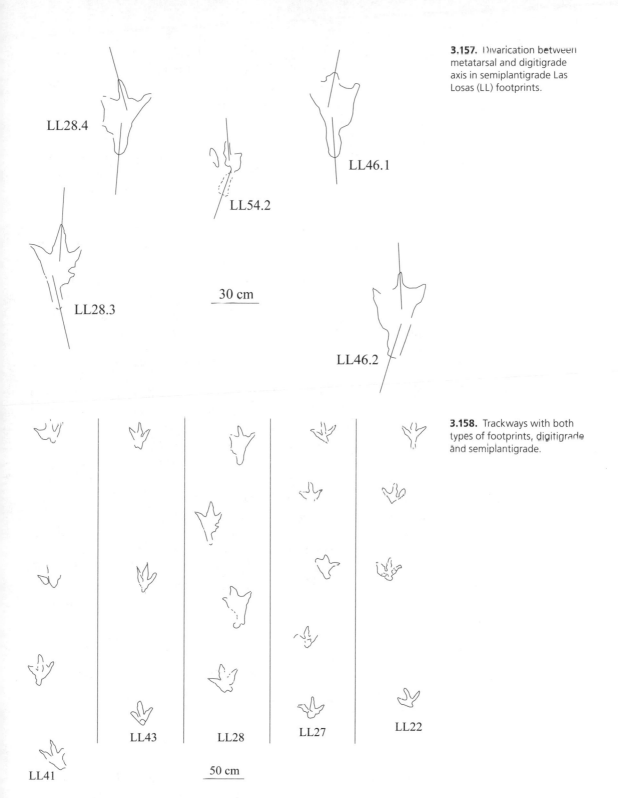

3.157. Divarication between metatarsal and digitigrade axis in semiplantigrade Las Losas (LL) footprints.

LL28.4

LL54.2

LL46.1

LL28.3

30 cm

LL46.2

3.158. Trackways with both types of footprints, digitigrade and semiplantigrade.

LL43

LL28

LL27

LL22

LL41

50 cm

course of unusual behaviors of the trackmakers. In contrast, Romero-Molina et al. (2003) concluded that at Las Losas the dinosaurs left semi-plantigrade footprints so that the autopod would not sink so deeply in low-viscosity mud. The pressure of the dinosaur's weight would not be concentrated on the acropodial portion of the foot but rather spread across the entire autopod.

The depth of the footprints in Las Losas changes. Some trackways consist entirely of deep footprints and others entirely of shallow prints. Some trackways contain both deep and shallow prints, and some individual footprints (LL46.5 and LL46.6), vary in depth between their digital and metatarsal portions. The nature of the mud across the surface must have changed from one point to another. Consequently, the depth of the footprints, the presence or absence of mud collapse structures, and the difference in the way of walking (digitigrade versus semiplantigrade) are possibly related to changes in mud viscosity. Variations in these three features can therefore be used to distinguish wetter and drier patches of substrate at the time the tracks were made (Fig. 3.149D).

MARKS ALONG THE LENGTH OF DIGITS

There are both isolated footprints and footprints in trackways (Table 3.9) that show long, narrow grooves or striae running through the axis of the digits (Figs. 3.150, 3.152). The number of striae, from one to three, depends on the footprint examined. The striae are straight or slightly curved. There is only one stria for each digit. The striae are not flattened by the sole of the digits but appear on the impressions on the sole, reaching from the apex to the back of the digits. There may be collapse structures alongside the track wall and small piles of mud accumulated by the claws at the back (Fig. 3.155).

The sequence in which these structures formed was, first, impression of the sole of the digits; second, incision and accumulation of mud at the rear; and third, the collapse of print walls immediately after the formation of the grooves. The conclusion is that the claws made these structures. The claw marks converge proximally, following the axes of the digit impressions, which means that the claws came together at the time the foot left the ground (Fig. 3.155). This movement of the digits is common in certain birds and depends on the angle between the metapod and zeugopod (Romero-Molina et al., 2003). The digits close as the zeugopod approaches the metapod.

Footprint Distribution

Semiplantigrade and digitigrade footprints are scattered across the track-site surface (Fig. 3.149). The semiplantigrade footprints do not all occur together and were made by dinosaurs moving in different directions at different speeds. Careful examination of Las Losas suggests that the distribution of different track features is related to changes in behavior of the mud.

The marked variation of footprints in the same trackway allows a distinction to be made between normal footprints and irregular footprints. Normal footprints are digitigrade, the outlines of which are not deformed and which show neither grooves nor collapse of mud in their shafts or walls. Semiplantigrade prints are included among irregular footprints. When groups of irregular footprints are enclosed within a zone, it is observed that there are several sectors with this type of footprint (Fig. 3.149C). Similarly, there are other areas of the site characterized by normal digitigrade footprints. If numerous individual dinosaurs traveling in different directions left behind distinct sectors containing normal or irregular footprints, the variation in footprint shapes is unlikely to be due to differences in the trackmakers but rather to differences in the substrate. The most likely explanation is that the mud was much softer (less viscosity and consistency) in those patches containing irregular footprints, presumably because they were wetter.

LL44 (Figs. 3.149A, 3.153) illustrates some of the previous irregularities, where the complete trackway is deep and distorted, so the ground must have been quite soft. In addition, several footprints have slide marks made before the foot entered the mud. These marks are more than 20 cm long. The distance between different anatomic parts of the foot (start of the shaft, hallux entry point, start of the digitigrade print) is variable, showing that the foot slipped forward in the mud more in some footprints than in others. The path followed is curved, indicating an abrupt change of 90° in the direction of travel (Fig. 3.153). Finally, the size of the steps is variable, from 62 to 155 cm. The stride and speed also vary. Blanco et al. (2000) suggested that the dinosaur was trying to get out of the muddy area.

The directions of travel of the dinosaurs look chaotic, with no order whatsoever (Fig. 3.149). A rose diagram (Fig. 3.149B) shows four peak trackway orientations: N119E (average between N128-108E), N171E, N129W, and N73W. The trackways that do not fit into these four peaks are not straight but curved. None of the previously described trackway features shows any association with direction of travel. Trackways in the N108-128E direction may have been made when the mud conditions were similar throughout the site because all their footprints are normal. Despite the relatively orderly distribution of the trackways (only four maxims; Fig. 3.149B), there is no criterion for assuming that it was due to herd behavior or natural barriers.

Table 3.11. Correlation between the outcrops of Viera, Torres, and Aguirrezabala (1984) and the tracksites of Pérez-Lorente (2001a).

Viera, Torres, and Aguirrezabala (1984)			Pérez-Lorente (2001a)	Abbreviation
Nivel 1	Capa 2	Outcrop 4	Umbría del portillo	UP
		Outcrop 3		
		Outcrop 2		
		Outcrop 1		
	Capa 1	Outcrop 2	Peñaportillo	PP
		Outcrop 1		

Main Site Features

The footprint density is relatively high (0.9 footprints/m^2), almost one footprint for each square meter, which is one of the highest densities in medium-size tracksites. As in other sites, only theropod dinosaur tracks are found. In fact, given the size of the tracksite, other ichnotaxa should be included, and it is surprising that despite the low viscosity and consistency of the mud, no ornithopod footprints were found.

There are several special aspects of Las Losas. The first is that it is the only site of more than 250 m^2 that contains only one type (theropod) of dinosaur footprint. There are no shape differences or gaps in the size distribution between the large and small digitigrade footprints. There are variations in shape that can be attributed more to the structures of the footprint than to the foot of the dinosaurs. Apparently the track-making dinosaurs at Las Losas had identical footprints, differing only in size. Las Losas is also the only site in La Rioja with a large number of trackways with semiplantigrade footprints, although some of the trackways contain both digitigrade and plantigrade footprints.

Peñaportillo and Umbria del Portillo

History

The site was found by Aguirrezabala, Viera, and Torres in 1982 in the course of their study of all the outcrops in San Vicente de Munilla (Viera and Aguirrezabala, 1982). They divided it into six outcrops located on two layers ("capa 1" and "capa 2" of the same "nivel 1" [stratigraphical unit]). The first data were published in Viera, Torres, and Aguirrezabala (1984). Subsequently, during summer research field and courses sponsored by the university and government of La Rioja, the highest outcrop (Peñaportillo) was extended over three field seasons (1991–1993). This was done to expose a new trackway with three footprints in the center and to establish the relationship between five parallel theropod trackways. The results were published in Casanovas et al. (1993c). The area, which today constitutes La Umbría del Portillo, was cleared during the summer of 1996 and the data published in Casanovas et al. (1998). Summary data about the sites and correlations between the descriptive terminology of

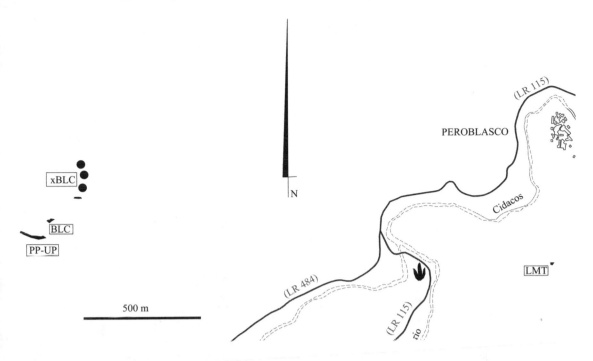

Location of site of Peñaportillo Umbría del Portillo (PP-UP) and neighboring outcrops of Barranco de la Canal (BLC and xBLC) and La Mata (LMT). Local roads are labeled LR 115 and LR 484.

Viera, Torres, and Aguirrezabala (1984) and later studies are detailed in Tables 2.4 and 3.11.

Description of Site

The Peñaportillo and La Umbría del Portillo tracksites (Fig. 3.159) are to the north of the town of Munilla, and access is by a road starting from the town's first houses. This road is usually in good condition, although there may be problems after the rainy season as a result of the mud, making it difficult for some cars. The slopes are steep but are easy to negotiate if the soil is dry.

The sites are on top of two overlapping strata. The oldest is Peña-portillo (Fig. 3.160), or capa 1, and above that is Umbría del Portillo, or capa 2. The study surface is the topographic surface, and they coincide on both slopes. The location lends itself to being cleared for discovering new footprints (Fig. 3.161). The sites of Peñaportillo, Umbría del Portillo, and La Canal could be joined together after clearing the slope, which would not be too difficult.

There are other studied outcrops around the periphery of Barranco de la Canal (BLC). Within a radius of 500 m there are four levels with six outcrops (BLC and xBLC; Fig. 3.159) described by Viera, Torres, and Aguirrezabala (1984), and between 800 and 1600 m away are the sites of San Vicente de Munilla, Munilla, Sobaquillo, and Malvaciervo. La Mata is 2200 m away. In addition, there are several other small occurrences of isolated footprints, or ones in small groups of prints. In one of the entrances to Peñaportillo, on the west side of tracksite there are two

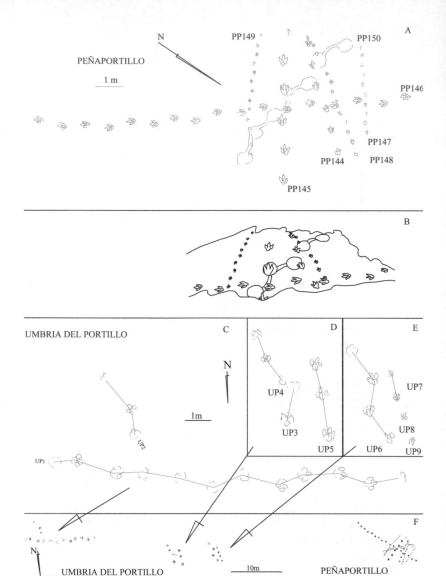

3.160. Maps of outcrops of Peñaportillo (PP) and Umbría del Portillo (UP). (A) Peñaportillo (PP), outcrop in its current state (Casanovas et al., 1993c). (B) Outcrop before digging (Viera, Torres, and Aguirrezabala, 1984); drawn from photograph. (C, D, E) Enlargement of trackways of outcrop of Umbría del Portillo (UP). (F) PP and UP distribution of footprints.

layers with footprints. One of the layers has mud cracks that the footprints have deformed; there are certain polygons of desiccation moved by the footprints. It is easy to find isolated footprints on the sides of the outcrops. This is one of the places where weathering and erosion would probably lead to the discovery of large numbers of new footprints.

At Peñaportillo (Fig. 3.162), 64 theropod and seven ornithopod footprints have been recorded, the latter accompanied by tail marks, while at Umbría del Portillo there are two theropod and 42 ornithopod footprints. The Peñaportillo footprints are completely different from the Umbría del Portillo ones. The Umbría del Portillo outcrop is not fully mapped, and only the parts we have studied are shown. Viera, Torres, and Aguirrezabala (1984) also cited and studied two large ornithopod trackways with a total of 11 footprints, which are not shown on the diagrams. The two

trackways are located in the area included in La Umbría del Portillo and Peñaportillo (Fig. 3.161). Remains of freshwater fish (*Hybodus, Lepidotes*), gastropods, bivalves, and bioturbation produced by invertebrates have also been found in the vicinity (Viera, Torres, and Aguirrezabala, 1984).

3.161. View of outcrops of Peñaportillo (PP), Umbría del Portillo (UP), and Barranco de la Canal (BLC).

Lithology

All the surfaces with tracks in Peñaportillo and Umbría del Portillo are on silty sandstone. The Umbría del Portillo footprints are younger because they are in a layer about 20 cm above Peñaportillo. The thickness of the strata varies from several decimeters to layers of a few centimeters. The layer thickness tends not to be constant and varies laterally, sometimes over distances of less than a meter. The color of the rocks is quite dark, probably as a result of being formed in a reducing environment.

Sandstone and shale layers interbedded with lenticular siliceous limestone and sandstone are sedimented on the Peñaportillo and Umbría del Portillo rocks. The limestone and more siliceous layers are more resistant to erosion and are prominent in the landscape. According to sedimentologists (Doublet, 2004), they are rocks of lacustrine origin. Sedimentary structures on the top of the strata are not visible in the Umbría del Portillo and Peñaportillo outcrops. The rocks of both outcrops are different in their structure and internal composition. At Peñaportillo there are fine sedimentary layers, while at La Umbría del Portillo there is no sedimentary lamination. Within the predominantly lacustrine

3.162. Peñaportillo cover interior.

environment, there must have been episodes of emergence above water level, at least at this location. The dinosaur footprints themselves suggest that the depth could not have been very substantial, at least when the rocks containing the footprints were formed. The mud cracks and possible pedogenesis of the level in La Umbría del Portillo indicate that the sediments occasionally were emergent.

Footprint Features in Mud

The footprints sink little into the mud, usually less than 5 cm at Peñaportillo and less than 10 cm at La Umbría del Portillo. Digital pads, claw structures, and digit and heel contours are well distinguished in many footprints. Only a single Peñaportillo (PP) trackway, PP150, has a circular footprint contour without any irregularities in either the outline or the interior. It could be said that almost all the footprints are stamps except those in PP150.

The marks in PP150 predate the other trackways crossing Peñaportillo, as seen in the interference of the tracks (theropod footprints are superposed). This means that no deposition of mud or sand sediments occurred in the area from the time PP150 was formed until the rest of the dinosaurs in Peñaportillo passed. The PP150 footprints do not prevent the subsequent footprints from being very well impressed (Fig. 3.160A). It is therefore assumed that the Peñaportillo study surface is the tracking surface for all the trackmakers (PP144, PP145, PP146, PP147, PP148 and

PP149), with PP150 being the first to pass by. There are several trackways overlapping the footprints of PP150 (Fig. 3.160A, B).

The fact that the outline of the footprints in PP150 is so round and that they lack interior details might suggest that they are undertracks. However, further examination of the tail drag marks, and the overlapping stamps eliminate that possibility—is not possible that the same surface may have a real footprint (more recent) over an undertrack (older one). On a visit to the site, the Anglo-Australian paleontologist Tony Thulborn suggested the round outline and the lack of anatomic details could be a thixotropic phenomenon. The disturbance caused by the foot on the ground liquefied the mud so that it flowed into the shaft of the footprint as the dinosaur lifted its foot. None of the footprints has pronounced extrusion rims. There are only small and narrow raised borders in parts of the Umbría del Portillo tracks.

Footprint Features (Ichnotypes)

Four different ichnotypes can be distinguished at Peñaportillo and two at La Umbría del Portillo. Of these, four are theropod ichnites and two ornithopod.

THEROPOD FOOTPRINTS

All these have relatively long (Fig. 3.163), separate, tapered digits, with several pad marks per digit (depressions on the sole, lateral constrictions of the digits). The heel is prominent and rounded and associated with the proximal part of digit IV, nearly aligned with the axis of toe III. The footprints in trackways PP146, PP145, and PP144 have these aforementioned features. However, trackway PP146 differs from the other two (Figs. 3.160, 3.163) in that the foot is narrower, with $[l - a]/a = 0.35$ compared with 0.32 and 0.30; the distal part of digit IV is further away from digit III than digit II; and digit III is shorter (III/l = 0.59 compared with 0.65 and 0.70). They could be considered as two large theropod ichnotypes.

PP147, PP148, and PP149 (Figs. 3.160, 3.163) are much smaller footprints. The length of the foot is between 13 and 15 cm, and it is longer than it is wide. The pads and claws can also be seen, but less distinctly than in the large prints. Probably the layered structure of the sediment was better able to withstand the weight of these dinosaurs and was consequently less deformed. It cannot be said they belong to same species or to different species of trackmakers from those that left PP145 and PP144.

There are two footprints in the Umbría del Portillo outcrop where the morphological features cannot be so clearly seen (Fig. 3.160E; UP7). These can only be said to be theropod footprints.

ORNITHOPOD FOOTPRINTS

Two different ichnotypes can be distinguished. One encompasses all the footprints at Umbría del Portillo except for the UP7 trackway, and the

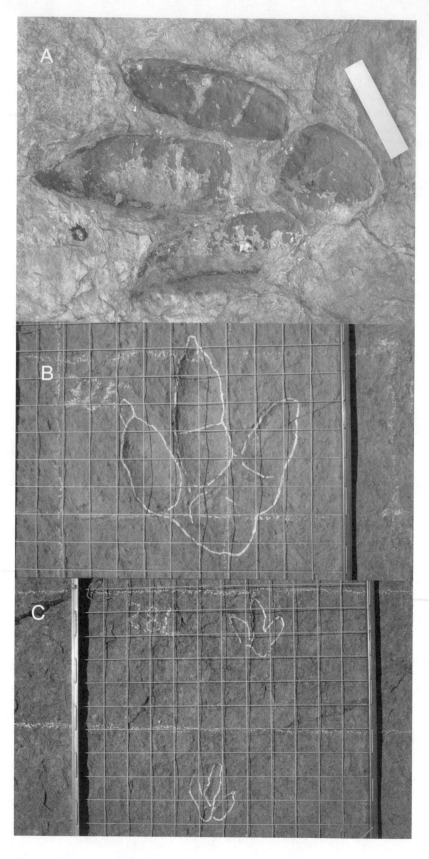

3.163. (A) Large theropod footprint PP146.13. (B) Large theropod footprint PP145.2. (C) Small theropod footprints PP149.6 and PP149.7. (Square net width.) A, Scale bar = 10 cm. B, C, Mesh width = 5 cm.

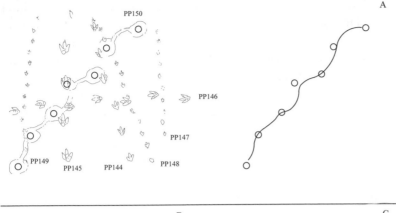

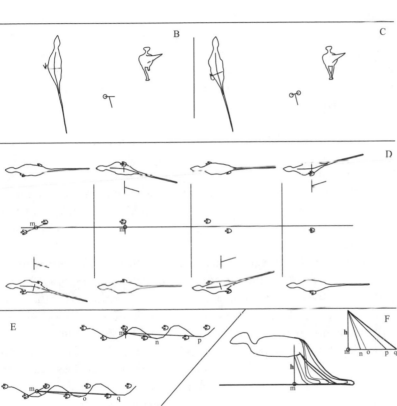

3.164. Tail of dinosaur (A) Schematic reproduction of trackway PP150. (B and top of D) Position of tail, feet and acetabular axis of dinosaur in aviform walk. (C and bottom of D) Position of tail, feet and acetabular axis in sauriform walk. (E) Plot of center of pelvic girdle (m) and possible support points of tail in aviform walk (n, p) and sauriform (o, q). (F) Deduced positions for tail using triangle and adapting it to possible dinosaur.

other applies to trackway PP150. Trackway PP150 has rounded footprints (Figs. 3.160A, B; 3.164A), lacking any details in the outline or interior of the footprint that would allow recognition of morphological features of the trackmaker autopod. These tracks were concluded to be ornithopod footprints as a result of the relationship between the length and width, and because ornithopod feet have shorter and less separated digits than theropods. It is not usual to find theropod footprints where the digits cannot be recognized. It is easy to consider ornithopod feet as more appropriate for leaving this type of footprint.

Previously, we have seen that there is an indirect argument supporting this theory, which is the wide mark and flattened base of the trace of the tail, which is not typical of theropod dinosaurs (see Pérez-Lorente and Herrero, 2007). Polish paleontologist Gerard Gierlinski says (oral comm.) that this could be the trackway of a thyreophoran with a bipedal gait (a stegosaur). This hypothesis about this trackway was defended during International Symposium on Dinosaurs and Other Vertebrates Palaeoichnology (Fumanya, Spain, October 2005). This would not be the first bipedal thyreophoran cited, as several articles have been written on the subject of trackways involving this type of bipedal dinosaur (Hadri et al., 2007).

There are no problems identifying the rest of the ornithopod footprints. They are large, with three rounded digits and a single pad in front of the rounded back end. Viera, Torres, and Aguirrezabala (1984) attributed them to *Iguanodon*, but the prints would be included in the ichnogenus *Brachyiguanodonipus* according to Díaz-Martínez et al. (2009). The average length per trackway is between 38 and 52 cm, while the width ranges from 39 to 49 cm. As in many large ornithopod footprints, they have a quite negative orientation and low estimated travel speed (between 2.8 and 4.1 km/h).

Track Distribution According to Ichnotypes

The separation of ichnotypes between the two outcrops is clear. At Peñaportillo, in the bottom layer with laminated sedimentary structures, there are several types of theropod footprints and possible thyreophoran footprints. At La Umbría del Portillo, the massive layer, large ornithopod footprints predominate.

At Peñaportillo there are six trackways with theropod footprints: three large (PP144, PP145, PP146) and three small (PP147, PP148, PP149). The features and travel direction of PP146 are different from the other two large theropod trackways (PP144, PP145), which are nearly parallel and perpendicular to the first. The small theropod trackways (PP147, PP148, PP149) are also roughly parallel along their path and are also parallel to the two large theropod trails. It is not known whether the five dinosaurs passed by at the same time, although they came after PP150 and, at least the small ones, came before PP146. The tracksite was dug and extended to find more surface area to analyze this association (Fig. 3.160A, B). However, the dig could not be continued as a result of the risk of destroying the tracks. The further the digging progressed, the more the upper layer became more strongly bonded to the bottom. At first glance, the area observed cannot be said with certainty to be a theropod family group, but as Pérez-Lorente et al. (2001) have noted, this possibility cannot be denied either.

At La Umbría del Portillo there are several ornithopod trackways of large animals of the same morphotype, as well as two theropod footprints among the remaining prints. All the trackways (including those not mapped) go in one of two directions. The speed of travel is not fast in all cases. These probably represent part of a herd of ornithopods.

The separation between the types of tracks according to the layer/ type of sediment is apparent. Sandstone sediments with fine lamination characteristic of moving water (Peñaportillo) are associated with the theropod footprints, while more fine-grained sediments, lacking internal sedimentary structures, typical of standing water (Umbría del Portillo), are associated with the ornithopod footprints. In the same series of layers, although a few decimeters below Peñaportillo there are unidentified dinosaur footprints that deformed the mud polygons of an emergent layer.

TAIL MARK IN PP150

Even though the type of dinosaur that made this trackway is uncertain, the information that is available pertains to three support points for the animal: both feet and the tail. These data allow further interpretations of and new hypotheses for the type of gait to be developed. The basis established for making these inferences (Fig. 3.164) is as follows. The pelvic girdle can be thought of as a horizontal line running from one acetabulum (top of the femur) to the other, with the tail at the center. As the dinosaur moves, the line advances with a rocking motion transmitted to the tail. This rocking motion is reflected in the sinusoidal form of the tail drag. The movement of the tail may be due to two causes (Casanovas et al., 1993c). The movement may be a continuation of the posterior dorsal sector of the spine (sauriform walking) or alternatively the counterbalancing movement of the dinosaur's body while walking (aviform walking). In the first case the tail movement is a reflection of hip motion, while in the second it balances the sway of the body (Fig. 3.164B, C). In the first case, the tail would touch the ground at the points farthest from the midline (the more external parts of the trackway) when the dinosaur is supported by both feet. In the second, the tail would touch the ground at points farthest from the midline only when the dinosaur has one foot on the ground and the other is halfway through its stride (Fig. 3.164D).

Torcida et al. (2003) argue that support of the tail by the substrate only occurs in particular postures, such as when the dinosaur lowers its body toward the ground. However, if it is supposed that the trackmaker in PP150 did not adopt a particular posture during the formation of this trackway, the height of the hip can be calculated according to the proposed formulas of Thulborn (1990). First the height of the acetabulum (H) is calculated, which is 295 cm (the average length of the footprint is 50 cm). Then the distance between the center of the hip and the support of the tail is solved by invoking a right-angle triangle (Fig. 3.164E, F.) One side (H, acetabulum height) is the vertical line running from the hip to the ground (point m), and the other side is the distance between m and the point where the tail touches the ground. The hypotenuse is the length of the tail.

The vertical line is taken as the midline in the two positions, as proposed for each method of walking. One is in the middle of the pace, between two successive footprints (sauriform walking), and the other at the same position as a footprint, i.e., in the orthogonal projection of the middle pes point on the midline (aviform walking). Casanovas et al. (1993c) suppose that the tail position might have two possibilities in relation to pes position: sauriform or aviform. In sauriform walking, depending on the step (left–right or right–left), the support of the tail will be on the right or left of the sinusoid tail drag. In aviform walking, depending on the supported foot (right or left), the support of the tail is on the left or right of the sinusoid. Figure 3.164C illustrates the left–right step (sauriform) and the right foot support (aviform). The tail length governs whether the support is provided in the first, second, third, or fourth option (Fig. 3.164F). If the support is near, the inclination of the caudal appendage is anatomically inappropriate, and if it is far, the length is excessive. Having drawn the models, the consistent result was for the trackmaker to be sauriform and for the length of the tail to be 470 cm.

After preparing the previous work (Casanovas et al., 1993c), Gierlinski (pers. comm.) pointed out the possibility that the trackmaker was not an ornithopod but a bipedally walking stegosaur with its spine bent. If this is the case, the number of uncertainties become greater because in addition to the above variables (type of walking, distance of support of the tail), the slope of the spine would have to be considered.

General Information and Main Site Features

The density of footprints is low (0.2 prints/m^2), about 2 every 10 m^2. However, as in almost all the outcrops, there are many interesting features. The large theropod trackways have valuable properties for studying the anatomy of the foot of the trackmakers. The sharpness of the digital pads and the precision of the contour lines of these footprints are features guaranteeing accuracy in the reconstruction of the sole of the autopods.

It has been suggested, with appropriate reservations, that the two large theropod trackways (PP144, PP145) and the three small ones (PP147, PP148, PP149) may belong to a family group. It is safe to argue that the three small footprint trackways belong to a group. It is more difficult to believe that three dinosaurs of this size walking in the same area and in parallel are not related individuals. However, the relationship between the five individuals cannot be established from data in this outcrop.

It is one of the three locations in the Iberian Peninsula where ornithopod tail marks accompanying footprints have been cited. The others are in sandstone in the Urbión Group in the site of Regumiel in Burgos province, bordering La Rioja to the west (Torcida et al., 2003), and 2TT of Corral del Totico (Jiménez Vela and Pérez-Lorente, 2006–2007). There, the tail mark is associated with tridactyl footprints, with rounded digits and some forefoot marks. If Gierlinski's suggestion is correct, it would probably be the only stegosaur trackway with a tail mark. However, the type of walking style and length of the appendage would still be unknown.

Conservation

As shown in previous sections, some sites have been given protective covers (private sponsors), e.g., Los Cayos, El Villar-Poyales, and Valdecevillo. In addition, the La Rioja government commissioned a conservation project for a site where both the rocks and the protective covering were exposed to atmospheric conditions. Peñaportillo was chosen because the site is ideal in terms of exposure to rain, snow, diurnal and annual temperature variations, and wind, as well as for the visual impact produced. The result was a construction (Fig. 3.165) that has so far protected the footprints from the direct action of water and snow (and therefore frost) and has prevented massive colonization of the rock by plants. The structure is transparent, with a translucent sheet roof. The roof has had to be repaired only once, after strong winds. Peñaportillo is in the upper part of the watershed between the two ravines.

3.165. Covering in Peñaportillo. Behind cover is reproduction of bipedal stegosaurus, proposed by Gerard Gierlinski (in foreground of picture) as trackmaker of PP150 trackway. Photograph by K. Sabath.

History

Barranco de la Canal

The Barranco de la Canal (BLC) site, along with those of Peñaportillo, Umbría del Portillo, and the XBLC sites (Fig. 3.166), was discovered by a team from the Aranzadi Science Society (Aguirrezabala, Viera, and Torres) in 1982. At that time, the Barranco de la Canal was divided into two separate outcrops by a cultivated plot of land. A large ornithopod trackway (Fig. 3.167) was observed under the plot on one side while appearing on

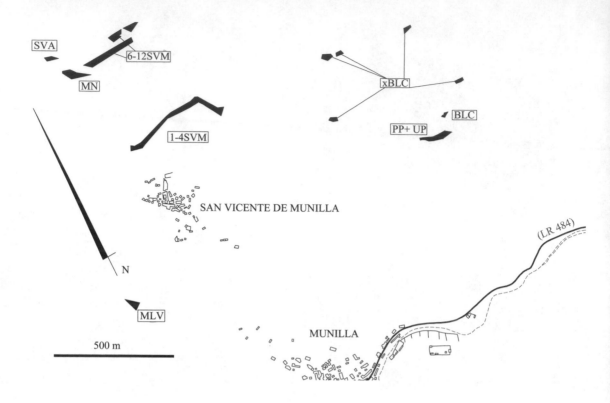

3.166. Location of Barranco de la Canal (BLC) and neighboring tracksites: other La Canal outcrops (xBLC), El Sobaquillo (SVA), Munilla (MN), San Vicente de Munilla (1–4SVM, 6–12SVM), Peñaportillo and Umbría del Portillo (PP + UP), and Malvaciervo (MLV). Local road is labeled LR 484.

the other side. During two digging campaigns in 1992 and 1993 (Pérez-Lorente, 1993b), the part of the plot overlapping the trackway was removed. The new unit that it revealed was later studied and the results published by Casanovas et al. (1985). Viera, Torres, and Aguirrezabala (1984) found four trackways (Table 3.12), a number that increased to seven after clearing during 1992–1993.

Description of Site

The site is on the lower slopes of the Barranco de la Canal. It is probably in the same bed as La Umbría del Portillo but occurs about 70 m to the northeast of the latter. The sites of tracks and other fossils cited nearby are

Table 3.12. Correlation between the trackways of Viera, Torres, and Aguirrezabala (1984) and Casanovas et al. (1995).

Viera, Torres, and Aguirrezabala (1984)			Casanovas et al. (1995)		
Trackway	**Footprints**	**Type**	**Trackways and isolated footprints**	**Footprints**	**Type**
A	36	Iguanodontid	BLC1	29 (31)	Ornithopod
B	2	Iguanodontid	BLC2	6	Ornithopod
C	2	Iguanodontid	BLC3	5	Ornithopod
D	2	Iguanodontid	BLC5	3	Ornithopod
E	2	Iguanodontid	BLC7	5	Ornithopod
			BLC4	3	Ornithopod
			BLC6	3	Ornithopod
			Isolated	6	Ornithopod
			Isolated	4	Theropod

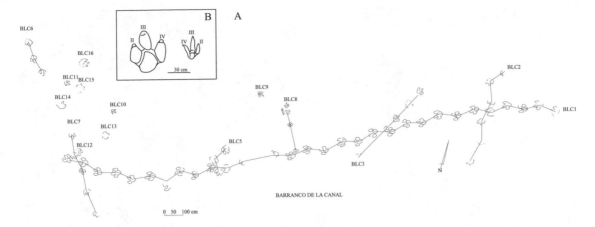

3.167. (A) Map of footprints and trackways at Barranco de la Canal. (B) Generalized shapes of ornithopod and theropod footprints at BLC.

similar to those described previously. The tracks are on a sloping layer and easy to investigate; the orientation is N120E and the dip 20NE. Access is by foot from the same road leading to the outcrops Peñaportillo and La Umbría del Portillo.

Lithology

The Barranco de la Canal layer is probably the same as that of La Umbría del Portillo, although this cannot be guaranteed, as various similar-looking layers overlap each other despite their varying thicknesses. This thickness varies for all layers and depends on where it is measured.

The lithology is clayed sandstone with a lot of shale and clearly visible fine sand–size phyllosilicate material. The quartz grain content is much less than in the Peñaportillo layer but similar to La Umbría del Portillo. The thickness of the layer at the site is about 20 cm. The stratification surface is irregular, with coarse granules occurring throughout; burrows of less than 1 cm in diameter are also seen. The granules and some burrows are also evident in the base of the ornithopod footprints, so that bioturbation came after the footprints were made.

On a more siliceous level than Barranco de la Canal and immediately above it, in the upper (western) part of the site, is a theropod footprint. In the lower (northeast) part, several layers above there are two other theropod footprints. Viera, Torres, and Aguirrezabala (1984) included the entire unit consisting of Peñaportillo, Umbría del Portillo, and Barranco de la Canal in their "nivel 1." At this level there are at least four layers with footprints. The thickness of "nivel 1" is a little less than 1 m.

Footprint Features in Mud

The footprint depths vary, depending on the trackway in which they occur, but all footprints in the same trackway have a similar depth across the site. The deepest prints are those of BLC1, with a maximum of 10 cm. The rest of the footprints are shallower and sometimes difficult to see.

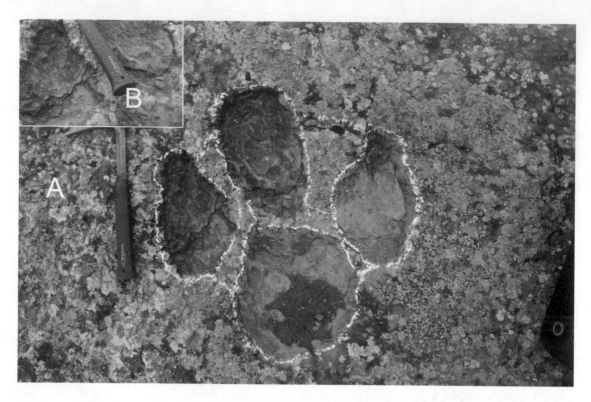

3.168. Right footprint BLC1.28. (A) Digital and heel pads are clearly visible. In distal part of toes II and IV, nail/hoof prints can be seen. (B) Close-up of proximal II–III sector to see nondeformed filling of toes.

Because all trackways intersect, it can be concluded that the dinosaurs did not all pass this place at the same time. The difference in depth of some footprints, even at the places where trackways cross, means that not all animals passed at the same time. In the 3600 m² surface (Umbría del Portillo–Barranco de la Canal), the ornithopod prints are more abundant than theropod prints, and sauropod footprints do not occur. It is possible that the environment provided better conditions for ornithopods than for other dinosaurs.

Several footprints in trackway BLC1 have pad marks filled with laminated sandstone without deformations (Fig. 3.168). The sandstone filling followed the formation of the footprints. The tracks clearly show the digital pads, their separation, and the shape of the claws, especially in the lateral digits. These are assumed to be real tracks, with the area of study being the tracking surface. There is evidence of bioturbation (irregularities in the top bed, and horizontal and vertical burrows) on the real tracks and throughout the tracking surface, although this has hardly altered the footprints.

Footprint Features (Ichnotypes)

There are four theropod footprints (Fig. 3.167) with distinct outlines, as well as subdivisions between the digital pads. Also, the independent, elongated, tapering digits are clearly separated from each other; their size (30–39 cm) means they are classified as large theropod prints (sensu Thulborn, 1990).

Table 3.13. Pace lengths and stride lengths at BLC1 trackway.

Footprint number BLC1.	1	2	3	4	5	6	7	8	9	10	11	12	13	14	15	16	17	18	19	20	21	22	23	24	25	26	27	28	29	30	31
Pace length: left–right	78		96		102		98		99		95		105		97							82		77		100		103		92	
Pace length: right–left		87		83		96		90		90		90		92		91							87		106		91		88		
	–		+	–		+	+	–	+	–	+	–	+	+	+	–	–						+	–	+	–	–	+	–	+	–
Stride length – left	175			176		182		190		186		181		191		186						180		167		194		192		178	
Stride length – right		168			176		196		186		187		180		192		185						170		162		182		199		
	–					+	–		–		+		–		+		–						+		+		+		+		–

Note: The relation between left–right and right–left pace lengths is greater than one at the beginning of the trackway and smaller at the end. The relation between strides is more complex.

Ornithopod footprints dominate the site. Their length ranges from 26 to 52 cm, and so they are classified as large ornithopod prints. Digit impressions are large and rounded, but somewhat elongated along the length of each digit; a rear pad constitutes the heel. Interdigital angle II^III is less than III^IV, as is normal in bipedal dinosaur footprints; digit III is closer to digit II than to digit IV. The rotation of the footprints is highly inward or negative (−20°). Many digits, especially those at the sides of the print, are slightly flared at the rounded tip, and sometimes with a depression in the sole, produced by rounded claws (Figs. 3.167, 3.168). They are classified in the *Brachyguanodonipus* ichnogenus of Díaz-Martínez et al. (2009). The trackway has alternating long and short steps, more or less according to their position in the trackway (Table 3.13). The left–right paces are shorter in the first part of the trackway, while right–left paces are shorter in the second. The estimated travel speed is quite slow (1.8 to 2.7 km/h). No sauropod footprints were found at the site.

Ichnotype Footprint Distribution

Theropod footprints occur in layers of different lithology than ornithopod prints. There is a predominance of theropod footprints and a lack of ornithopod prints (except the possible thyreophorans) in the fine laminated sandy sediment of Peñaportillo. In contrast, at La Umbría del Portillo and Barranco de la Canal, ornithopod (*Brachyiguanodonipus*) footprints predominate, with a lack of theropod footprints in an environment of abundant shale deposition and fine- and medium-grained phyllosilicates, apparently massive, or perhaps totally bioturbated.

La Umbría del Portillo and Barranco de la Canal are probably in the same study surface: they have a similar lithological composition and external rock appearance. The environment in these places would have been muddy and relatively compact when the dinosaurs passed by because the footprints penetrated only a little into the mud.

No herd behavior can be proposed for the trackmakers in the Umbría del Portillo–Barranco de la Canal area.

Trackway BLC1

Trackway BLC1 has the most ornithopod footprints in La Rioja: Twenty-nine footprints from a possible thirty-one were found, with two intermediate ones missing. This trackway is 27 m long, one of the longest in La Rioja. The longest trackway is ANDI (in Cuesta de Andorra), which is 32 m long and has 21 of its 25 footprints preserved. In fact, VA13 (Barranco de Valdecevillo), a sauropod trackway, would be the longest, at 54 m, if the intermediate footprints between its initial and final parts were found. The trackway with the largest number of footprints is at La Pellejera (1LP17), a theropod trackway of 28 m, which contains 34 of the 41 footprints it would

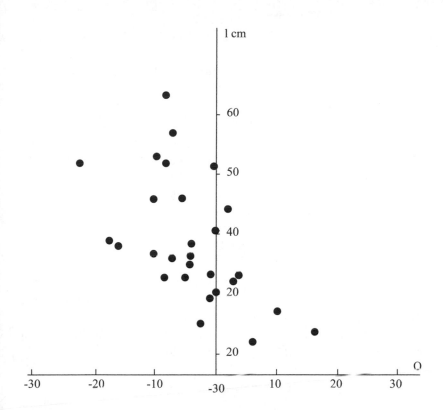

1 cm

60

50

40

20

20

-30   -20   -10   -30   10   20   30   0

3.169. Relationship between footprint rotation/orientation (positive indicates outward with respect to trackmaker's direction of travel; negative indicates inward with respect to direction of travel) and footprint length (l, in centimeters) in tracks attributed to ornithopods. From Casanovas et al. (1995a).

have were it complete. Another long ornithopod trackway La Rioja is 2TT16 (El Totico, 18 m long, with 24 of its 29 footprints preserved).

In trackway BLC1, like almost all long trackways, the path is not straight but bends during its itinerary. Looking at the footprints, it can be seen that the dinosaur took longer steps when leading off from its right than from its left foot in the first part (BLC1.1 to BLC1.17) of the trackway, while at the end (BLC1.22 to BLC1.27) the opposite happens. The difference in left stride length from the right may be used as an argument to infer that the dinosaurs leaving these tracks were lame. If the difference is not marked, it may be attributed to laterality (i.e., right-handed or left-handed dinosaurs); however, the problem is when one section of a trackway has longer steps than others. In the first part of BLC1, the dinosaur turns left, while at the end it turns right. Looking at the strides, it can be seen that the distance between the left and right feet varies in no apparent order along the trackway. Two conclusions can be drawn from this. First, the variation in the dinosaur path does not occur because of a change in stride length but rather pace length. Second, because all long biped trackways in La Rioja have undulating paths, the variation in length of the steps detected at short distances should not imply lameness. Variations from straight lines in dinosaur paths may be due to laterality (right- or left-handed dinosaurs) and subsequent course corrections. Conclusions could probably be drawn by examining longer trackways with pace and stride changes throughout the whole "irregular" trajectory.

Negative Rotation

A highly negative rotation of the footprints in BLC1 is apparent. Because the tracks are also relatively large, the variation in the rotation of the ornithopod footprints was examined as a function of footprint size (Casanovas et al., 1995a). Trackways in La Rioja were used, as well as those in other parts of the world, where sufficient information for the study was known at the time. Morphometric data were used from 51 trackways totaling more than 400 footprints. Footprint rotation/orientation was plotted in relation to footprint length (Fig. 3.169); increasing foot length also increased the negative value of orientation (Fig. 3.169). Casanovas et al. (1995a) proposed an anatomic explanation for this trend; a negative footprint orientation was hypothesized to be a mechanism enabling a biped animal with a horizontal back to place its center of gravity above the foot, especially when walking slowly. The observed relationship has continued to hold as new data for ornithopod trackway have accumulated. Pérez-Lorente (1996a) analyzed the same relationship in theropod trackways, and although there seems to be a similar effect, the relationship between print size and orientation is not as strong as for ornithopods.

Ornithopod Size and Speed

It is a paradox that the estimated speed traveled in BLC1 is so low when from the data available the length of the footprint (l) and stride (z) are so great. The velocity is obtained via formulas (Alexander, 1976; Demathieu, 1986; Thulborn, 1990) based on the relationship between the height of the acetabulum (H) and stride length (z). Because the height of the acetabulum (H) is inferred from pes length (l), a relationship was also sought between stride (z) and pes length (l). It was seen that a linear relationship between stride length and the foot measurement of the dinosaur was observed not only in the above-mentioned sample of 51 ornithopod trackways (Casanovas et al., 1995a) but also in theropod trackways (Pérez-Lorente, 1996a:45). The speeds estimated from the 51 trackways were plotted against footprint length (Fig. 3.170), with a vertical line connecting the speed values according to the Alexander and Demathieu formulas. For large footprints the Alexander speed is higher, and for lower values the Demathieu value is higher.

Casanovas et al. (1995a) concluded that the lower the foot length, the higher the ratio of stride/footprint length, along with the speed. This led to the expectation that larger ornithopod dinosaurs would be slower than smaller ones. He also concluded that if the actual speed estimates are plotted as a function of footprint length (l), the minimum speed (1.2 to 3.5 km/h) actually occurs when l is about 50 cm. For higher and lower values of l, the velocity increases. Also, if a theoretical calculation of the variation in velocity is made as a function of l, the result is similar.

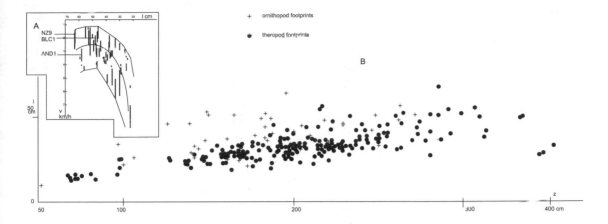

+ ornithopod footprints

• theropod footprints

3.170. (A) Relationship between velocity (v) and length of ornithopod footprints. Mean values are on parabola. Projection of values of AND1 (La Cuesta de Andorra, section 3.9), NZ9 (Navalsaz, section 3.10), and BLC1 are shown. (B) Relationship between footprint length and stride length from data (Pérez-Lorente's team) of ornithopod and theropod footprints of La Rioja.

Extremiana and Lanchares (1995; cf. Casanovas et al., 1995a) using the Mathematica program (version 2.2 for Macintosh), studied the variation of the height of the acetabulum as a function of l and calculated the theoretical variation of the velocity from variations in l and z. They plotted the change in velocity as a function of l and obtained a parabola. According to Alexander and Thulborn, the minimum speed is 4.8 km/h for an ornithopod foot 0.21 m long, and according to Demathieu, the minimum speed is 3.4 km/h for an ornithopod foot 33 cm long. According to these proposed formulas, small and large ornithopod trackmakers go faster than those of intermediate size. It should be noted that the small ones reach a high speed but the large ones do not. Finding trackways of ornithopods with larger footprints might support this trend. Because the large ornithopod footprint length does not exceed 65 cm in trackways, the travel speed is low.

In the theropod footprint analysis of Pérez-Lorente (1996a), the same plot was constructed, with the mean speed value curve again giving a parabola. However, it was not as marked as in the case of ornithopod prints. In this case, the reverse was not true: there were small theropod footprints showing both slow and fast walking.

General Information

The density of tracks in the site is low (0.34 footprints/m²), equivalent to a little over three tracks in every 10 m². This density seems to be real, although, we do not know if there were smaller or shallower footprints destroyed by bioturbation affecting the whole strata. The environment of the site is promising, however, for the number of footprints seen on the rock outcrops between the debris—all over the side of the ravine—and for the dip in the layers in this slope. As mentioned for Peñaportillo, there may be a continuous outcropping between La Umbría del Portillo and Barranco de la Canal under the formerly cultivated plots and debris.

El Sobaquillo

History

The El Sobaquillo (SVA) site was discovered in 1993 by one of the members of the team (Arturo Fernández Ortega) while digging the site at Munilla, located about 100 m to the south. There are two trackways, one of which we (Casanovas et al., 1995i) considered spectacular. It contained 19 sauropod footprints, which were the largest in La Rioja to occur in a continuous and clean trail. It seemed that the footprints were not deformed by either the mud or the movement of the animal's feet. The possibilities of deducing a quadruped walking style (primitive alternate pace, alternate pace, or amble; see Casamiquela et al., 1987:plate 8) were analyzed from the trackway data, which was published by Casanovas et al. (1997a). El Sobaquillo is close to a series of outcrops containing theropod and ornithopod footprints studied by Viera and Aguirrezabala (1982), in San Vicente de Munilla (SVM).

Description of Tracksite

The study surface is on top of a layer with a strike of N128E and dip of 31NE, which is clear of vegetation and relatively easy to find. The rock outcrop between the two plots was once cultivated. Access is difficult and is only by foot. No road or path passes by the site, and it is quite far from the main road, so you have to walk about 30 minutes to get there. In addition, the ground has strong slopes and is colonized by shrubs of spiky gorse and the sticky leaves of rock roses. It is therefore unsuitable for many people and it is not recommended for a casual visit.

3.171. Location of El Sobaquillo tracksite (SVA). Tracksites of Munilla (MN), San Vicente de Munilla 1 to 4 (1–4SVM), San Vicente de Munilla 6 to 12 (SVM), Malvaciervo (MLV), and two as yet unstudied tracksites (footprint symbol), are also indicated.

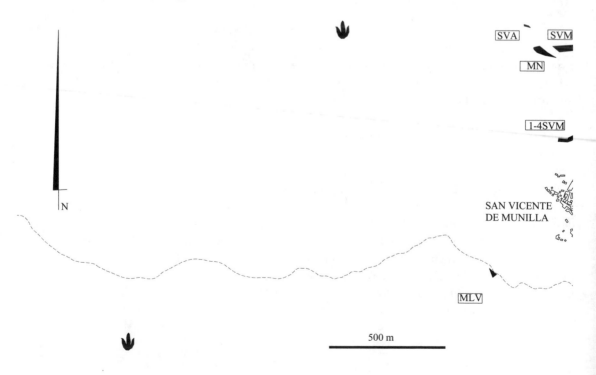

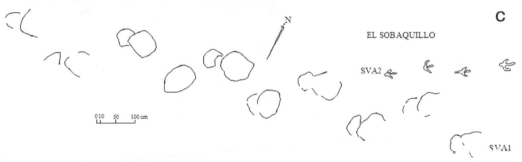

EL SOBAQUILLO

SVA2

SVA1

010 50 100 cm

C

N

El Sobaquillo is in the same layer that Viera and Aguirrezabala (1982) refer to as "nivel 4" in the study they made of sites in this area. These authors say they found unionids with closed valves and several occurrences of mud cracks in this layer. The dip of the layers is slightly more than the slope of the land, so there are no extensive outcrops. The trackway could probably be extended by digging, but the ground is steep, and a large amount of earth would have to be manually removed to expand the trackway by even 1 m.

The site (Fig. 3.171) is 100 m from the site of Munilla and the more southerly group of San Vicente de Munilla (6svm–12svm) outcrops, as well as 500 m from another group in San Vicente de Munilla (1svm–4svm). It is also 1000 m from Malvaciervo and 1600 m from Peñaportillo. The first site to the west is Las Mortajeras, which is located at 2800 m. There are two occurrences of unstudied footprints (Fig. 3.171) at a distance of 780 and 2100 m.

Lithology

The rock is a dark sandy siltstone. The beds are thin-bedded, 10 to 20 cm thick, with parallel lamination. According to the official geological map of Spain (Hernández et al., 1990), El Sobaquillo is at the top of the Enciso Group, corresponding to the Upper Aptian–Lower Albian period, according to Doublet (2004). These fine-grained, somewhat sandy beds are probably storm deposits in shallow water. However, there are clear mud cracks (Fig. 3.172) on this site that suggest the emergence of the

3.172. (A) Photograph of sauropod manus-pes set, mud cracks, and SVA1.6. (B) Features of pes penetration into ground in footprint SVA2.3; collapse of mud in toe IV and in proximal zone of toe III; possible hallux print and exit mark of three toes through central distal hole. (C) Map of trackways SVA1 and SVA2.

location. They are considered to be coastal lake sites in the terminology of Doublet (2004). The footprints occur in two layers, the lower s_1 level with an irregular top and the upper s_2 level with a smooth top broken by mud cracks.

Footprint Features in the Mud

Features in the two trackways are different. First, the surface is different: while the SVA1 sauropod trackway is on the surface with mud cracks (s_2), the theropod trackway SVA2 is on the lower surface (s_1). The footprints in SVA2 are highly deformed (Fig. 3.172). They are different from each other in the shape of the digit impressions, the angles between them, and their relative positions. In SVA2.1 digits II and III are close to each other; in SVA2.2 and SVA2.4 digits II and IV are widely separated and may show toe mark retroversion; finally, in SVA2.3 all three digits completely enter the mud and exit together through the center (digit III mark). In this track, there is also a collapse of mud in the mark of proximal part of digit III after the foot exits. Retroversion (Boutakiout et al., 2006) is the structure produced by backward collapse (W phase and/or K phase) of the interdigital II–III and III–IV zone in footprints with forward slide (T phase) of the pes. The final feature shows two wide and rounded "hypexes."

The prints in SVA1 were made by a quadruped, and their footprints deform the mud cracks (s_2). The footprints in theropod trackway SVA2 do not affect the SVA1 trackway and are therefore deduced to have come first. Also, SVA2 does not continue through s_2 and is therefore supposed to predate the deposition of sediments s_2.

The SVA1 footprints sink very little in the mud. The base of the print is flat and possibly a little deeper at the front. SVA1.6 is the deepest print, with a relief of 15 cm between the top of the extrusion rim and the deepest part of the footprint. In several footprints, the feet push the mud crack polygons forward, deforming the forefoot impression. It is clear, therefore, that the footprints were made after the mud dried because the polygons in the interior of the footsteps are destroyed. It cannot be deduced whether the study surface is the tracking surface, as there are no marks for the digits or claws. There may be a fine layer (s_3) above the actual surface s_2 that was the tracking surface and is the reason why the footprints sink little into the mud of layer s_2.

The forefoot marks are slightly shallower than or as deep as those of the hind feet. In one of the footprints (SVA1.7), the hind foot is superimposed on the forefoot and erases it completely. In the first footprints, only the forefeet and the front part of the hind feet are seen clearly.

Footprint Distribution by Structure

The theropod tracks in the lower layer have structures typical of soft mud, while the sauropod tracks in the upper layer sink little into the mud. Although the bed is eroded, the depth of the sauropod footprints increases from the start of the trackways onward. SVA1.6 seems to be the deepest; however, the footprints that follow in the trackway do not provide enough information about whether this increasing depth is a trend maintained in the latter footsteps. The s_1 and s_2 levels are in some of the higher strata of the Enciso Group—higher than the Barranco de la Canal, Peñaportillo, or La Umbría del Portillo sites.

Footprint Features (Ichnotypes)

SAUROPOD FOOTPRINTS

SVA1 is intermediate between a narrow- and wide-gauge trackway (Farlow, 1992). The position of the forefoot prints and their separation from the midline (trackway deviation) is slightly higher than for the hind foot impressions. The pes print length is 76 cm and the width 73 cm, while the length of the manus marks is 33 cm with a width of 49 cm. They are the largest sauropod dinosaur footprints in La Rioja.

The Farlow (1992) criteria for trackway gauge, based on the separation of the forefeet and hind feet marks from the midline, are commonly used in the classification of quadruped trackways. In addition, the heteropody criteria (relative size of manus and pes prints; Lockley, Farlow, and Meyer, 1994) have been added. Finally, other morphological classification features are required, such as the shape of the forefoot, hind foot, and toes and claws, if present.

The Farlow criteria can be expressed numerically in several ways: Ar − a/2, or the difference between the amplitude of trackway and half the width of the autopod (Casanovas et al., 1997b) (Table 2.1); Ar/a, or the relative width of the track (Casanovas, Pérez-Lorente, et al., 1991); the trackway ratio; or TR = (sw/ow × 100), with sw indicating side width (corresponding to width of the track perpendicular to the long axis to the trackway) and ow indicating overall width (low, corresponding with external trackway width; Lr in Table 2.1) (Romano, Whyte, and Jackson, 2007; Whyte, Romano, and Eldvidge, 2007). The three measurements are used to establish the degree of separation of the prints. The first two are also used in relationship with the midline; in the third, the width of the footprint is also considered. This therefore relates the amplitude and width of the trackway. Finally, another two criteria (heteropody and morphology) are well established and are followed in La Rioja studies.

SVA1 is a trackway where the hind feet do not touch the midline and thus must be classified as *Brontopodus* Farlow, Pittman, and Hawthorne, 1989. This is quantified such that the forefoot trackway amplitude (41 cm) is similar to that of the hind feet (39 cm); the relative trackway width (Ar/a) is much greater for the forefeet (0.84) than for the hind feet (0.52); and the midline separation (Ar − a/2) is also greater for the forefeet (17 cm) than for the hind feet (3 cm). The ratios Ar/a and Ar − a/2 indicate that the hind feet touch the midline and the forefeet are farther apart (see also Farlow, 1992). The heteropody ratio is between ⅓ and ⅕—that is, it is between the Texas and Colorado types of Lockley et al. (1994). The forefoot marks are wider than they are long. If these marks are not distorted by interference from the hind foot prints, then they are closed and semilunar, similar to those of the more derived sauropods, such as Camarasauromorpha or even higher groups (Wright, 2005).

On this trackway, the deduction has been made of progression in ambling gait for the sauropod. The calculations are based on the following. According to the glenoacetabular distance of quadruped dinosaurs and the type of walking (amble, normal pace, serpentine progression), certain autopods, depending on the type of gait, support themselves on the ground at the same time; others rest on the ground. When ambling, the limbs on each side support the body alternately; during primitive alternating pace (serpentine progression), the two diagonal limbs support the body; and three limbs support the body when using the alternating (normal) pace (Leonardi, 1987).

The glenoacetabular distance is deduced directly from the trackways if the type of walking is known by two procedures. First are the Demathieu figures (Leonardi, 1987:plate 7) and the measurement ratios of Farlow, Pittman, and Hawthorne (1989). Second, the glenoacetabular distance in sauropod dinosaurs may be similar or somewhat larger than the length of the rear limb (Casanovas et al., 1997a). Four formulas may be used to calculate the height (H) of the rear limb in sauropods (l, pes length; a, pes width). According to Alexander (1976), H = 4l; to Lockley, Houck, and Prince (1986), H = 4a; to Ishigaki (1986), H = 3.6l; and to Thulborn (1990), H = 5.5l. Agreement between one of the three lengths derived for the acetabulum and one of the four limb heights will provide a consistent answer to the type of walking and height.

The calculations providing the most consistent values are 337 to 321 cm (Demathieu, 1987b; Farlow, Pittman, and Hawthorne, 1989) for the glenoacetabular distance; 320 cm for the height of the limb (Alexander, 1976); and walking type as ambling. The travel speed for limb heights of 320 cm ranges between 3.6 and 4.1 km/h.

THEROPOD FOOTPRINTS

These tracks were highly deformed by the condition of the mud, which must have had a low viscosity and consistency. The footprints differ greatly among themselves (Fig. 3.172), and the measurements do not conform to the feet of the trackmaker. Footprint length varies from 25 to 40 cm and the width from 24 to 33 cm. The interdigital angles are not reliable because of the shape of the digits, which in some cases shows toe mark retroversion. The variations are not attributable to a particular cause because in SVA2.1, for example, digit III is quite long (perhaps produced during the K phase), but in SVA2.3 (Fig. 3.172) there is an elongated projection in the heel. This is formed either by the sliding of the foot during the T phase or by the impression of the metatarsus. The trackway is narrow. The foot definitely has relatively long, separate digits, which is consistent with theropod footprints.

The acetabulum is estimated to be about 125 cm above the ground for a theropod digitigrade foot 25 cm long (Thulborn, 1990). The estimated walking speed is between 5 and 7 km/h, which is relatively high and consistent (Kuban, 1989; Pérez-Lorente, 1993a) with the anomaly of the semiplantigrade tracks (or with the impression of the metatarsus).

General Information

Of all the sites in this general area, this is the only one with sauropod footprints. The nearest sites to this with sauropod footprints are more than 7 km away in the Oncala Group and about 8.5 km away (Barranco de Valdecevillo) in the Enciso Group. This can be considered as the youngest site with sauropod footprints in the Cameros Basin, or at least the one closest to the Oliván Group base. The Barranco de Valdebrajés, Los Cayos E, and El Sobaquillo sites are the youngest of the Enciso Group.

The footprint density (0.07) is low, both here and at the other nearby outcrops, and is barely equivalent to one footprint every 10 m². Numerous other exposures in the same area also contain footprints. The El Sobaquillo layer corresponds to the youngest in the "nivel 4" of Viera and Aguirrezabala (1982) containing footprints. Outcrops in the surrounding area contain tridactyl footprints, mostly theropod but some ornithopod prints as well. The outcrops are in poor condition and generally contain isolated footprints, i.e., with few trackways. No footprints have been detected in younger Enciso Group layers on this site.

El Sobaquillo is being eroded by climatic conditions. This decomposition is made easier by the outcrop conditions: the layers flake easily and contain a lot of moss and lichen. It is far from human contact and even contact with cattle. However, its restoration would require a major effort in getting people there and removing material, and probably other sites with easier access should take precedence.

History

San Martín de Jubera (SM) was found by R. Ezquerra in 1992. It was worked the same year, and three outcrops over three layers were found. Although mentioned by Pérez-Lorente (1993a), a description of the site was not published until the study of Casanovas et al. (1995b).

Description of Location

San Martín de Jubera (Fig. 3.173) is north of the Cameros Basin. The geological situation is similar to that of several other sites, including Soto 1 and Soto 2 (1ST, 2ST), San Vicente de Robres, and others located on the northern edge of the basin. All of them occur in limestone-dominated formations attributed to the Oncala Group. This stratigraphic allocation has not always been thus; some of these sites have occasionally been

San Martín
de Jubera

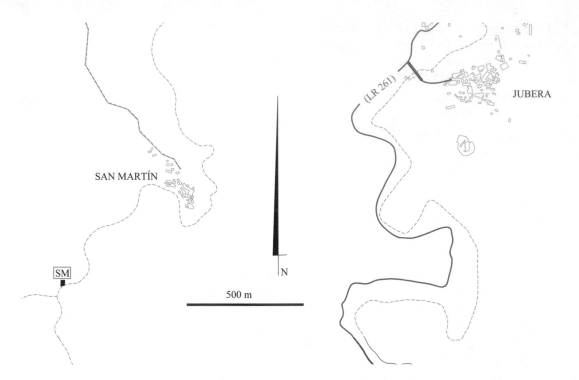

3.173. San Martín (SM) site location. Local road is labeled LR 261.

included in the Enciso Group (1ST, 2ST, San Vicente de Robres) or the Urbión Group (San Martín de Jubera). The age of San Martín de Jubera was clarified by reviewing microfossils and the stratigraphic position of the rocks, and it has been classified as Berriasian (Doublet, 2004).

The nearest localities known to have footprints are the San Vicente de Robles tracksite, about 5.5 km distant, and the Soto 2 tracksite (2ST) at 8 km, both outside the map of the location (Fig. 3.173). There is no modern road leading to the site. However, it can be reached by walking along an old bridle path from the town of San Martín, which can be reached by a main road.

The San Martín tracksite consists of three outcrops, 1SM, 2SM, and 3SM. The three are contiguous, although the composition of the rock, the structures, and the footprints are different for each. 1SM is on the north side of the path, 2SM is reached from the path by going down the ravine, and 3SM is adjacent to a stream in the ravine. Natural casts of footprints were found from other surrounding layers on the same site. On the path between San Martín and the outcrops, layers appear with irregular and poorly shaped footprints.

Lithology

The exposures consist of sandstone, shale, and limestone. The two lower sites (1SM, 2SM) are in limestone, while the upper (3SM) is in sandstone–siltstone with thin siltstone layers. The siltstone and sandstone tend to be thinly bedded. The laminae, especially those containing more sand, show cross-bedding. The study surface of 3SM shows ripples deformed by

the footprints. In neighboring beds there are mud cracks, consistent with the environment where the footprints were made. According to Doublet (2004), the rocks belong to the Upper Leza Formation, deposited in lake, marsh, and river settings. Calcareous sediments, pedogenesis, fluvial sedimentary structures, and drying and karstification processes are well pronounced in the period for the upper part of the Leza Formation.

3.174. Footprint 3SM6.2, an overtrack. Ripples in track layer were deformed by dinosaur's foot. This and other overtracks at site consist of nondeformed pieces of upper sandy bed, which tightly adhere to ripple surface.

Footprint Structures and Distribution

The two lower outcrops (1SM, 2SM) have no special mud structures. Their tracks are shallow (usually less than 5 cm) and have hardly any extrusion rims.

3SM STRUCTURES

The structures in 3SM are striking because the tracks were printed on soft sediment, with clearly visible laminae. In addition, many of the footprints stand out on the study surface as untrue casts (overtracks) (Fig. 3.174). Most of the study area has ripples, which are deformed in the outline and interior of the footprints. In addition to the ripples, some overlying

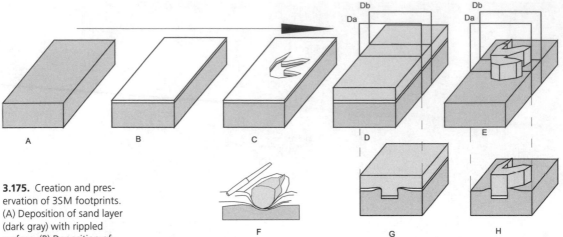

3.175. Creation and preservation of 3SM footprints. (A) Deposition of sand layer (dark gray) with rippled surface. (B) Deposition of sandy siltstone (white) over initial sand layer. (C) Impression of dinosaur footprints. (D) Deposition of another sand layer (light gray); Da and Db are two vertical planes for visualizing internal structures. (E, F) Creation of section through toe cast (Fig. 3.176). Deformed layers occur beneath toe mark and sand-filled toe mark. This corresponds to what would be observed in vertical plane Db in (E). (G) Vertical section Da as it would be observed before modern erosion. Upper sandstone layer (light gray) contacts lower sandstone layer (dark gray), becoming tightly adherent in extrusion rims of prints. (H) Aspect of overtracks as presently seen.

3.176. Section through toe cast. Crests of deformed ripples occur beneath toe mark. Marginal folds can be seen in silty sandstone to right and left (arrows). Sandy cast of toe mark and claws is outlined in center. Above this is an overlying sandy layer, separated from toe cast by an erosional surface (marked by pencil). In this ichnite, cast does not firmly attach to upper sandstone layer.

silty layers are also deformed. The shafts of the tracks are filled with sand (Figs. 3.175, 3.176).

The sequence of events responsible for the formation and present exposure of the footprints is illustrated in Fig. 3.175. First, there was deposition of a sand layer (dark gray) with a rippled surface (Fig. 3.175A). This was followed by deposition of a sandy siltstone (white layer) (Fig. 3.175B). The dinosaurs then trod on the sandy siltstone, leaving their footprints (Fig. 3.175C). The ripples were deformed around and in the walls of the shafts. They were crushed at the base of the shafts, i.e., there are no ripples at the bottom of the footprints. The sandy siltstone layer was bent downward (Fig. 3.176) and thinned before being crushed and disappearing. Extrusion rims were formed. Afterward, a sandy layer (light gray) filled the footprint shafts, creating casts (Figs. 3.175F, 3.176). The thickness of the sandstone and siltstone layers varies. This means that the same sandstone or siltstone level may be wider or narrower from one part to another. There are erosive bases at higher levels above the ripples. The sandstone in the most recent level (light gray) comes into contact with the sandstone of the lower level (dark gray) at the edges of the footprints.

At the edges of the footprints there are extrusion rims that raise the lower sandstone layer above the footprints and the surrounding rock. At the print margins, the siltstone layers thin, at some points so much that the siltstone layers disappear. Consequently, the two sandstone levels have the potential to meet and adhere at the edge of the tracks. Put another way, between the two sandstone levels is a siltstone level that thins out or disappears at the edge of the tracks.

The overtracks, whether isolated or in trackways, are structures formed in sediments after the passing of the dinosaurs. During diagenesis, sandstone at the top of the lower level and at the bottom of the upper level become welded at the extrusion rims of the footprints. These are therefore special anchor points of the upper level in the lower level with a higher resistance to erosion; these are the points where the upper level enters the lower (cast) and they weld (extrusion rims). The upper sandstone

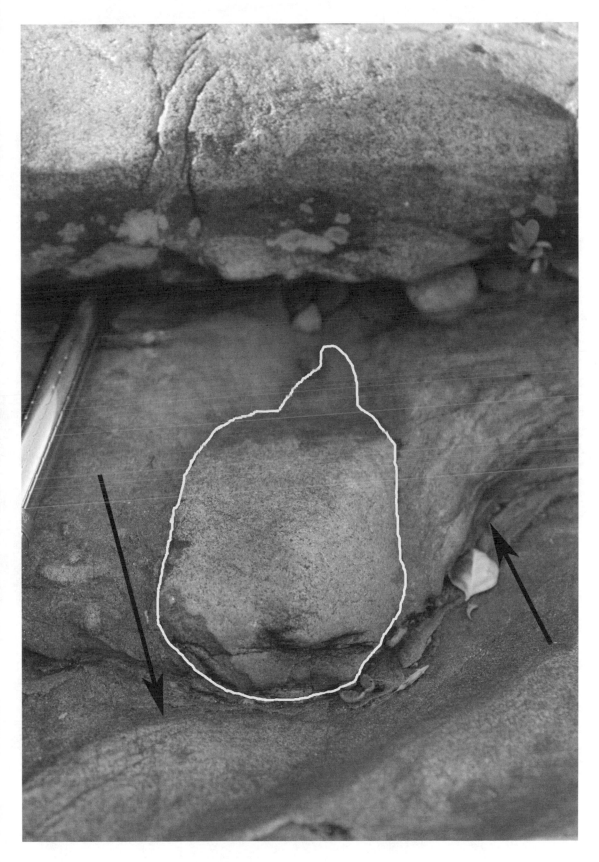

levels have otherwise separated from the lower ones at all points as a result of intense weathering of the intermediate siltstone.

The structure of a complete footprint before modern erosion (Fig. 3.175G) consisted of a bed of sand (dark gray) with deformed ripples and undertracks at its top; a very thin deformed layer of sandy silt (white) whose top was the tracking surface, i.e., with true footprints; and a sandy bed (light gray) that filled the shafts (natural casts) and came in contact with the lower sandy bed in the extrusion rims. Depending on their resistance to erosion, in some footprints a variety of structures are visible: pad marks in the bottom of the true footprints, flattened ripples in the bottom of undertracks, the sand filling of the footprint shafts, and overtracks.

Footprint Features (Ichnotypes)

SAUROPOD FOOTPRINTS

All sauropod footprints occur in the 1SM outcrop (Fig. 3.177). The forefoot marks are crescent shaped and probably open (i.e., Neosauropoda; Wright, 2005). The number of footprints is too small to permit reconstruction of the original form of the manus prints, which were deformed as the pedes pushed the mud forward, deforming the manus prints. The marks of the hind feet are oblong, with their lengths parallel to the direction of motion. These are the smallest sauropod footprints in La Rioja. The hind foot marks vary in length between 24 and 34 cm, and the forefoot print lengths between 18 and 26 cm.

The forefeet are placed in front of the hind feet, with the anteroposterior axis parallel to that the hind feet. There are three parallel forefoot–hind foot alignments, two of which may be part of the same trackway (1SM1, 1SM3). According to Casanovas et al. (1995b), the two associations are parallel with the position of the forefoot–hind foot sets alternating in fashion. The distances between manus/pes sets in each of the sequences 1SM1 and 1SM^2 can be considered as strides, and so sequence 1SM^2 would be the left part of another trackway. The previously mentioned authors thus interpreted track (1SM1–1SM3) from a very wide trackway, which they classified as part of the ichnogenus *Brontopodus*, following the Farlow (1992) criteria.

ORNITHOPOD FOOTPRINTS

There are two footprints assigned to this ichnogroup in 1SM and seven in 2SM. The latter outcrop has only ornithopod footprints (Fig. 3.177). They have three digits close together and a pad for each. The 1SM footprints are probably deformed by the sauropod footprints. The 2SM tracks have a rhomboid pad mark for digit III, with the side toes slightly longer than they are wide. In the distal part there is a transverse depression across the foot, like that of a wide claw. The length of the prints is between 33 and 36 cm, and they are slightly longer than they are wide.

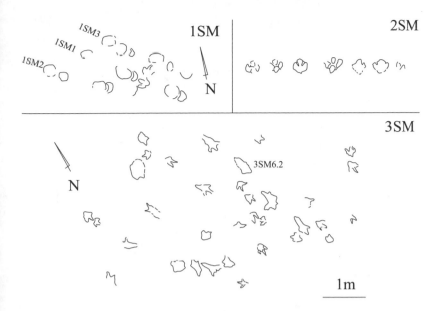

Because the digits are united by a central depression, Casanovas et al. (1995b) assigned them to the ichnogenus *Hadrosaurichnoides*. At the time this caused few problems because it was assumed that San Martín de Jubera was in the Enciso Group (i.e., in the same group in which the footprint taxon had been defined – the Era del Peladillo site). San Martín de Jubera, however is older, although it also falls within the Lower Cretaceous.

THEROPOD FOOTPRINTS

Theropod footprints were found only in 3SM. The general mud structure features are described above.

No footprint can be considered as a stamp in order to define their features and to find a correlation with an ichnogenus. Most are preserved as deformed sediment (subtracks) or as deformed sediment and the remains of casts, or they are overtracks. Digitigrade tridactyl and semiplantigrade tetradactyl footprints are distinguished.

The length of the digitigrade footprints is between 20 and 31 cm. There are some larger ones, but this is because the heel area is elongated, probably as a result of the impression of part of the metatarsus. The semiplantigrades are between 50 and 55 cm in length, with the mark of the hallux present in almost all of them.

Only four trackways can be recognized among the total of 35 tracks: two semiplantigrade and two digitigrade trackways. Virtually all the semiplantigrade footprints are also overtracks. Although the hallux can be seen, it is not possible to measure the length of the digitigrade footprints to make calculations about the speed. Nor is it possible to determine whether these footprints show the divarication between the metatarsal shaft and the acropod axis.

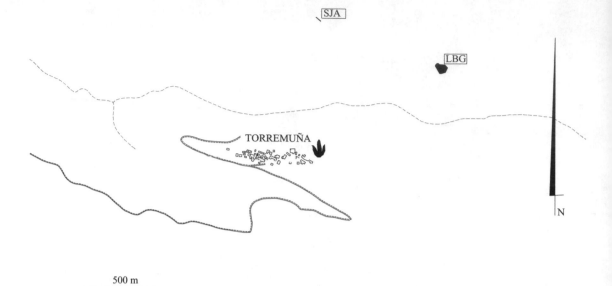

3.178. Location of El Contadero (CT) tracksite and neighboring occurrences of footprints. LBG, La Barguilla; SJA, Santa Juliana.

Footprint Distribution According to Ichnotypes

It is probably misleading to assign different dinosaurs to different outcrops. In La Rioja there are examples where the footprints are usually grouped according to ichnotypes in one part or another of large sites. If the large sites in La Rioja with footprints of various types are subdivided into smaller sectors, then the groups of footprints in each fragment, such as those of San Martín de Jubera, would be mostly of the same type. Therefore, there are insufficient criteria here to propose dinosaur selection on the basis of bedding distribution.

General Information

The track density is around 0.5 per square meter, equivalent to one every 2 m². The environment of the site has tops of layers with many irregularities, sometimes surrounded by extrusion rims, which are certainly more footprints. Casts have been found in the vicinity of San Martín de Jubera and in the bottom of some of the strata. Outcrops 2SM and 3SM are in good condition. However, at 1SM fragments are breaking off, destroying part of the footprints. The location is difficult to access for transporting material to restore the site and far from any accommodations for teams to work on it.

El Contadero

History

Most discoveries of footprint sites in La Rioja are made by people not professionally involved with paleontology. These people inform organizations or persons devoted to studying footprints. However, El Contadero

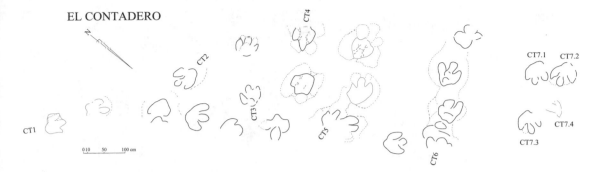

EL CONTADERO

010   50   100 cm

CT1 CT2 CT3 CT4 CT5 CT6 CT7.1 CT7.2 CT7.3 CT7.4

3.179. Map of footprints of El Contadero (modified from Pérez-Lorente, Romero-Molina, and Pereda, 2000).

(CT) was found in 1998 by David Quintana, a forest ranger and geologist, who was also responsible for discoveries at other sites, including footprint shafts (San Román), natural casts (slope of the Nido del Cuervo), and bones and teeth (Trevijano). The El Contadero site was cleared in the spring of 1999 and published by Pérez-Lorente, Romero-Molina, and Pereda (2000).

Description of Location

Access is difficult because the site is far from a main road (Fig. 3.178). The easiest way to reach it is by a path that crosses a small pine forest. The trail follows at a distance of 300 m from a road going from Torremuña to the peak of Nido del Cuervo located west of the figure. The closest other footprint sites are La Barguilla (LBG, 1000 m) and Santa Juliana (1250 m) (Fig. 3.178). El Subaquillo is 10 km away. Isolated casts have been seen in the vicinity of Torremuña, one of which is from a large ornithopod.

Twenty-four footprints (Fig. 3.179) have been studied at the site, although there are two more not counted because of their poor condition. Digging would likely result in more being found, as there are suggestions of prints up to 50 m from the site. No other types of fossils have been seen in the vicinity of El Contadero.

Lithology

The bed with the footprints is a dark gray to black limestone with strike N110E and just over 25° dip to the north. The limestone is micritic in texture, and there are some small shell fragments. There are no visible sedimentary structures on the surface or inside the stratum. The original rock must have been mud rich in organic matter and saturated with water. The bed is about 15 cm thick and is in the upper part of a group of calcareous strata of similar composition and thickness (between 10 and 40 cm).

Footprint Structures in Mud

The dinosaurs passed through this location when the mud was still soft. Footprint depth is variable, between 2 and 11 cm along the same trackway (Fig. 3.180), with some footprints punching completely through the bed.

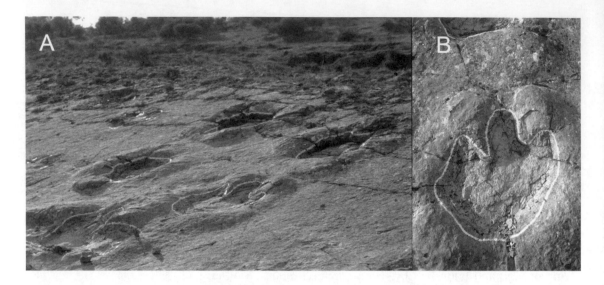

3.180. El Contadero. (A) General appearance of tracksite. Most obvious footprints belong to tracks CT4 and CT5 (Fig. 3.179). Note mud extrusion rims around prints. (B) Footprint CT5.2.

The original thickness of the layer must have been greater because the original rock was limestone mud with abundant organic matter, which can compact by more than 30% during diagenesis. Most of the footprint shafts are surrounded by a mound produced by the extrusion of the mud (Figs. 3.180, 3.181). The height of the extrusion edges is related to the penetration of the feet beneath the tracking surface.

In several of the ichnites there are mud structures extruded on both sides of and between the digits. The footprints (Fig. 3.181) currently have the following structures: actual outline of the footprint stamp in some sectors; extrusion rims at the sides; drop-shaped lobes of mud pushed forward between digits II/III and III/IV (two lobes per footprint), with the apical parts of the drops pointing to the footprint hypex; and inward collapse of part of the extruded mud into the footprint shaft. There is usually no extrusion rim in the rear; in other words, the extruded mud does not form a complete ring around the footprint.

To form these structures, the following must have occurred during the T phase. The foot was inclined relative to the substrate while starting to be supported at its very rear. A small mound of mud formed in front and toward the sides of the heel pad. The foot pushed forward at the same time as it was inserted into the ground, so that the mud was not extruded behind the heel. The foot entered the substrate at an angle (Fig. 3.182A), with the sole being gradually supported. The side digits were gradually supported (Fig. 3.182B). Mud accumulated on the sides ahead of the supported part of the foot and in the gap between digits II and IV.

In the W phase, the foot was fully supported. The distal parts of the outer digits (II, IV) and the central digit (III) sank into the mud (Fig. 3.182C). The excess mud could not be accommodated within the shaft of the footprint, so it was pushed into open space—that is, into the interdigital spaces. Interdigital lobes of extruded mud were formed (Fig. 3.181, b1). In the K phase, or after the foot left the shaft of the footprint,

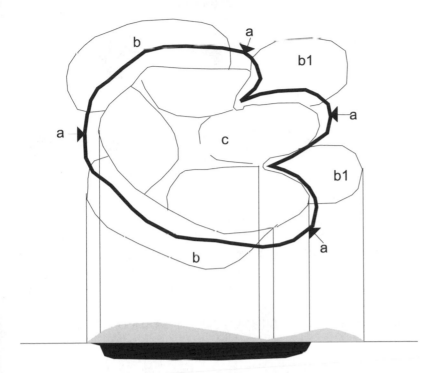

3.181. Features of footprint CT5.2 (Fig. 3.180). Outline of footprint (a) was modified by extrusions of mud along outer sides (b) of toes and between toes (b1). Present shaft of print is indicated by (c). Lower part of figure shows back-to-front distribution of surface heights and depths along print: depressions beneath rock surface are in black, and mud extrusion rims in gray. Modified from Pérez-Lorente, Romero-Molina, and Pereda (2000).

some of the mud fell back into the footprint, giving it its current structure (Figs. 3.180, 3.182D).

The environment in which these tracks were made was similar to that associated with theropod footprints studied by Gatesy et al. (1999) in the Fleming Fjord Formation of Greenland. However, there are notable differences, both in the shape of the digits and in the mechanism and relative depth of the footprints. For example, in El Contadero, the foot was first supported at the very rear of the heel with the distal part of digit III touching the ground only at the end of the support phase. After contact with the substrate, the foot slid forward in the Greenland example, while in El Contadero there was no such movement. Finally, the feet of the Greenland trackmakers were narrow toed, unlike those of the El Contadero dinosaurs, which had broader and relatively shorter digits. Even if the penetration in the mud was the same, the result would be different. Feet with broad digits close together have to move the mud to the outside of the track contour. In narrow-toed feet, the lateral displacement is less. The possibility of total collapse of mud walls is greater in the latter case.

Distribution of Footprints by Their Structures

The depth of the footprints is variable, ranging from less than 2 cm to more than 11 cm. The deepest marks are in an area in the lower part of the site. The depth of the ichnites is proportional to the height of the mud extrusion rim. There is interference between footprints; for example, the depth of the base of CT1.10, was altered by emplacement of CT6.1.

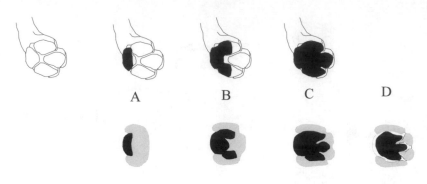

3.182. Sequence of events responsible for formation of El Contadero (CT) footprints. (Upper row) Part of dinosaur's foot was supported by substrate (black), but also mud pushed outward and forward. Sole of foot came down at an angle to mud surface, first touching at back of toe region of foot (A). Support of foot passed forward (B) until it was completely supported by substrate (C); mud continued to be pushed outward and forward (gray). As foot was lifted off ground, mud collapsed inward toward shaft, resulting in final form of footprint.

Footprints CT4.1, CT4.2, CT5.2, and CT5.3, which have the best mud extrusion structures between the digits, occupy the central part of the site.

It is likely that there were areas of the site where the mud was softer. No textural heterogeneity is seen in examination of the bedrock sections (e.g., linsen bedding of interspersed sand in the calcareous clay), which would make the surface firmer in places even if it was equally wet. Consequently it is more likely to have been a uniformly muddy area with shallow ponds, which had dry patches–perhaps between scattered pools–or areas less saturated with water.

Footprint Features (Ichnotypes)

The footprints show the marks of three broad digits, rounded at their ends. There are four wide pads, one for each digit and another in the center at the back of the print. The footprints are relatively large (the average length is 57 cm). The ratio $(l − a)/a$ is close to 0, indicating circular footprints in outline. The footprint rotation is negative (inward), and the width of the trackway ($Ar/a = 0.3$) is narrow. The ratio of z/l (approximately equal to 3) is a typical value for trackways made by dinosaurs with thick to very thick limbs. The estimated walking speed is very low (between 1 and 1.5 km/h), as calculated by Casanovas et al. (1995a), where l is footprint length, a is the footprint width, Ar is the trackway deviation, and z is the stride length; where $(l − a)/a$ is the relative footprint length; where Ar/a is the relative trackway deviation; and where z/l is the limb thickness. According to Díaz-Martínez et al. (2009), these prints would be classified as *Iguanodontipus*.

Footprint Ichnotype Distribution

As in almost all small footprint exposures (about 50 m²), the 24 footprints are from the same ichnogroup. The prints occur in seven trackways with no single or parallel direction. Similar footprints have been found up to 50 m away on the same layer. This set of data is consistent with the passing of a group of dinosaurs probably much larger than seven, but the tracks do not follow the same direction, which means that they were not behaving in herd-transit fashion when they left the tracks. It may be that all the footsteps in El Contadero were left by two or three dinosaurs of the same size. The difference in size of the footprints is less than that sometimes found in other examples of prints of the same trackway. The actual set of footprints (those in outcrops and concealed) is higher because similar footprints have been found in small areas of the same layer up to 50 m away.

General Information

The footprint density is 0.5/m², equivalent to one footprint every 2 m², which is low. The site is well conserved and seems to resist inclement weather. The site could be extended by digging at the base of the layer (northeast side) to perhaps more than 1 m of distance. As in other trackways studied in La Rioja, paradoxically, the walking speed calculated is small even though the dinosaurs were large.

History

La Barguilla (LBG) is a site (Fig. 3.183) that two brothers told me about. They were shepherds from Hornillos de Cameros named Juan Jose Santos and Antonio Santos. When asked what year they had learned that what they had found were dinosaur footprints, they were not sure. Both knew of quite a few places near the town that had footprints and that had been unreported before when they saw the sites of Enciso and Munilla, sometime after 1980. Walking through the countryside, the brothers had seen, for example, the footprints of Santisol, Santa Juliana, El Torrontón, La Pellejera, and others still unnamed. The number of tracks at these sites is more than a thousand. Documentation began in 2000 and is still continuing. Some of the discoveries remain to be mapped.

Location

The layer with footprints (Fig. 3.184) has a strike of N125E and dips gently (23N). The rock surface is smooth and easy to see, although some of the footprints are not clear. The site is 1 km from Hornillos de Cameros along the old path to Munilla, and one has to go on foot by this path because there is no access via a main road. The nearest other sites known to have footprints (Fig. 3.183) are Santa Juliana, 580 m away, and Santisol at 1500 m. There are also three points with unstudied footprints, the farthest of which is a little over 1500 m away. El Contadero, off the map, is about 1000 m away.

Lithology

The rock is a fine-grained gray-green sandy shale. It is a thin layer (about 15 cm) with faint mud cracks. The mud cracks are not deformed or crushed and so were formed after the tracks.

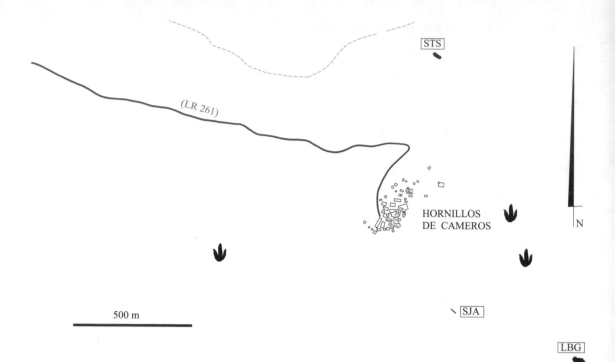

3.183. Location of La Barguilla (LBG) and neighboring sites of Santa Juliana (SJA) and Santisol (STS). Local road is labeled LR 261.

Footprint Structures

The morphology of the footprints varies widely across the tracksite. This partly reflects the fact that their markers belonged to two different groups of dinosaurs. However, there were also differences in the physical condition and composition of the sediments, and perhaps their water content, which influenced the depth of the footprints and the structures they contain. Some footprints show only the distal part of the digits. The mud cracks cut across the marks and are not distorted by them, which indicates that the cracks formed later (Thulborn, 1990) and that the tracks are true prints, as opposed to undertracks. Some footprints have mud extrusion structures around them and between the toes (Fig. 3.185), as at El Contadero. After withdrawal of the foot, the mud must have collapsed and significantly deformed the outline of some of the digits. There are two sets of interference between footprints. One of these is an overlap between footprints LBG2.21 and LBG3.12, where it is impossible to say which trackway is older. The second interference is where print LBG2.11 walks over an isolated theropod footprint. In this instance, the mud extruded from LBG2.11 covers part of the theropod footprint.

There are several semiplantigrade footprints (Figs, 3.184, 3.185) at La Barguilla in which extruded and collapsed mud closes the proximal part of the digit III mark. The collapse of mud is related to the depth to which the foot penetrated and the low consistency of the mud. The semiplantigrade walking mode was probably adopted by the dinosaur to negotiate this area of soft mud.

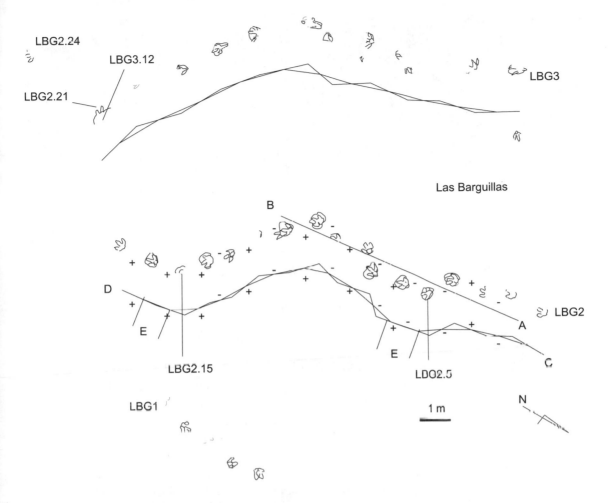

3.184. Map of footprints at La Barguilla. For tracks LBG2 and LBG3, diagrams of paces and trackway midlines are shown adjacent to footprints themselves. Line segment AB shows Leonardi's (1979, 1984) observation of footprints positioned ("as pegadas se mudan em grupos de três direita para esquerda e vice-versa"; Leonardi, 1979:505) on either side of line running through portion of trackway LBG2. Midline position in relation with CD line is abnormal in points (E) of change of pace sequence (Table 3.14). Note position of footprints LBG2.5 and LBG2.15 in curve of trackway and in Table 3.14.

Footprint Structure Distribution

The footprints with only digit impressions, or where the digits are deeper than the heel pad mark, are at the east side of the site (Fig. 3.184). All are part of trackway LBG1. Some authors, such as Currie (1983, 1995), assume that the water depth was greater in footprints of this type, in which the digit pads and the heel appear shallower. The part of the animal's foot that exerted more pressure on the ground made a deeper mark (i.e., the digits, as the dinosaur pushed forward). LBG1 has no mud extrusion rims, and the mud cracks go through the tracks without being deformed. The opposite is seen in the central trackway, LBG2, where the extrusion rims are on the sides and front of the footprints. This is more consistent with a foot that is first supported by the heel, pushes the mud forward, and then is supported on the toe pads. Again, the mud cracks are not deformed in this trackway. Finally, the trackway farthest to the west has semiplantigrade footprints with collapsed mud marks. In this case, the digit marks are narrower than in the previous trackway's footprints, so the penetration of the foot into the soft mud is easier for this dinosaur.

3.185. (Left) Footprint LBG2.5; length of white bar = 10 cm. (Right) Footprint LBG3.9; note collapse of mud in proximal part of digit III; mesh opening size 5 cm.

Many of the isolated footprints are shallow, so it might be assumed that there was more water on the site when they were made. This does not seem likely because there are no claw marks or any other signs of swimming. It is therefore likely that the mud was harder at that time.

Footprint Features (Ichnotypes)

ORNITHOPOD FOOTPRINTS

Many of the footprints in trackways LBG1 and LBG2 are complete. They have three rounded pads for the digits and one at the back for the heel. The feet are relatively large (34–36 cm in length) and slightly wider than long, with a ratio of relative footprint length $(1 − a)/a = −0.05$ and $−0.03$, thus classifying them as wide feet. The digit impressions themselves are likewise broad, and their marks are well separated from each other. The tips of the digits are rounded. The remaining foot measurements are typical for all biped dinosaurs: digit III is the longest, and the interdigital angle II^III (16°) is less than angle III^IV (34°). The pace and stride length in trackway LBG2 is irregular because of its sinuous nature (Fig. 3.184). The limb thickness ratio, z/l, indicates that the limbs were thick. (Table 2.1 provides a key of abbreviations.) Classifying the footprints in the proposed groups is risky. Given their size, they could be of the *Iguanodonipus* type, but there are many with the heel pad well marked and rounded.

THEROPOD FOOTPRINTS

There are two types of theropod footprints at La Barguilla, digitigrade and semiplantigrade. The semiplantigrade prints (Figs. 3.184, 3.185) occur in trackway LBG3, where they alternate with digitigrade prints. The average length of the digitigrade sector of the footprints is 30 cm, and 50 cm if the metatarsus mark is included. The metatarsal impression thus constitutes 40% of the overall print length. The divarication of the digits is wide but probably does not reflect their normal anatomic position. Thulborn (1990) says that the foot sinking into the mud causes the digits to spread apart more than in a shallow footstep. The digit marks are relatively long, tapered, and well separated. The pace and stride length vary greatly as the midline (or trackway) twists about, and they vary in the type of progression made (semiplantigrade–digitigrade–semiplantigrade), as also seen in the variation of the trackway deviation (Ar) and the pace angle (Ap). This case can be added to those well known in the literature (Kuban, 1989; Pérez-Lorente, 1993a; Sarjeant et al., 2002) in which semi-plantigrade walking displays greater pace and stride length and therefore an apparent increased velocity than in the digitigrade mode. The speed is abnormally high (6–8 km/h), which is common in this type of trackway. There are no digital pad marks, with only LBG3.5 showing a small, rounded form in the heel. The isolated footprints are poorly preserved and cannot be classified. They may have been left by more than one type of dinosaur.

Footprint Distribution by Ichnotypes

The midline of trackways LBG2 and LBG3 are curved rather than straight. LBG2 also describes a sinusoidal path similar to that of a *Sousaichnum* trackway described by Leonardi (1984) over the interval, in that the interval LBG2.1 to LBG2.10 (Fig. 3.184) seems to have the ornithopod footprints placed three to the right, three to the left, and three to the right. Currie (1983, 1995) reported similarly curved ornithopod trackways at the Peace River Canyon tracksites. In the present book, sinusoidal trackways are also documented from the Valdeté, Cuesta de Andorra, and Barranco de la Canal tracksites. This sinusoidal pattern is not exclusive to ornithopod trackways; it is also associated with theropod trackways and the forelimbs of sauropods. In addition to the aforementioned studies there are also descriptions of sigmoidal trackways for the Suttles Quarry (Delair and Lander, 1973), Valdeté (Lockley et al., 1994), and Iouaridène (Nouri, 2007), among others.

In analyzing the cases cited and comparing them with those of La Barguilla, Pérez-Lorente (2003c) found that there are two types of sigmoid trackways. Those with a large amplitude and short wavelength are the more irregular trackways, while those with a short amplitude and long wavelength are more regular. There is also a difference in length between the left and right steps in sigmoidal trackways (Table 3.14) that suggests a course correction by the dinosaurs. The validity of this assumption is supported by the long routes, which are only slightly sinuous. The variation of curvature of the midline (turning points) determines whether the left–right steps are greater than the right–left, and vice versa. Changes in dinosaur progression due to external causes (different water depth; reaction to the movement of neighboring animals) are more likely to account for irregular trackways and those with a large amplitude and short wavelength.

Natural Barriers and Behavior

The quality of footprints varies from one trackway to another. In LBG1, the digit impressions are good, but the heel pad is shallower and less distinct. The LBG2 trackway has mud extrusion rim structures similar to

Table 3.14. Sequence of longer (+) and shorter (–) paces at trackway LBG2.

Right pes Left pes	Right–Left (cm)	Left–Right (cm)	Paces: P(1) to P(16)		P(x+1)-P(x) cm
LBG2.17	111		16		
LBG2.16		102	15	+	9
LBG2.15	80		14	+	22
LBG2.14		67	13	+	13
LBG2.13	100		12	–	33
LBG2.12		95	11	+	5
LBG2.11	97		10	–	2
LBG2.10		82	9	+	5
LBG2.9	100		8	–	18
LBG2.8		69	7	+	31
LBG2.7	95		6	–	26
LBG2.6		77	5	+	18
LBG2.5	91		4	–	14
LBG2.4		106	3	–	15
LBG2.3	70		2	+	36
LBG2.2		105	1	–	35
LBG2.1					

Note: Thick lines coincide with inflection points at midline.

those seen at El Contadero, although much less pronounced. The LBG3 trackway, although it is of theropod origin with metatarsus impressions, is also curved. The midlines of LBG2 and the initial part of LBG3 (Fig. 3.184) are somewhat parallel.

According to the structures seen, the softest mud was around LBG3 and the firmest around LBG1. The acetabulum heights derived in each case (without including the metatarsus) are 243 cm for LBG1, 258 cm for LBG2, and 148 cm for LBG3. This poses an alternative if it is assumed that the mud had the same physical characteristics: the feet in LBG1 might have exerted less pressure on the ground than in LBG2, and these less than in LBG3. The theropod dinosaur footprints, which sometimes left an impression of the metatarsus, effectively indicate that the water depth was less (because the foot sank deeper into the ground). The sinuous path of the trackway is not due to the animal being dragged by the current because the feet were well supported.

It cannot be argued that the parallelism between the trackways LBG2 and LBG3 is due to two dinosaurs walking together because they do not belong to the same ichnogroup. The speed deduced is also quite different: for the ornithopods it is between 2.3 and 3.2 km/h, and for the theropod it is between 5.6 and 7.9 km/h. It seems likely, therefore, that the LBG2 and LBG3 sinusoid trajectories are due to the ornithopod and theropod reacting to changes in water depth or displaying some other type of behavior whose interpretation has not been deduced.

LBG3 has several of the typical features of semiplantigrade trackways because the track maker can choose the semiplantigrade or digitigrade mode (Sarjeant et al., 2002), and because the speed of progress is abnormally fast (Kuban, 1989; Pérez-Lorente, 1993b; Sarjeant et al., 2002).

General Information

The footprint density on the site is 0.1 per square meter, which is low (equivalent to one footprint in every 10 m²). This figure seems low considering the long interval of time over which footprints could have been left (the longer the range of time, the larger the number of dinosaurs that pass). Likewise, by counting the three trackways and isolated footprints, the number of dinosaurs deduced to have passed and left some footprint in this tracksite is at least nine. There are footprints from at least two different points in time (overlapping footprints), all of them made before the mud cracks. There also are isolated prints that must be part of six trackways with no visible prints. There are no footprints made after the drying of the mud.

The surface of La Barguilla has tectonic fractures (joints). The footprints on the edges of the site exposed for longer have been erased. With respect to the durability of the footprints, I can say that it is likely that the siliceous composition of the rock provides sufficient strength to maintain the footprints.

Main Features of Site

La Barguilla is interesting for the curved and sinusoidal nature of two of the trackways, the possible variation of the height of the water layer, the difference in structures between footprints with wide digits and those with thin ones, and the possible parallelism of trackways. What is the influence of limping, laterality, or trackway wavelength on the control of the pace length? In the analysis of trackway LBG2, other trackways have been considered, leading to suggestions of different forms of behavior. Winding trackways with different wavelengths and amplitudes have been distinguished to justify the anomaly of the trackway cited. LBG2 is the only ornithopod trackway in La Rioja with a short wavelength and relatively large amplitude. It is possible that in La Rioja, because many of the outcrops with footprints are small, there are common patterns of dinosaur behavior being displayed, although they are not recognizable.

Although it is one of 11 sites with footprints with the metatarsus mark in La Rioja, the footprints are not well enough preserved to provide other useful data. At LBG3, digitigrade tracks alternate with semiplantigrade ones, indicating that this mode of walking was optional for the semiplantigrade trackmaker.

History

Santisol

The Santisol (STS) site was also discovered by Juan Jose and Antonio Santos and was cleared and mapped in July 2000. Parts of the sites now known as 1HR, 2HR, 3HR, 4HR, and 5HR were discovered in the outskirts of Hornillos de Cameros (Viera and Torres, 1996).

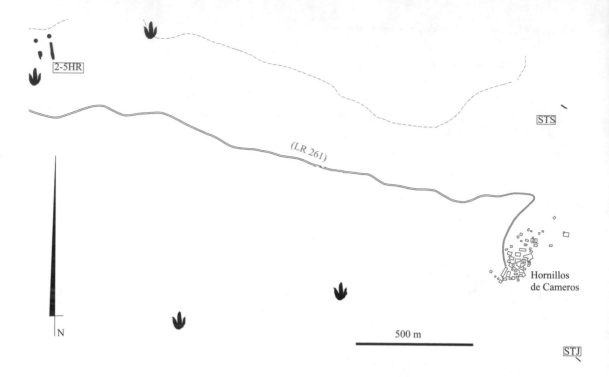

3.186. Location of Santisol (STS) and neighboring sites, Hornillos 2 to 5 (2–5HR) and Santa Juliana (STJ). Other sites in vicinity with footprints have been reported but have not yet been described. Local road is labeled LR 261.

Description of Locality

The location (Fig. 3.186) is a little over 2 km from the Hornillos de Cameros road (700 m in a straight line), near a dolmen, on high ground from which the arrangement of the strata and location of several sites can be seen well. The trackways have few footprints because the outcrop is long and narrow (Fig. 3.187). The layer strikes in the direction N110E with a smooth dip of 18°N. Santisol is next to a mountain road accessible by an ATV. The nearest sites with footprints are about 900 m away and have not yet been studied. Santa Juliana is 1 km away, and the sites 2HR to 5HR are 2 km away. This sites of La Barguilla and El Contadero are also within 2 km from Santisol.

Viera and Torres (1996) cite finds of fossils of plants, invertebrates, and vertebrates in the surrounding area. The fossils are situated in sections (Spanish *tramos*) in a column covering unspecified land in the Urbión and Enciso Groups, with no indication of the boundary between the two. The fossil remains can be located because two *tramos* contain two known tracksites: in *tramo* 44 we find 5HR, and in *tramo* 40 we find the lowest of the Hornillos de Cameros sites (1HR). The vertebrates (*Lepidotes* sp., *Microdon* sp., *Hybodus* sp., *Goniopholis* sp., *Pholidosaurus* sp., Chelonia indet.) are all below *tramo* 29, probably in lacustrine limestone. The invertebrates *Eomiodon cuneatus* Sowerby, 1816, and *Viviparus* (=*Paludina*) cf. *fluviorum* are between 1HR and 5HR. No other invertebrate fossils other than traces have been found nearby.

The area is rich in dinosaur footprints. In the vicinity of Santisol, the dip of the layers is contrary to the slopes, so that there are few outcrops

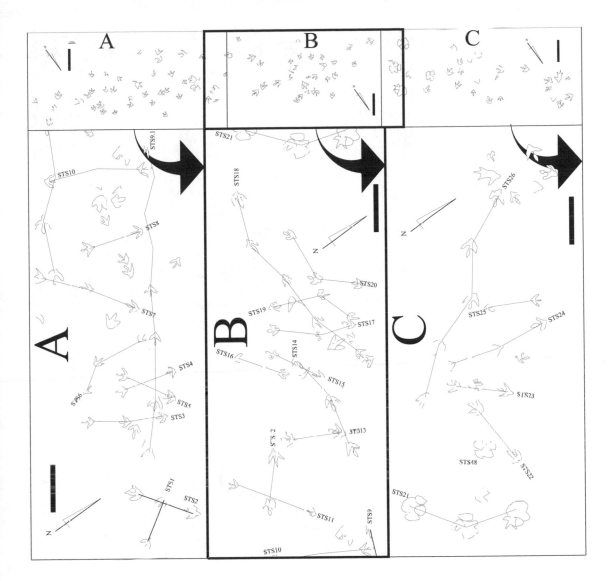

3.187. Map of footprints. (Top row) Entire site, demarcated into three sectors. (Bottom) Expansion of footprints for each sector.

with stratification surfaces. It is also an area of old periglacial influence, with very gentle relief, and therefore with lots of loose debris that hinders direct observation. Almost any rock outcrop where something more than a meter of stratification surface area can be seen contains footprints, both molds (top) and casts (wall). Viera and Torres (1996) cite an area with casts at the base of a bed, and we know the location of two large casts (a footprint about 60–80 cm in length) of prints outside the areas mentioned. There is also a small cast on display in the Hornillos de Cameros town hall. All the large casts are ornithopod footprints.

Lithology

The footprints are on a 15 to 20 cm layer of yellow-brown sandy shales. The footprint bed is on the top of several levels with the same style and color, separated by interspersed levels of less than 1 cm. The site is laterally

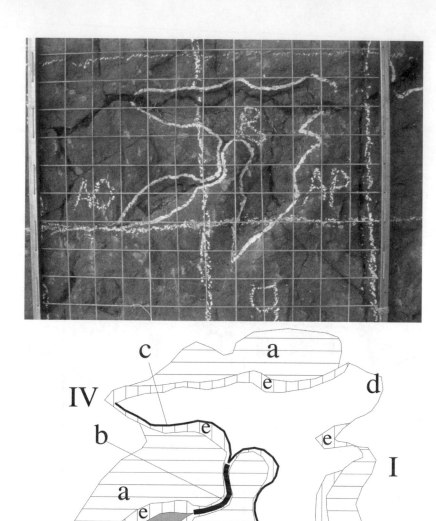

discontinuous, so it seems that the unit is lens shaped, i.e., a sandy silt sedimentary accumulation with footprints between other kinds of sediments. There are no ripples or mud cracks. There seems to be parallel lamination in very fine laminites.

Footprint-Mud Structures

Most of the footprints at Santisol have both simple and complex mud collapse structures (Figs. 3.188, 3.189). There are also footprints where the structures are simple, with no collapse of mud into the footprint.

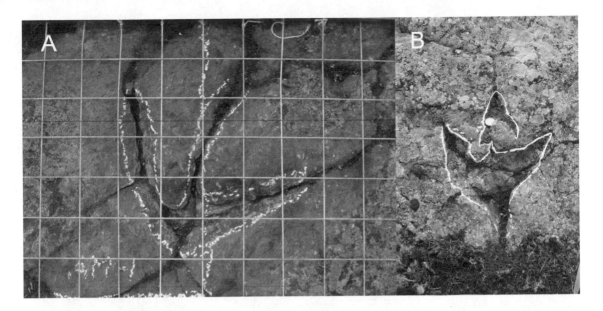

3.189. (A) STS11.2; collapse of footprint walls. (B) STS20.1; thrust structures of mud toward digit III hole; area of hypexes collapsed and mark of metatarsus as it passes. Mesh width = 5 cm.

SIMPLE COLLAPSE STRUCTURES

Some tridactyl footprints show thin, well-separated digits (Fig. 3.189A). The walls of the footprints are convex toward the interior of the holes left by the digits. This thinness is thought to be due to the mud walls collapsing. The final collapse occurs after the foot has exited the print (Pérez-Lorente, 2001a). In STS11.2 (Fig. 3.189A), the structures indicate that the foot did not penetrate so deeply that its upper surface was beneath the top of the mud. Nor was there total collapse after foot withdrawal, as in other cases in La Rioja (e.g., Virgen del Prado, La Senoba, Era del Peladillo [1PL]), where the digital shaft is left as a series of three converging lines.

COMPLEX STRUCTURES IN SANTISOL

Many of the tracks at this site (Fig. 3.189B) have similar structures to those described by Gatesy et al. (1999), where the foot sank completely into the mud. The footprint outline in this type of dinosaur ichnite (Fig. 3.188) has a rear projection, a middle part with well-defined digits II and IV, and an eyelet-shaped anterior portion separated from the impressions of digits II and IV. Several factors need to be considered in the formation of these structures: total penetration of the foot in the mud (Gatesy et al., 1999); the movement of the metapodium and acropodium, which is not parasagittal (Pérez-Lorente, 1993a); and the successive phases of the footprint's interaction with the substrate (Thulborn and Wade, 1989).

MOVEMENT OF FOOT

To explain the structures of these complex forms (Figs. 3.188–3.191), we must first consider the autopod structure of a bipedal dinosaur as it makes digitigrade and semiplantigrade footprints. In semiplantigrade footprints of dinosaurs, the axis of the metatarsal and the digital portions of the print

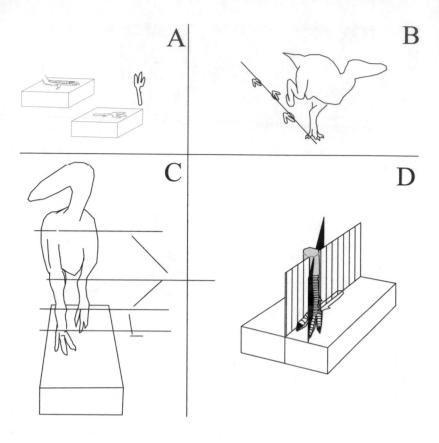

3.190. Different angles between acropodium and metapodium in bipedal dinosaurs. (A) In semiplantigrade footprints, digital portion of print points slightly inward, rather than being aligned with metatarsal impression (see Fig. 3.157). (B) Similarly, in digitigrade prints, footprint angles inward (negative rotation; viewed from front in C). (D) Thus, metatarsus may have moved in slightly different vertical plane (striped) than did digital part of foot (black).

are not aligned and form an angle (Figs. 3.188, 3.190A) other than 180° (Pérez-Lorente, 1993a). It is also known that the pigeon-toed formation or negative orientation (sensu Leonardi, 1987) of digitigrade footprints involves directing the front part of the foot forward toward the interior of the track, but with the metatarsus raised off the ground (Fig. 3.190B, C).

I have interpreted (Pérez-Lorente, 1993c) that the axis of the digitigrade foot points toward the inside of the track in both the digitigrade and semiplantigrade print modes. I also indicate the asymmetry of the condyle of the metatarsal III in theropod dinosaurs, suggesting that the axis of the metatarsus and the axis of the digitigrade foot moved in different planes (Fig. 3.190D). The feet of bipedal dinosaurs point toward the inside of the trackway without the dinosaur's body rotating around a vertical axis.

If the acropodium was indeed angled medially with respect to the metapodium, then when the foot of a bipedal dinosaur penetrates the ground (T phase), the metatarsus would not follow the path indicated by the axis of digit III. The metatarsus would instead move forward but along its own straight path, toward the outside of the track, i.e., toward digit IV (Fig. 3.191C). As the foot slides forward into the mud, it has to do so in articulation with the metatarsus. The digitigrade foot inside the mud therefore moves in a direction oblique to the axis of digit III. The foot slides forward (and therefore also toward the lateral side of digit IV,

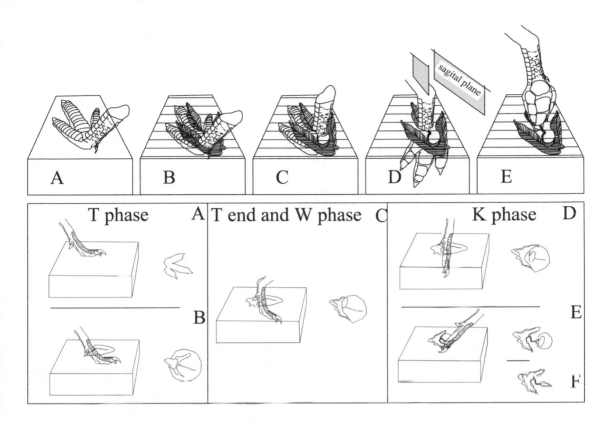

because the rotation of the feet is negative). The mud pushes forward and accumulates over the hypex; the digit II mark is wider than the digit IV mark. Once the foot is completely beneath the surface of the substrate, the mud collapsed into the three digit marks partially occludes them. Incisions of the bottom front part of digits II and IV (Figs. 3.188e, 3.188c, 3.189B) are evident. Next is the formation of side extrusion rims, then the mark of the hallux (Figs. 3.191A, 3.191B). Once the foot is fully supported (W phase), as the dinosaur moved forward, the metatarsus would become more vertical and therefore would be placed above the first joint of digit III (Fig. 3.191D).

Next occurs collapse of the mud; it falls into the hollows of the side digits and the gap left by digit III. Rounded lobes, or retroversion structures, are formed by the collapse of mud back into the shaft of the footprint in the place where the hypex made a mark. The metatarsus passes through the mud accumulated over the hypexes (Figs. 3.187, 3.190C, 3.190D) and leaves an incision (Figs. 3.188b, 3.191D), thus marking its path. The metatarsus starts its path in the inner proximal part of the digit IV mark and advances, rotating toward the center of the foot. This movement is consistent with the arrangement of the condylar slide surface of the joint of the third metatarsal distal part (Pérez-Lorente, 1993c). When the foot leaves the ground (K phase), the resistance of the acropod to moving inside the mud would force the metatarsus to move even farther toward the center of the footprint (Fig. 3.191D).

3.191. Interaction between moving left bipedal dinosaur footprint and substrate, and resulting footprint features. Foot contacts ground in T phase (A) and penetrates soft substrate (B). As T phase transitions into W phase (C), position of metatarsus slips outward toward digit IV (C). Once animal is fully in W phase and its weight is over its foot, the metatarsus becomes more vertical (D). As the foot is lifted off substrate, three toes gather together and exit, leaving a common eyelet-shaped opening at the front end of the footprint (E, F).

The metatarsus axis advances until it is on the central part of the footprint. This movement may influence the rotation of the metatarsus and the resistance of the mud to the movement of the foot (Figs. 3.188b, 3.191D). With the digits joined, the foot exits through the front of the footprint (Figs. 3.188f, 191E). A hole is formed, like a buttonhole, in the most advanced part of the footprint. This small hole is separated from the rear depression by an incision scar, which is sometimes sinuous (Figs. 3.188b, 3.191E, 3.191F). The hole is wide as a result of two sections of the autopod—the acropod and part of the metapod—passing through. The mud around the front of the hole falls into it, leaving a buttonhole shape. This may also have a central incision, or one side open that runs along the base of the side walls of the buttonhole shape (Figs 3.188f, 3.191E, 3.191F). The bottom of the eyelet is flat and parallel to the tracking surface in many footprints. This involves movement of the mud filling, closing the gap caused by the departure of the digits. The mud under the surface is probably more fluid and will prevent the footprint from showing signs of it passing, or the foot movement itself may cause a thixotropic flowing.

Distribution of Footprints by Structures

The ornithopod footprints are not collapse structures and are deformed by theropod footprints. Theropod footprints with collapse structures, on the other hand, occur all over the site, both isolated and as part of trackways. The direction of travel is variable. The shallowest theropod footprints, without any collapse features, were made later, as deduced from the interference points. They are few and apparently isolated.

It does not seem appropriate for theropod footprints to be on a site with soft mud: the cautious dinosaur hypothesis of García-Ramos, Piñuela, and Lires (2002a), proposes that theropod dinosaurs avoided soft ground, while herbivorous dinosaurs were equally mobile on both soft and solid footing. In this case, the rock with footprints is not on a level expanse but is wedged toward the sides. It may be the filling for a narrow, slightly lower area like a pool or the bottom of a canal. It is also possible that the cautious dinosaur hypothesis is not always true.

Footprint Features (Ichnotypes)

Most of the footprints of Santisol are tridactyl, with Pérez-Lorente (2003c) classifying them as theropod, ornithopod, or unidentified prints. The theropod footprints have independent, relatively long, tapered digits. Those with collapse structures provide no more usable biomorphic data. The average length of the footprints is between 22 and 36 cm, with a width between 27 and 46 cm, so they are wider than they are long. Those that penetrate more deeply in the mud also leave a hallux mark. As in the site at Virgen del Prado, the narrow digits are not considered to be biomorphic features of the trackmaker's acropodium. It is surprising that in this type of digit (typical of theropod footprints) the width of the footprint is greater than the length of the ichnite. This is probably due to the opening of the digits and the foot sinking into the mud in the W phase (Thulborn and Wade, 1989), while this opening is probably also responsible for the high divarications (up to 86° [STS3]) seen in prints at this site.

There are strong tridactyl digit marks that are tapered, with several pads for the digits and no hallux seen. Few footprints of this kind, where the contour line is complete, have been found. They measure more than 37 cm in length and are longer than they are wide. They would be classified as *Megalosauripus* Lockley, Meyer, and Santos, 1998.

Trackway STS21 and the isolated footprint STS48 (Fig. 3.187) are from feet with wide, rounded digits. Each digit has a pad, and there is another, similar pad on the heel. According to Díaz-Martínez et al. (2009) they would be included in *Brachyiguanodonipus*.

Footprint Distribution by Ichnotypes

It is possible, although unlikely, that the deformed footprints belong to the same type of theropod dinosaur. The size of the footprints (between 22 and 36 cm), and therefore the animals that left them, is variable. The calculated height of the acetabulum is 110 cm for the small ones and 170 cm for the large ones. The number of individuals that passed through the site cannot be calculated, nor whether the surrounding areas of tracksite were soft clay. According to the cautious dinosaur hypothesis, the site was not suitable for theropods. The site is also too small to infer whether the ornithopod footprints are from one or two individuals, and whether they walked alone.

General Information

At 0.9 prints per square meter, the track density is intermediate, equivalent to almost one footprint per square meter. Most of the footprints were made over a short interval of time when the mud was soft. This involves a concentration of dinosaurs in the area at a given time, but this fact does not permit us to draw any major conclusions.

Santisol could be extended a few square meters, which would take it to its north–northeast limit. Such a dig should be coordinated with the protection of the site as a result of the exposure to the elements (sunlight; diurnal and seasonal temperature changes; rain, snow, and frost).

The problem of hardening of the mud in a place that does not appear to have any emersion is being addressed. In theory, the first theropod footprints were made on mud of low viscosity (deep footprints) and average consistency (no mudflow but swelling of the walls). The ornithopod marks should therefore be printed on harder ground (shallow footprints) but are earlier than the theropod prints. All the footprints are real, and the study surface is the tracking surface. There are no mud cracks or signs of colonization by plants; therefore, the surface was apparently never above water level. Can we argue that the theropod feet (narrow toes) had more penetrating power than the ornithopod ones (massive feet) in the same type of ground? Does this mean more pressure on the soles of the theropod feet?

Main Features of Site

The complex structures of the footprints in Santisol have been studied for the first time at this point. After the study of Santisol, tridactyl footprints that have a front eyelet in the digit III position were examined more closely. This led to similar structures being interpreted at Era del Peladillo 2 (2PL), other sites like Las Mortajeras (not listed here, and located between Santisol and El Sobaquillo), and other points in other countries and environments.

In the study of footprints, it is important to distinguish the structures they show and not to devalue footprints that are not stamps. New definitions regarding the types of structures and surfaces being examined are necessary. An interesting problem for me is the description and interpretation of the footprints with structures involving the physical features of the mud, autopod penetration into the substrate, and the movement of the foot. I have not studied whether the relationships between the metatarsus position (rotated about its longitudinal axis?), the ankle joint movement, and the simple hinge joint at the ankle of dinosaurs (cf. Brusate, 2012) can be compatible.

La Pellejera

History

In 2001, Juan Jose Santos showed us this site, which he had discovered several years earlier. The place was then a rocky hillside with many loose stones, among which could be seen some footprints. The predominant structures were large holes, both shallow and deep. It was not necessary to dig to expose the clean surface, but we dug to remove items causing the changes and decay, such as loose stones and clay.

The first clearing was done during the summer of 2002, and data were collected in April 2003. The preliminary results were published by Requeta et al. (2006–2007), but La Pellejera (LP) remains under study today.

The number of footprints mentioned in 2006–2007 was 717. At the end of 2008, 780 tracks had been recorded, and currently 825 are known. In addition, during the summer of 2009, an area that had not been studied at the top of the site was cleared. The number of trackways has changed since 2006–2007 because some have been merged after finding intermediate footprints. Five new trackways have also been found. No date has been set for publishing new findings because the process of data acquisition is ongoing.

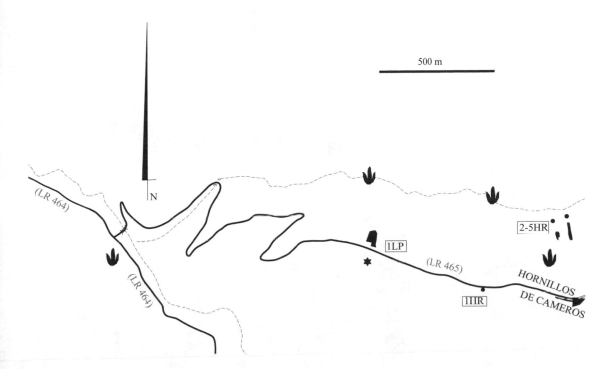

Location of La Pellejera Tracksite

3.192. Location of La Pellejera tracksite (1LP) and Hornillos 1 to 5 (1–5HR). Star shows
location of dinosaur model in
3.193. Local roads are labeled
LR 261 and LR 464.

La Pellejera is along the road (LR 465) leading to Hornillos de Cameros
(Fig. 3.192). It is easy to find, not only because of the signs indicating
La Rioja sites but also because there is a reproduction of a quadrupedal
ornithopod dinosaur walking in the semiplantigrade mode (Fig. 3.193).
The exposure layer has a strike of N100E and dip of 18N. The gentle
slope and wide area of expansion favors observation and study conditions
(Fig. 3.194).

The closest studied site is 1HR (Hornillos de Cameros 1 tracksite),
which is about 500 m away. At 800 m are 2HR, 3HR, 4HR, and 5HR, while
at some distance away are Santisol (3 km) and La Cela (5 km). There are
more outcrops with footprints in the immediate area (Fig. 3.192). A circle
of 1500 m centered on La Pellejera would contain 10 exposures with tracks
(1LP, 1–5HR) and four not studied (one off the map in Fig. 3.192).

Near La Pellejera are the fossil sites (containing shells and traces of
invertebrates; and vertebrates) cited in Santisol (Viera and Torres, 1996).
The vertebrates are at stratigraphic levels below LP. Alvarez et al. (2000),
Moratalla (1993), Moratalla et al. (1988), Moratalla, Sanz, and Jiménez
(1997b), Pérez-Lorente (2003c), and Viera and Torres (1996) deal with the
studied sites in the surrounding area.

Lithology

The footprints are located in silty sandstone (Fig. 3.196). There are several
layers of sandstone of varying thickness (10–40 cm), separated by parallel, finely laminated siltstone beds. Sandstones are generally massive,

3.193. Quadrupedal orthopod dinosaur with semiplantigrade walking. This is a scale replica of the trackmaker that left the 1LP15 trackway.

although laminites can be seen on the edges of some footprints. The sandstone–siltstone pair are considered the repetitive sedimentary constituents of the La Pellejera site. The footprints were made on the top of a sandstone level.

The sandstone levels sometimes amalgamate, and several layers merge laterally into a single bed with a massive texture (Fig. 3.196). One of these levels has small ripples on the top. The sand layers vary in thickness so that some are discontinuous. Despite this complexity, long trackways of true footprints (e.g., 1LP17) provide evidence that the study surface is isochronous throughout most of the site (Fig. 3.197).

SEDIMENTARY AND ICHNOLOGICAL SURFACES

The footprints occur in sedimentary discontinuity surfaces. These surfaces are classified according to the previously defined concepts: pretrack and posttrack surfaces (Gatesy, 2003), tracking surface (Fornós et al., 2002), and study surface (Requeta et al., 2006–2007). The study surface was defined by Requeta et al. (2006–2007) at the site to indicate that the 1LP surface is not isochronous. The age difference in different zones of La Pellejera according to tracksite sector may or may not be significant. Neither the constituent beds (different strata) nor the bottom surface

3.194. General view of site. All-terrain vehicle (box) provides scale. Part of site cleaned (1LP) in 2006 is bounded by black line. Direction and dip of study surface are same as slope.

of the footprints are isochronous (not all dinosaurs walked on the same stratigraphic surface).

At La Pellejera, most of the exposed study surface is produced by water that erodes some points but fills them at others; it forms natural cast in footprint shafts before erosion. True footprints and stamps above the study surface that occur as long trackways are recognized.

Most of the La Pellejera study surface is an isochronic synsedimentary surface of erosion (TC2; Fig. 3.198). Under the surface, the sandstones

Table 3.15. Relation between footprints and stratigraphic surfaces in La Pellejera tracksite.

Footprint generation	TC4 surface	TC3	TC2	TC2/TC1
6	a (b, c)	(d [b, c])	(d [b, c])	(d [b, c])
5	—	a (b, c)	(d [b, c])	(d [b, c])
4	—	—	b	(d [b,c])
3	—			c
2	—	—		a
1	—	—		a

Note: Footprint structures visible in sections A–M (Fig. 3.198) and other observations: a, stamp or true footprint; b, natural cast; c, subtrack, with the filling deformed because the foot goes through the upper bed; d, undertrack, or prints under the tracking surface. In brackets are other features. The footprint generation is the same as Fig. 3.198.

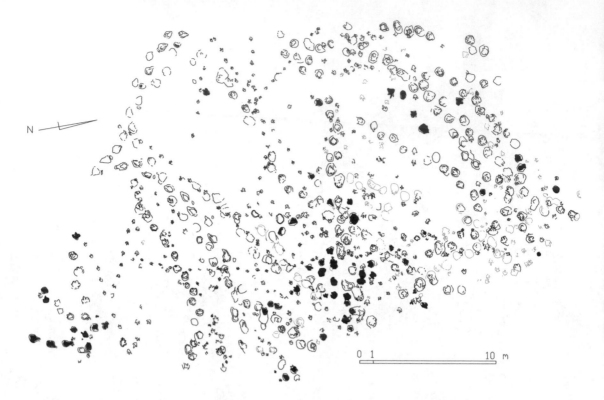

3.195. Map of footprints of 1LP published in 2007.

in contact with it are not necessarily of the same age. The footprints beneath such a surface may not be coeval. On the study surface there are also subtracks and undertracks, while above it are subtracks and casts of younger footprints than the surface of erosion; i.e., the TC2 surface forms part of the stamps, true footprints, subtracks, and undertracks. Subtracks and casts exist as well, but separate (Table 3.15).

The stratigraphic surfaces visible in 1LP involved in the formation of footprints are in relation to the following sedimentary processes (Fig. 3.198). First was deposition of the sandy bed and formation of the TC1 surface in the top of the sedimentary material (Fig. 3.198A). Passage of dinosaurs probably occurred twice, first recorded in mud of low viscosity and low coherence (collapse structures; Fig. 3. 198B), then recorded in slightly viscous but consistent mud, resulting in subtracks (Fig. 3.198B) and footprints without collapse (Fig. 3.198B). Next was generation of the TC2 surface and general erosion of TC1. Only true remnants of TC1 are identified at the bottom of the shafts in some of the previous footprints (Fig. 3.198L). The erosion removes earlier sediment above and fills the footprint shafts before TC2, forming casts (Fig. 3.198C). Dinosaurs passed on the TC2 surface after it occurred (Fig. 3.198D). Next was deposition of shale (very thin laminites) and sandstone (cast on TC2 footprints; Fig. 3.198E), formation of TC3 (Fig. 3.198F), and passage of dinosaurs (Fig. 3.198G). True footprints are evident on TC3 and undertracks (and/or subtracks) on TC2. Next was deposition of shale (very thin laminites) and sandstone, permitting formation of TC4 (Fig. 3.198H), then passage of

dinosaurs (Fig. 3.198I). True footprints are evident on TC4 and under-tracks (and/or subtracks) on TC3 and/or TC2.

Thus, the only remains of the original surface of the top of TC1 must surely be at the base of the hollows of the first footprint generation. TC2 occupies most of the study surface and is also the tracking surface for some footprints (e.g., the shallow theropod footprints). TC3 and TC4 are recognized under the subtracks of the last two generations of footprints and in some remains (Fig. 3. 199).

Several problems still remain. One is the surface of erosion. It may be that between TC1, TC2, TC3, and TC4 are other undetected erosion surfaces in the site, or perhaps a fusion of two. If we confuse the erosion surfaces, then we run the risk of changing the relative time of printing of some footprints. Second is the lateral continuity of the levels. It is known that there are discontinuous levels before TC2 that wedge and disappear laterally. This problem has not been resolved with post-TC2 levels. The footprints that TC2 shows filled with more modern material are assumed to be after TC3, when they have a single level of sandstone inside, and after TC4 if they contain two levels, although this need not always be true.

3.196. Sandstone levels at La Pellejera. Footprints at this tracksite occur in multiple surfaces of overlapping beds. Individual layers are so similar that surfaces containing true footprints, subtracks, and undertracks cannot be correlated across the site. Study surface goes from the top of one bed to another. Note variability in thickness of beds and how they disappear by pinching out beneath other beds.

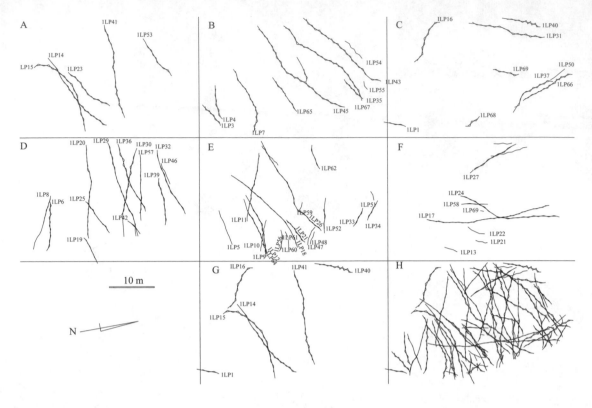

3.197. Orientation of trackways at 1LP (cf. Fig. 3.195). Beginning of each trackway is indicated by location of its label. For site as whole (H), dinosaurs moved in a variety of directions. (A, B, C, G) Trackways consisting of deep footprints. (D, E, F) Trackways of shallow prints. (A, D) Eastbound trackways. (B, E) Westbound trackways. (C, D) Trackways headed in directions other than east or west. (G) Trackways consisting of semiplantigrade footprints.

Footprint Distribution by Structures

DIFFERENT GENERATIONS OF FOOTPRINTS

BEFORE TC2 The first tracks that appear to have formed at this site were made on the TC1 surface; they have been grouped in the first generation (Fig. 3.198). They differ because the mud viscosity was low and the consistency intermediate. The theropod claw prints cut through the bed and leave their marks at the front part of the print. An example is trackway 1LP7 (Fig. 3.200). Almost all of the footprints in this trackway have similar claw marks, with three parallel grooves that cut down to 20 cm into the bed. It is likely that the dinosaur had long, separate nails capable of leaving deep parallel incisions and grooves. There is a visible collapse of mudflow in one structure and a closing of the grooves. After collapsing, the mud hardened slightly.

Other footprints made before the general erosion record a different interaction between the feet of dinosaurs and the sediment. In these, the mud was already slightly hard at the time the footprint formed (Fig. 3.198B). Its viscosity was higher and its consistency lower. Many of these footprints have vertical walls with distinguishable sedimentary laminae cut through. These have been called second-generation footprints.

First- and second-generation footprints are partially or totally filled with undeformed finely laminated sediment. The filling presumably took place at the same time as the general erosion event (TC2). In the walls of some footprints, there is erosion, as if a pool of water had filled a part of

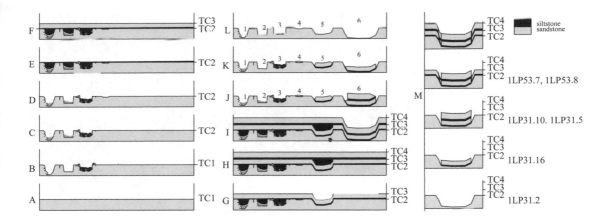

the shaft (Fig. 3.201). It is likely that part of the walls of the footprints were eroded before or during the filling of shafts.

Finally, there are marks with deformed subtracks that also predate the TC2 erosion surface (Fig. 3.198B); these are third-generation footprints. The top of the shaft can be filled with sandy sediments syngenetic with the formation of TC2.

Before the general erosion producing the TC2 surface, or perhaps in the same stage of erosion, the existing footprints were filled with sandy and silty sediment (Fig. 3.198C) with fine parallel laminite structures. It has not been possible to deduce whether the third-generation footprints come after the silty sediment filling or not.

DIRECTLY ON TC2 (FOURTH GENERATION) There are theropod stamps and other true theropod footprints with distinguishable pad and claw marks (Fig. 3.203). The theropod stamps (Figs. 3.198D, 3.204) are made on TC2, overlapping the earlier footprints. These have been called fourth-generation footprints. The size of these footprints is variable (average length from 20 to 54 cm), and they are not deep. Some footprints have welded parts of the upper level (post-TC2) that protrudes. Some have digits thinned by the collapse of mud and/or grooves produced by the drag of the claws in the K phase. There are no ornithopod footprints. The fourth-generation footprints are the shallowest. Their depth is usually less than 5 cm, and these prints are different because they are the only ones with recognizable, well-pronounced digits, pad marks, and sharp claws.

AFTER TC3 Above TC2, the deposition continues. Following this, there is another stage of footprints, the fifth generation. Until now they have been distinguished because we can recognize TC3 and TC2 under the sole of the footprint, or because another, later, footprint is superimposed on them (Figs. 3.198G, 3.199, 3.204), so they are like a sandwich between TC1 and level TC3.

AFTER TC4 There are also footprints identified as having been made after TC4 that deform all surfaces (Figs. 3.198I, 3.199, 3.204), which are

3.198. Types and generation of La Pellejera footprints. (A) First tracking surface, top of sandstone layer. (B) True footprints and subtracks at TC1. (C) Second (TC2) tracking surface; partial erosion of TC1, and formation of casts (D) True footprint and stamps on TC2. (E) Formation of casts in previous shafts. (F) Third tracking surface, TC3. (G) Formation of true footprints (subtracks not indicated), undertracks, and dead surfaces. (H) Fourth tracking surface, TC4. (I) formation of true footprints (TC4) as well as undertracks, subtracks, and dead surfaces (TC3, TC2, TC1). (J, K) Erosion maintaining filling of some shafts. (L) Intensive erosion of casts and subtracks; actual top of lower sandstone layer (TC1 + TC2). (M) Styles of preservation, and examples of post-TC4 footprints. See Table 3.15.

3.199. (Top) Pretrack and posttrack surfaces TC2 and TC3, and posttrack TC4 surface with post-TC4 footprints (1LP53.7 [left] and 1LP53.8 [right]). (Bottom) Schematic drawing.

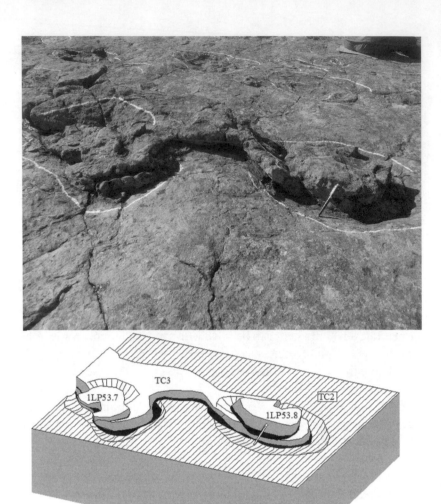

called sixth-generation footprints. The footprints detected after TC4 seem to be subtracks and undertracks. It is not certain whether the sixth-generation footprints are direct structures.

Subtracks are detected on the study surface (TC2), which, according to their erosion status, have remnants of between one and four superior levels. The subtrack filling may be silt and/or sand from TC3, TC4, or both. Generally speaking, they are recognizable and distinguishable from one another (post-TC3 and post-TC4 subtracks), but one must be cautious when assigning them to one or another generation. Sometimes the filling of the shafts is eroded so that there are gaps (Figs. 3.198K, 3.198L). In these cases, it is difficult or impossible to know to which generation the footprint belongs. Other times, part of the filling can erode, so a sixth-generation footprint may lack some of the filling (Figs. 3.198K, 3.198M); the bottom of the footprints can be any of the surfaces referred to.

3.200. Footprint 1LP7.17, left pes print (beneath screen, toe marks directed to right). This first-generation print is characterized by narrow toe and claw marks and mud collapse structures. To right, fourth-generation footprint 1LP6.1 (left foot) is impressed near toe end of print 1LP7.17. Mesh width = 5 cm.

OTHER STRUCTURES

DEEP FOOTPRINTS to determine which generation the deep unfilled footprints are from, the whole trackway they belong to must be examined. The deep footprints are either filled (Fig. 3.198I), half-filled, or empty (Figs. 3.198B–L), and it is normal to find all types in the same trackway. On the top of TC1 and TC2 are weaker sedimentary levels prone to erosion of the rock, so that removal of the fillings (casts and subtracks) is normal. Any of the deep tracks can be free of their contents and leave a gap, usually formed by TC1 and TC2 (Fig. 3.198L).

Sixth-generation footprints are also empty and form deep holes (Figs. 3.198L, 3.198M). It is difficult to know which generation an isolated, deep footprint is.

SHALLOW FOOTPRINTS These are post-TC2 footprints, some of whose filling has not eroded. 1LP4 is a trackway created after the TC2 surface. In this trackway there are both shallow footprints (2–10 cm depth) and very shallow ones (0 cm or overtracks). In the shallow footprints the cast or filling has disappeared; in the overtracks a part of the filling sandstone is conserved, protruding above TC2.

DEAD ZONE The fifth- and sixth-generation footprints (post-TC3 and post-TC4) generally provide marks that are reproductions of the contours of the feet. The two post-TC2 sandy levels respond rigidly within a more plastic context. The post-TC2 sandy layers break through the outline of the tridactyl footprints. The undertracks produced on the TC2 surface bend the surface stratification (Figs. 3.198G–I, M, 3.199, 3.204). Footprints

3.201. Semiplantigrade left footprint 1LP15.12p. (A) Oblique view. (B) Overhead view. Erosion occurred at argillaceous level at base of wall of footprint, while infilling produced undeformed laminites in bottom of footprint shaft. Both erosion and filling were probably syngenetic with creation of TC2 surface. (A) Oblique view shows erosion of wall base. (B) Vertical view shows sandy undeformed laminites in inner part of footprint, truncated by eroded shaft wall (top). Manus print 1LP515.12m is seen at lower corner (B). Mesh width = 5 cm.

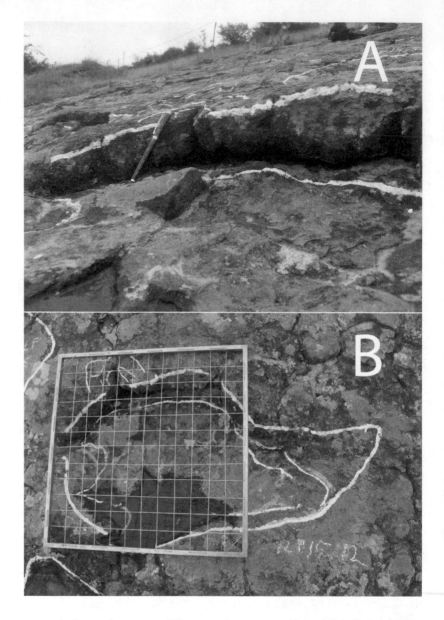

1LP43.(20)21 and 1LP31.10 (Fig. 3.204) are examples of this behavior. In 1PL53.7 and 1PL53.8 (Fig. 3.199), the deformation of the lower level is seen, probably reaching its breaking point in other subtracks.

An interesting observation is the occurrence of ornithopod footprints with acuminate digit subtracks. If the TC4 dead zone is eroded, the tridactyl ornithopod true footprints (post-TC4) show tapered digits (theropod morphology) at the TC3 level. This implies that the brittle behavior of the dead zone may leave angular fragments under the feet with rounded toes. The formation of undertracks with tapered digits is therefore possible under footprints with rounded toes.

EXTRUSION RIMS Some of the deep footprints have mud extrusion rims surrounding them (W phase). The height of the rim is variable but

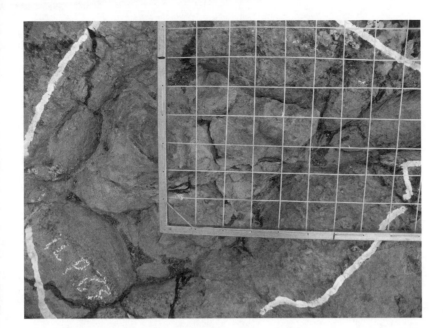

3.202. 1LP23.3 left subtrack; shaft occupied by sediments deformed by foot. Mesh width = 5 cm.

3.203. Theropod footprints. (A) 1LP12.6, right footprint with marks from pads and claws. (B) Isolated footprint 1LP129; partially eroded sandy cast of digit III. (C) Left footprint 1LP30.10; narrowing due to collapse of mud in digit III; eroded remains of silt filling in digits II and IV. (D) Left footprint 1LP12.5, incision left by digit III claw drag. Mesh width = 5 cm

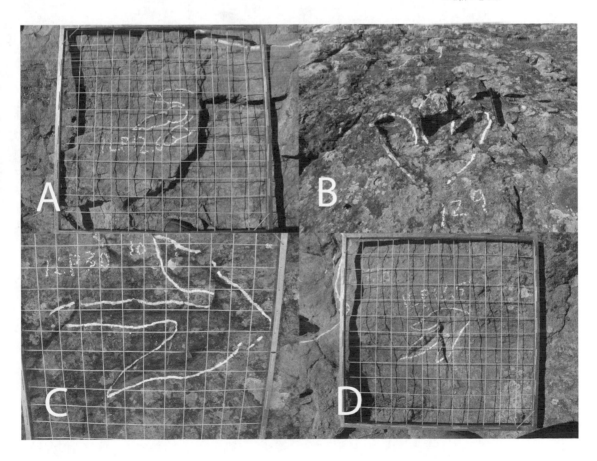

3.204. Footprints of three different generations. (Top) Photograph of the drawing area in which there are also other overlapping footprints, 1LP31.9 and 1LP43.(19)20 (upper left). (Bottom) Schematic drawing of 1LP182.7 (on TC2 surface) and superposition of 1LP31.10 (on TC4 surface) onto 1LP43.(20)21 (on TC3 surface).

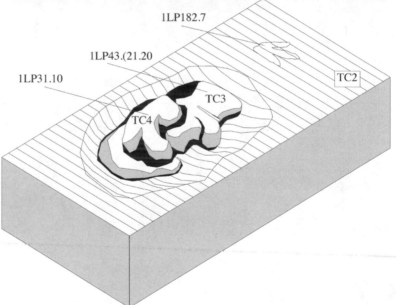

not very pronounced (Fig. 3.205). Mud deposits are also seen in the front part of trackway 1LP15, probably as a result of the forward movement of the foot during the T phase.

Footprint Distribution by Structures

According to the structures, the first classification of footprints in La Pellejera would be either deep or shallow. It has been seen that shallow footprints with direct structures, both large and small, are produced on one sedimentary level (TC2) and are fourth generation. However, deep

3.205. Subtrack next to tracksite. Deformed ripples can be seen in extrusion rim, created earlier than TC1. Deformed sediment is present in shaft.

and shallow tracks are equally distributed throughout the site. There are no areas where one type of footprint or another dominates.

However, there is a change in the footprints over time or vertical TC surfaces. The relative position according to certain structures in time with respect to TC2 seems established. For example, the deep trackways with sedimentary filling (first and second generation) not deformed by footprints are older than TC2; shallow footprints with direct structures (fourth generation) are later than TC2 and before TC3; deep footprints with dead zones are after TC3 and TC4 (fifth and sixth generation). Third-generation footprints have deformed sediment inside but are destroyed by TC2, so they may be formed in the same period, although later, i.e., while the TC2 erosion phase was happening, when passing dinosaurs also left tracks.

Footprint Features (Ichnotypes)

For each trackway, one footprint has been selected for illustration (Fig. 3.206), either a print with the most complete outline (theropod) or the one best showing the trackway features to which it belongs.

THEROPOD FOOTPRINTS

Prints with long, separate, and tapered digits and pad marks are included in this group. The heel may be prominent but relatively narrow, bilobed (1LP33.2), and with or without a central pad mark.

3.206. Representative footprints from 1LP trackways arranged in order of increasing trackway number. Prints are oriented with respect to north.

1LP1.4 — os
1LP3.4 — t
1LP4.7 — o
1LP5.9 — t
1LP6.5
1LP7.17 — ts
1LP8.5 — t
1LP9.3 — t
1LP10.9 — t
1LP11.2 — t
1LP12.6 — t
1LP13.3 — t
1LP14.13 — o

1LP15.8p — os
1LP15.8m
1LP16.3m
1LP16.3p — os
1LP17.14 — t
1LP18.14 — t
1LP19.7 — t
1LP20.6 — t
1LP21.1 — t
1LP22.5 — t
1LP23.5 — os
1LP24.6 — t
1LP25.11 — t
1LP26.5 — t
1LP27.5 — t

1LP28.9 — t
1LP29.11 — t
1LP30.16 — t
1LP31.16 — os
1LP32.6
1LP33.2 — t
1LP34.1 — t
1LP35.9 — os
1LP36.2 — t
1LP37.2 — os
1LP38.2 — t
1LP39.1 — t
1LP40.11 — os

1LP41.6 — os
1LP41.15
1LP42.2 — t
1LP43.10 — os
1LP44.3 — o
1LP45.8 — t
1LP46.1 — ts
1LP47.2 — t
1LP48.1 — t
1LP49.2
1LP50.1 — o

1LP51.2 — t
1LP52.5
1LP53.5
1LP54.7 — os
1LP55.1 — o
1LP56.1 — t
1LP57.4
1LP58.1 — t
1LP59.2 — t
1LP60.1 — t
1LP61.1 — t
1LP62.1 — t

1LP63.5 — t
1LP64.1 — t
1LP65.1 — os
1LP66.16 — o
1LP67.7 — t
1LP68.1 — o
1LP69.5 — o
1LP182.4 — t
1LP (36.15) 183.6
1LP184.5 — t
1LP185.2 — t

t theropod footprints
o ornithopod footprints

0   50   100 cm

ts semiplantigrade theropod footprints
os semiplantigrade ornithopod footprints

N

All small theropod footprints (pes length 20 to 25 cm; Table 3.16) have a prominent heel. In many of them (Fig. 3.206), the mark of a relatively thin pad can be discerned. Most theropod footprints that are longer than an average of 40 cm, however, have a broad heel (1LP9.3, 1LP32.6, 1LP33.2, 1LP36.2, 1LP39.1, 1LP44.3, 1LP49.2, 1LP57.4).

A systematic study of the footprints has not been done, but there are likely to be at least two theropod ichnogenera at the site. Most intermediate footprints have the same features as the small ones.

ORNITHOPOD FOOTPRINTS

All wide footprints without thin, separate digit marks have been provisionally allocated (Requeta et al., 2006–2007) to the ornithopod ichnotaxon. This may not be true in some cases.

Some trackways with wide footprints have sharp claw marks. All have mud collapse structures, so they must have occurred on TC1 and before TC2. The footprints are large, and they include trackways 1LP3, 1LP7, 1LP45, and 1LP67.

The more modern trackways with wide footprints are on surfaces TC3 and TC4, which behave elastically. It has not been possible to check whether the highest surfaces with these footprints have direct structures. It is therefore problematic to assign them to one type of dinosaur or another, even if the digit mark outlines are clear. The same line in the subtracks may be misleading. As stated previously, it may be that there are tapered copies of rounded toes above due to the brittle behavior of the lower levels trampled upon. Examples of these trackways are 1LP31, 1LP37, and 1LP43.

Footprints in trackways 1LP35, 1LP40, and 1LP41 have no examples with sharp claw marks. The digit marks show only indications of a broad and rounded pad. There is therefore no reason to remove the footprints from the ornithopod group to which they were assigned (Requeta et al., 2006–2007).

Finally, there is a group of footprints with metatarsus marks, some with hind foot and forefoot marks. These are trackways 1LP1, 1LP15, 1LP16, 1LP41, and 1LP40. They all have a wide front (digital region) that is usually rounded and an elongated rear part (metatarsal region) with no hallux mark. Those with digit marks are mesaxonic, with three rounded front toes. The metatarsus marks show a slight deviation from the axis in relation to the axis of the digitigrade foot. Some trackways have digitigrade and semiplantigrade footprints (1LP40, 1LP41). They are classified as semiplantigrade ornithopod footprints.

1LP15 and 1LP16 also have forefoot marks (Fig. 3.206). These are small and semicircular, with their major axis oblique to the trackway. The anteroposterior axis (minor) has a clear positive (outward) orientation (crossing the midline, behind the track). There are no separate digit marks. The phalanges and metacarpal bones would probably be vertical and encompassed in a skin structure.

Table 3.16. Measurements at La Pellejera trackways.

Trackway	l	a	O	Ar	Lr	P	Ap	z	H	z/H	v1	v2	(l − a)/a	Ar/a	z/l	Footprint depth	Ichnotype	No. of footprints	Direction of trackway
1LP1	54	50	−9	14	69	124	150	235	334	0.7	2.8	3.6	0.07	0.29	4	21-S	os	5	N205
1LP3	53	45	−10	18	80	83	128	149	355	0.4	1.3	2.2	0.19	0.4	3	13-S	t	8	N243
1LP4	55	56	−17	18	88	120	142	225	370	0.6	2.4	3.3	−0.43	0.34	4	12-S	o	7	N256–281
1LP5	33	23	−6	4	30	95	172	186	157	1.2	4.7	4.2	0.11	0.19	5	3-N	t	8 (9)	N258
1LP6	34	28	6	20		97	131	176	162	1.1	4.1	3.9	0.24	0.72	5	3-N	t	9	N99
1LP7	55	49	−5	14	76	88	144	162	369	0.4	1.4	2.4	0.14	0.29	2.9	11-S	ts	19	N297–253
1LP8	29	23	−10	2	27	85	163	163	142	1.1	4.2	3.8	−0.13	0.12	5.6	2-N	t	14	N115
1LP9	51	38	11	10	62	125	160	235	229	1.0	4.4	4.3	0.05	0.34	5	5-N/S	t	10	N274–250
1LP10	25	17	5	2	24	65	173	131	118	1.1	3.6	3.4	0.21	0.14	5.0	2-N	t	9 (11)	N290
1LP11	30	21	4	2.5		85	171	170	145	1.2	4.4	4.0	0.41	0.10	5.0	1-N	t	13 (19)	N293
1LP12	24	17	–	5	27	75	162	148	112	1.3	4.7	3.9	−0.01	0.29	5.9	1-N	t	11 (7)	N76
1LP13	49	30	7	3	31	98	172	197	220	0.9	3.5	3.7	−0.32		6.2	4-N	t	4	N208
1LP14	63	53	−9	19	85	98	131	176	427	0.4	1.3	2.4	0.2	0.3	2.9	10-S	o	20 (21)	N56
1LP15	58	47	−13	24	88	127	137	236	373	0.6	2.6	3.4	0.26	0.5	4.1	12-S	os	17 + 8m	N9–61–89
1LP16	51	50		26	103	117	145	212	327	4.3	2.5	3.3	−0.11	0.6	4.3	15-	os	11 + 2m	N147–118
1LP17	23	18	−2	6	28	72	167	141	106	1.3	4.5	3.8	0.09	0.4	6.0	2-N	t	35 (41)	N18–8–0–354
1LP18	32	22	6	4	27	95	171	190	155	1	4.0	4.3	0.45	0.2	5.6	2-N	t	19 (24)	N257–232
1LP19	24	18	−7	6	31	92		163	114	1.4	5.6	4.3	0.32	0.28		1-N	t	6 (8)	N71
1LP20	29	22	−1	8	37	88	155	179	143	1.2	4.9	4.2	0.36	0.34	5.4	2-N	t	23 (24)	N99
1LP21	34	25	9	14	58	96	142	191	161	1.2	4.7	4.2	0.37	0.55	4.2	1-N	t	5	N242
1LP22	32	27	1	5		105	171	202	153	1.3	5.5	4.6	0.16	0.2	5.6	1-N	t	5	N205–223
1LP23	53	52	−8	18	86	99	134	183	395	0.5	1.5	2.6	0.04	0.3	2.6	11-S	os	12 (15)	N55
1LP24	34	29	2	10	50	119	155	233	163	1.4	6.6	5.1	0.23	0.3	4.5	3-N/S	t	22 (24)	N39–9
1LP25	20	15				81		140	95	1.5	5.2	4.0	0.31			1-N	t	5 (12)	N63
1LP26	30	20	−4	7	35	93	161	179	145	1.2	5.0	4.2	0.41	0.34	5.4	2-N	t	22 (27)	N231–260–246
1LP27	31	25	11	6	56	166	162	302	148	2	11.3	6.9	0.31	0.4	5.0	2-N	t	8 (9)	N333
1LP28	29	23	−6	8	43	83	152	149	139	1.1	3.7	3.5	0.32	0.3	5.3	1-N	t	14 (22)	N295
1LP29	35	29	−5	8	40	112	161	214	166	1.3	5.5	4.6	0.2	0.3	6.4	2-N	t	16 (20)	N83
1LP30	41	38	−2	20	59	149	149	267	202	1.3	6.4	5.2	0.09	0.52	6.5	4-S	t	13 (17)	N115–104
1LP31	44	48	−6	17	76	109	138	203	290	0.7	2.7	3.3	−0.1	0.4	4.8	11-S	os	15 (17)	N202
1LP32	54	42	−1	12	58	140	162	275	244	1.1	5.2	5.0	0.3	0.5	5.1	4-S	t	12	N76
1LP33	40	35	7	14	59	117	150	225	186	1.2	5.3	4.6	0.10	0.4	5.6	3-N	t	5	N308
1LP34	39	34	−9	11	47	97	155	189	198	0.8	3.7	3.7	0.2	0.3	4.8	3-N	t	6	N308
1LP35	44	48	−9	19	80	101	136	187	201	1.0	3.1	3.5	−0.0	0.4	4.8	9-S	os	19 (26)	N244–221–231
1LP36	42	28	8	22		120	138	231	194	1.4	4.5	4.3	0.4	0.3	6.3	3-N/S	t	13 (19)	N84–70

1LP37	48	50	-15	31	93	107	97	201	309	1.0	2.2	3.0	-0.05	0.6	3.9	9-S	os	10 (12)	N155
1LP38	28	24	1	2	26	75		148	136	1.1	3.8	3.5	0.4	0.4	5.3	1-N	t	6	N301
1LP39	43	32	-2	11	57	113	158	224	198	1.0	4.9	4.5	0.3	0.2	4.7	3-S	t	6 (10)	N92
1LP40	51	54	-5	15	88	89	137	157	340	0.4	1.4	2.4	-0.1	0.4	3.1	13-S	os	7 (12)	N205
1LP41	47	46		19	83	116	138	213	296	0.7	2.3	3.5	0.01		4.7	9-S	os	19 (20)	N89
1LP42	35		10	15	50	194	162	381	165	2.3	9	8.3				3-S	t	3	N54
1LP43	54	59	-10	15	55	103	140	201	352	0.5	2.7	3.0	-0.02	0.2	3.7	8-S	os	23 (24)	N230
1LP44	38	22	1	7	42	96	162	182	179	1.0	1.8	2.8	0.9		4.8	2-N	o	5 (6)	N296
1LP45	53	46	-2	16	80	104	140	195	330	0.5	2.1	3	0.18	0.3	3.5	9-S	t	23 (28)	N212–225–249
1LP46	37	25				62		192	181	1.0	4.1	4.0	0.4		5.1	3-S	ts	4 (6)	N77
1LP47	30	29				131		196	154	1.3			0.01		6.5	4-N	t	4 (5)	N306
1LP48		26				91		186								1-N	t	3 (4)	N299
1LP50	25	26				122			158				-0.1			4-S	o	3 (5)	N153
1LP51	45	29	-9	7	41	100	168	199	205	0.8	3.3	3.9	0.57			2-N	t	3	N260
1LP52	46	35	10	16	41	137	156	300	210		7.4	5.8	0.32			3-N	t	5 (6)	N283
1LP53	48	45	-8	14	80	99	151	191	319	0.5	2.1	2.9	0.3	0.36	4.3	7-S	o	10 (11)	N65
1LP54	62	57	-2	18	84	96	143	175	374	0.5	1.5	2.5	0.14	0.38	3.0	6-S	o	7	N237
1LP55	46	55		25	100	103	116	175	210	0.8	3.0	2.4	-0.1	0.4	3.8	5-S	o	3	N217
1LP56	27	21				127		231	135	1.7	8.0	5.6	0.22			2-N	t	4 (5)	N286
1LP57	42	30				95	180	157	191	0.8	2.7	3.1	0.28		4	2-S	t	8 (15)	N100
1LP58	44	42	13			137		275	321	0.9	3.9	4.3				4-N	t	3	N11
1LP59	28	27	0	7		97	161	193	137	1.4	5.8	4.6	0.05	0.27	5.5	3-N	t	3	N300
1LP60	25	24				169		127	127				0.15			1-N	t	3	N278
1LP61	40	29				144		185	185				0.4			2-N	t	2	N255
1LP62	43	32				106			197				0.5			3-N	t	3 (5)	N255
1LP63	29	27				66							0.04			2-N	t	3 (5)	N25
1LP64	31	25	3	5	35	69	164	129	159	0.8	2.5	3.1	0.18	0.17	4.1	3-¿N?	t	4	N257
1LP65	45	53		10	70	87	157	174	298	0.6	2	2.8	-0.16	0.18	3.9	7-S	os	8 (10)	N248
1LP66	55	55		10	74	87	146	179	371	0.5	1.7	2.6	0.0	0.17	3.2	8-S	o	10 (17)	N161
1LP67	52	47				95	140	170	347	0.5	1.6	2.5	0.09	0.35	3.2	8-S	t	7 (8)	N241
1LP68	50	46	-23	26	113	106	120	186	337	0.5	1.9	2.8	0.07	0.6	3.7	8-S	o	5	N154
1LP69	50	55	-6	13	80	86	149	166	335	0.5	1.6	2.5	-0.08	0.25	3.3	6-S	o	5	N202
1LP216																		5 (6)	
1LP217																		8 (10)	
1LP218																		3	
1LP219																		5 (10)	

Note: Abbreviations as in Table 2.1. N, shaft empty; S, shaft full of sediment.

Footprint Distribution by Ichnotypes

The La Pellejera trackways have been categorized as footprints that are deep or shallow (Fig. 3.197). In addition, trackmaker direction is assessed. The name of the trackway is written at the beginning of each one. Figure 3.197 shows the prevalence (49 trackways) of animals moving either to the east or the west. A direction perpendicular to the east–west axis is less common (18 trackways).

Among the east–west trackways heading westward, there is a group that starts on the east edge of the site but does not continue (Fig. 3.197E). The direction of travel goes from N280E to N320E. All the footprints in this group are shallow. No conclusions have been drawn from this evidence.

The footprints with metatarsus marks (Fig. 3.197G) have been separated. Their directions of travel do not coincide.

There is no indication of any natural barrier or herd behavior.

General Information

The footprint density is relatively high – at almost one footprint per square meter. This information, however, is not significant because there are at least six generations of footprints separated in time at La Pellejera. If there were no doubts about the age of the footprints, a calculation in stages could be done. This is currently impossible because there are many tracks of an unknown generation.

There are many footprints in the site surroundings. It is thought that about 20% of the footprint surface tracksite has been cleaned. The surfaces probably maintain the same ichnological richness under the old cultivated land in neighboring fields.

1LP is quite fragile. The layers are parallel to the surface and break up as a result of the levels with more shale content. Frost, plants, and diurnal temperature variation in the rock are the most active destructive agents in the tracksite. Dinosaur paleoichnological works and courses in La Rioja have been going on since 2003 to preserve the site.

Main Features of Site

1LP has been an ichnologically active place for a long time. There are at least six different stages of generation of dinosaur footprints, separated by sedimentation and hardening of the levels. There are 3 stages before TC2; the fourth between TC2 and TC3; the fifth between TC3 and TC4; and the sixth after TC4.

The rigid and plastic behavior of the levels does not depend on time or superimposition in La Pellejera. The TC3 level is broken under the feet of the dinosaurs in the track formation stage, while under TC2 the behavior is plastic. The same can be said for the track formation stage on TC4 (TC4 is broken, TC2 folds, and TC3 breaks or bends). It can be assumed that the water content at lower levels (under TC2) was higher than in the upper ones. According to the depth of the footprints, the sandy level beneath TC2 was initially plastic (large tracks with collapse structures), then rigid (shallow footprints), and finally once again plastic (deep tracks formed on surfaces TC3 and TC4). It is likely that hydration levels were different from and opposite to the normal, in that some deep levels were more saturated than other surface ones.

The area where the study surface is observed covers approximately 4400 m². This means that if the track density is unchanged, about 4000 footprints could be revealed on the site.

The trackways are relatively long and uninterrupted. This is normal in almost all large sites in La Rioja. We can therefore assume that they were places of transit. There is no evidence of dinosaurs wandering aimlessly; they seem to follow a definite route without interruptions through the site.

Trackways of the same ichnotype, however, are not parallel in a way that might indicate herds or groups of few dinosaurs. It seems that in 1LP the dinosaurs were solitary animals or the opposite: gregarious animals spread over a large area.

The concentration of theropod footprints on the relatively rigid TC2 surface can be explained by the cautious dinosaur hypothesis. Theropod dinosaurs could move with agility over this relatively firm sediment, so this would be a poor place for an herbivorous dinosaur to protect itself.

Research is ongoing at this site for more accurate information about the passing stages of the dinosaurs, the footprint structures, their ichnotypes, and the placing in time of all trackways. Another issue raised by the study is the two quadruped semiplantigrade ornithopod trackways. More studies are needed regarding the size and relative placement of the parts of the foot (digits and metatarsus), the placing with respect to the body, and the relationship with the forefoot marks.

History

Soto 1 and Soto 2

This section describes two small places of interest among the sites of La Rioja, Soto 1 (1ST) and Soto 2 (2ST), numbered chronologically as they were discovered. The most interesting feature of 1ST is that it has the first trackway with the most rapid estimated trackmaker speed published in La Rioja. 2ST is of interest as the first place with footprints of a herd of sauropods discovered in the Iberian Peninsula.

In both cases, the finds were made accidentally by two different groups of naturalists not engaged in paleontology or any other branch of geology; one was a group of graduate biologists and the other a group of amateurs. 1ST received its name in 1989 and 2ST in 1990.

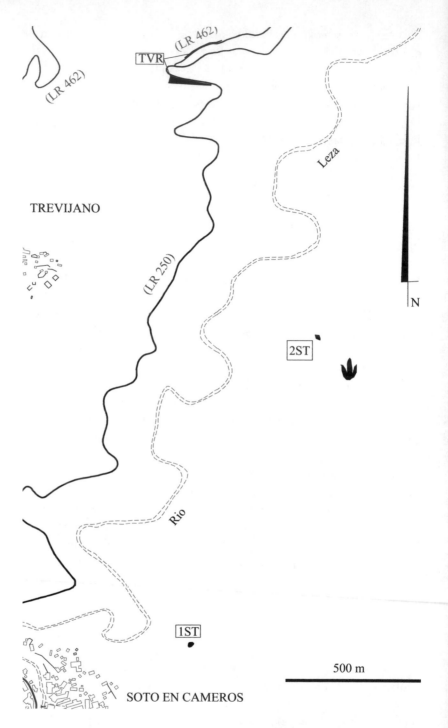

3.207. Location of Soto 1 and 2 (1ST and 2ST) and neighboring sites: Trevijano (TRV) and an undescribed outcrop indicated by footprint symbol. Local roads are labeled LR 250 and LR 462.

The age and stratigraphic position of the rocks that contain them (the Leza Limestone Formation or lower Leza Formation) have been under discussion for a long time. They were included in the upper, or modern (Aptian–Albian), part of the Enciso Group, but Doublet's (2004) review places the Leza Formation in the Lower Oncala Group, which brings it back to the Berriasian age. Suárez-González et al. (2011) say the Leza Formation is in the Enciso Group of Barremian–Aptian age.

A

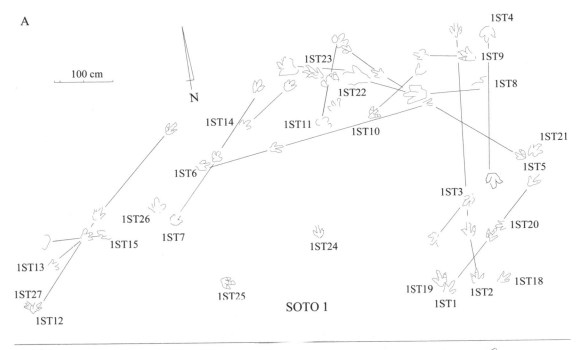

100 cm

N

1ST4
1ST9
1ST23
1ST8
1ST22
1ST14
1ST11
1ST10
1ST21
1ST6
1ST5
1ST3
1ST26
1ST7
1ST15
1ST20
1ST24
1ST13
1ST27
1ST25
1ST19
1ST2
1ST18
1ST1
1ST12

SOTO 1

B

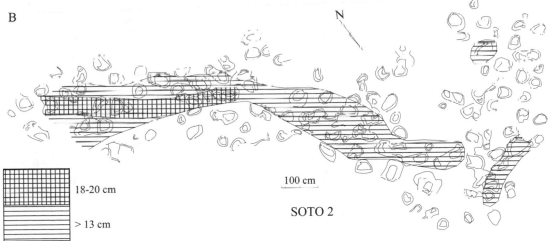

N

18-20 cm
> 13 cm

100 cm

SOTO 2

Geological mapping of the area of 1ST and 2STm tracksites indicates that the stratigraphic position of Leza Formation is below (older than) the Urbión Group, and therefore is included in the Oncala Group (older than the Enciso Group).

Description of Location

1ST and 2ST are on an old bridle path connecting two neighboring villages: Zenzano, a remote village in the mountains, and Soto de Cameros, larger and formerly a place of trade and industry. The path runs along one side of the slope of a deep canyon carved out of Weald Facies limestone (the Leza Formation). The area is a bird sanctuary, many of which nest in the canyon wall rocks. Along the bottom runs a river whose flow is

3.208. (A) Footprints of eastern part of 1ST. (B) Tracksite 2ST; maximum depth of footprints: squares 18–22 cm; horizontal lines, 13–18 cm; no lines, less than 13 cm. Footprints continue northward in lower right corner but with undefined outlines.

seasonally intermittent. The water disappears shortly after passing the village of Soto to emerge a few miles farther below.

1ST is approximately 500 m north of Soto (Fig. 3.207) in a layer of southward-dipping limestone. Walking further north, the beds become horizontal, and 1.5 km from Soto is 2ST, occurring in a layer of sandstone–siltstone, interbedded with limestone of the same formation but older.

1ST is exposed in an outcrop divided into two parts separated by a covered area. The larger portion (Fig. 3.208A) contains most of the footprints (49), while the smaller one has six. 2ST (Fig. 3.208B) is part of the path surface, in the sense that you go over the footprints if you walk along it.

Nearby there are several sites with studied footprints. The closest are Trevijano 1 and 2 (TVR) and Soto 3 (3ST). Located further away are La Ilaga and Camino a Treguajantes. Unstudied casts of footprints have also been located at various points, one of which is indicated on the location map in Fig. 3.207.

Some 2300 m north of Soto there is a fossil study site with various types of dinosaur teeth (theropod and ornithopod), badly eroded vertebrae, and eggshells. There are also dinosaur bones yet to be studied in another site to the south, just under 1500 m from the village.

The limestone of the Leza Formation is of continental origin. Alonso and Mas (1993) found two interbedded limestone layers containing marine algae (*Salpingoporella cemi* and *S. dinarica*), attributed to the Barremian–Aptian. Doublet (2004) reclassified the same fossils, concluding that they are *S. urdaganasi* formed in a mesosaline lake environment and of an older age.

The difference in age and location of the Leza limestone formation in the canyon is under discussion. The latest data on the subject are from Doublet (2004), so the temporal correlation of this author is being followed. According to his interpretation, all the footprint sites mentioned in this description, including the nearby places, must be considered Berriasian (Oncala Group). Until recently, we assumed them to be Barremian lower Albian (Enciso Group).

Lithology

1ST and 2ST are in the Leza Formation, dominated by limestone. The track sites are located in the Leza River canyon. The river flows about 200 m below 2ST. A total of 300 m of the canyon wall in 2ST are rocks of the Oncala Group. Lacustrine limestone with interbedded shale and some sand predominate. The limestone is dark—almost black when freshly cut. Minor amounts of sandstone are present; this rock is dark green-gray but never black. Both types of rock are well stratified. In the vicinity of the sites there are laminites and spherical structures of algae (ooliths, oncolites, pisolites), mud cracks, and ripples in some layers of limestone of the same formation.

Site 1ST is in a level of gray limestone alternating with marl and other limestone levels of varying thickness. The top is rough, probably as

3.209. Variation in depth of footprints at site 2ST. North is to right.

a result of the interference of ripples and erosion. There is no roughness within the footprints, nor are they flattened. There are mud cracks in lower layers.

Site 2ST is in limestone with algal structures (laminites) but with no appreciable ripples or mud cracks.

Footprint Mud Structures

The footprints in 1ST are shallow, with digital structures (pads, claws, heel) generally well marked. The ends of the toes are well defined, with the depth of the marks less than 3 cm in all cases.

At 2ST, the footprints are deeper, and there are structures related to the pressure of the sauropod feet, extrusion and pushing of mud, and flexibility of the substrate. Because of the pressure of the feet, many of the tracks bend the top layer without breaking it. There is no visible mudflow in the extrusion rims. The mud displaced by the volume occupied by the shaft of the foot must flow under the laminites.

The depth of the footprints varies according to the sector in 2ST (Figs. 3.208B, 3.209), leading to an inference about which were the most and least viscous parts of the site. To the north, the depth of the footprints is less and their width greater. No footprint has been noted at the northern part of the tracksite because the shafts become more and more shallow and the footprint outlines cannot be accurately drawn.

The variation in footprint depth indicates that the tracking surface was more or less plastic. This variation depends on location within the outcrop. This must be a consequence of variable physical properties of the sediment across the site. If the upper layer bent but did not break, the variation in depth would depend on the viscosity of the mud in a layer below the tracking surface or on the variation of the resistance to bending of that surface.

In the 2ST zone, the deeper the footprints, the more vertical are their walls. At these ichnites, it is likely that the laminites are bent in the extrusion rims and are cut into the walls of the footprints.

All the tracks are shallow in 1ST. At 2ST, all the footprints are sauropod and of variable depth.

Footprint Features (Ichnotypes)

TRIDACTYL FOOTPRINTS

All the footprints in 1ST are tridactyl, and the trackways are biped. Casanovas et al. (1990a) classified them as theropod, ornithopod, and unidentified footprints (Table 3.17).

The theropod footprints are classified by Casanovas et al. (1990a) according to foot size in Carnosauria and Coelurosauria (Thulborn, 1990). The length of the theropod dinosaur footprints ranges from 20 to 43 cm, but it seems there is no morphological separation at 25 cm. There are no footprints between 29 and 42 cm. The 1ST8 trackway in this field is remarkable for its size (43 cm). Casanovas et al. (1990a) assigned large theropod footprints to the superfamily Tyrannosauroidea (with digital pads) and Megalosauroidea (without pads).

Nowadays, distinguishing between the morphosuperfamilies mentioned by the existence of pads cannot be used. In the 1ST trackways, sometimes the pads are visible in the footprints; sometimes they are not (Fig. 3.210). The heel is well marked in many footprints; many of them are bilobed. According to current nomenclature, large footprints may be associated with *Megalosauripus*, despite 1ST8 not having clear morphological features. If less than 29 cm, they cannot be included in a group or separated by size. In these trackways, the ratio z/l, which gives a measure of the thickness of the extremities (Pérez-Lorente, 2001a), is in accordance with fast trackways and with data from dinosaurs with long, thin limbs (cf. Haubold, 1971).

There are also ornithopod footprints in 1ST, which are distinguished by having digits with a single, relatively broad pad with a rounded end. They are attributed to the ichnofamily Iguanodontidae by Casanovas et al. (1990a). All ornithopod footprints in 1ST are less than 28 cm in length, i.e., they are not large ornithopods.

SAUROPOD FOOTPRINTS

Site 2ST has sauropod footprints chaotically arranged (Fig. 3.208B). Only in some cases can the mark of a forefoot be related with the hind foot, and there are no digit or claw marks in any of the footprints. The forefoot

Table 3.17. Measurements of trackways at Soto 1 (1ST) track site.

Trackway	l	a	O	Ar	Lr	z	H	P	Ap	II	III	IV	II^III	III^IV	z/H	v2	v1	(l − a)/a	z/l	Ar/a	Footprint type
1ST1	28	22	0	0	21	234	136	117	180	14	19	13	25	30	1.7	5.6	8	0.28	9	0	t
1ST2	26	23					129	81		11	17	12	7	33				0.15			t
1ST3	24	17					113	86			15	10	27	23				0.50			t
1ST4	27	28					132			14	16	15	27	37				−0.04			t
1ST5	20	17					93	69		7	12	9	38	25				0.16			t
1ST6	24	18				269	113	127		11	14	8	21	35	2.4	7	13	0.31	11		t
1ST7	22	20	−2	2	22	269	114	134	176	11	12	5	35	28	2.3	7	12	0.13	12	0.13	o
1ST8	43	34	−2	7	51	214	197	112	165	17	22	8	6	27	1.1	4	4	0.26	5	0.21	t
1ST9		23						76		11	16	1	26	30							t
1ST10	25	16						110		12		1	22	42							t
1ST11	27	28	9	2	30	141	175	71	176	11	12	1	29	44	0.8	3	2.6	0.00	5	0.07	o
1ST12	25	19	14	4	31	378	120	190	176	12	13	2	22	17	3.1	9.7	21	0.34	15	0.20	t
1ST13	21	21					108	77		8	10	9	40	43				−0.06			o
1ST14	21	21					100	111		11	12	1	14	14				0.13			t
1ST15																					n
1ST16	24	25						94		11	14		49	33				−0.04			o
1ST17	24	27	16	4	32	153	125	77	168	14	19	13	33	45	1.2	4	4	−0.06	6	0.14	o

Note: Abbreviations as in Table 2.1.

3.210. 1ST trackways.
o, ornithopod; t, theropod;
n, unidentified footprints.

marks appear to be relatively closed, more similar to evolved sauropods (Wright, 2005). The shape of the foot is oval but elongated. There is no evidence to indicate that the feet were wider at the front. The width of the hind foot print is similar to that of the forefoot. The axis of the forefoot is a continuation of the hind foot axis. The heteropody has not been calculated because in the forefoot–hind foot pairs, the forefoot marks are deformed by the mud pushed forward by the hind foot. The footprints are similar to *Brontopodus* Farlow, Pittman, and Hawthorne, 1989.

Footprint Distribution by Ichnotype

Sites 1ST and 2ST are small, so the possibility that each contains several different ichnotypes is low (Casanovas et al., 1999). Also, the mud conditions were different, with 2ST softer than 1ST.

Site 2ST has only sauropod dinosaur tracks. This would be the best defense for herbivorous dinosaurs. At 1ST the footprints are bipedal dinosaurs. In this case, the physical state of the mud (shallow footprints, more viscous mud) can be used to explain the lack of sauropod footprints.

At 1ST the tridactyl footprints have both ornithopod and theropod features. The cautious dinosaur hypothesis separating the normal habitats of carnivores and herbivores cannot be applied to this site. However, the number of footprints and trackways at 1ST is predominantly theropod.

Site 2ST is a typical occurrence of sauropod footprints. There are many examples (several in Spain in Teruel and La Rioja) where the passage of these dinosaurs left chaotically arranged footprints, many

Table 3.18. Fastest dinosaur trackmakers in La Rioja.

Site/outcrop	Trackway	No. of footprints	Footprint length	z/H (relative stride)	Speed (km/h) according to:	
					Alexander (A) or Thulborn (T)	Demathieu
La Torre 6-B	1	6	31	3.4–4.6	33 T	12
La Virgen del Campo	LVC24	3	27	2.4	14 A	8
La Virgen del Campo	LVC137	5	34	2.2	12 A	8
La Era del Peladillo 3	3PL11	3	30	3.1	22 A	10
La Senoba	LS2	6	47	2.0	11 A	7
Las Losas	LL49	3	22	2.3	11 A	7
Barranco de Valdecevillo	VA3	4	30	2	10 A	7
Las Mortajeras	LM53	10	27	2.1	10 A	7

overlapping, where it is difficult to separate them into trackways. In these cases, the activity of the dinosaurs can be attributed to gregarious behavior involving the passing of herds or large groups of animals. Overlapping footprints and trackway interference is normal in this context. The footprint surface should continue around the current boundary of the track site. The limit of the tracksite is not the limit of the possible herd: the bed is not continuous to the west and south because it has broken and fallen to the base of the canyon wall of the Leza River; to the east and northeast the study surface extends beneath the hillside; and northward the footprints become shallower and more open, and the outline becomes more uncertain.

General Information

The footprint density in 1ST is 0.8 footprints/m² (intermediate), about one footprint per square meter. At 2ST it is high (2.4 footprints/m²), i.e., equivalent to over two prints in every square meter. The higher density in 2ST is explained by the passage of a herd of sauropods across the site and because the footprints are small (the length and width dimensions are 38 × 18 cm for the hind feet and 14 × 20 cm for the forefeet).

THE FASTEST THEROPOD DINOSAURS

The estimated speed of the theropod dinosaurs at 1ST is relatively high (Table 3.17). The only trackmaker not going fast is the one responsible for the large footprints in 1ST8. The velocity is low for the ornithopod footprints, in accordance with the general trend.

In the literature for the region, there is only one trackway cited that indicates faster travel than 1ST12, and that is at El Peladillo (La Torre outcrop 6-B; Viera and Torres, 1995b). In general, and not including the semiplantigrades, there are few trackways with estimated speed faster than 10 km/h in La Rioja (Table 3.18). These estimates are obtained from the formulas of Alexander (1976), except for La Torre 6B, where Thulborn's (1990) formula was used. When using the formula proposed by Demathieu (1986), most trackways do not reach 10 km/h. Applying the method of Thulborn (1990) for those whose z/H ratio is greater than 2 (trotting and running), the maximum speed by Alexander's (1976) formula is also less.

The above results must be considered with caution because several of them are for trackways with three footprints. The calculation in these cases has been done with only one stride. Only those trackways with more than five footprints can be considered valid, as then there are three or more strides: LVC137, LS2, LL49, and Las Mortajeras (LM) LM53. The four trackways 3PL11, LS2, LL49, and LM53 occur in places where the mud was very soft. In La Senoba, Las Losas, and Las Mortajeras, irregular trackways and mud collapse structures are typical. Perhaps many of the trackways with large strides in La Rioja are the result of a variant of dinosaur behavior, probably similar to that adopted in the semiplantigrade walk: a hyperextensive pace (Aguirrezabala and Viera, 1980; Kuban, 1989; Pérez-Lorente, 2001a).

The trackway at La Torre, however, is different. According to Viera and Torres (1995b), the footprints are shallow, and only the front parts of the toes are seen. At the same outcrop there are other footprints with the same depth but with the soles completely impressed and with shorter strides. The stride in the fast trackway of La Torre 6B is 560 cm. It is assumed that the weight of the dinosaur was not excessive.

The above authors obtained a speed of 37.4 km/h. Hip height (height of the acetabulum) is 128 cm following Alexander's (1976) protocol where hip height = 4 × footprint length. When using the allometric formulas of Thulborn (1990) for the calculation of H and the running speed, the result is 163 cm for the hip height and a speed of 33 km/h. As Henderson (2003) shows, the Alexander formula is more appropriate.

This speed exceeds that found by Day et al. (2004) of 29.2 km/h in a theropod trackway from the Ardley Quarry in England. This is between 30 and 43 km/h found by Farlow (1981, 1987) in trackways with a stride length of 538 to 660 cm.

Irby (1996) cites quicker trackways by providing data from different authors. Some only have three footprints, which is insufficient to provide reliable velocity values. We found eight trackways in the study with four or more footprints whose speed ranged from 17 to 42 km/h.

As with Irby (1996), the conclusion in La Rioja is similar. By 2003, 981 biped trackways had been studied (Pérez-Lorente, 2003b), of which 856 were theropod. The quickest of these are provided in Table 3.18. Only 10% exceeded the speed of 10 km/h, and only two of them had a Thulborn (1990) relative stride length more than 2.9 (3PL11 [z/H = 3.1]; La Torre 6B [z/H = 4.3]).

The proportion of theropod trackways of over 10 km/h in La Rioja is about 1%, while running is found in only one trackway with more than three footprints (La Torre 6B), i.e., 0.12%. Assuming that the percentage can be applied to all theropod trackway in La Rioja, there are two trackways with a relative stride (z/H) of 3. This can also be interpreted as theropod dinosaurs running 0.23% of the time throughout their lives.

The La Torre 6B dinosaur was not one of the largest theropods in La Rioja. The speed and body size should not pose problems in terms of the "strength indicator" (Farlow, Smith, and Robinson, 1995).

Main Features of Site

The ornithopod footprints have a speed in keeping with that found in La Rioja. The large theropod footprints (trackway 1ST8) also indicate normal walking.

Several theropod trackways in the site (1ST) are of high speed (12–21 km/h). The average length of footprint is between 23 and 25 cm. Because they are concentrated in this site, no theory can be ruled out for similar trackways. Perhaps small theropods ran away quickly after a large theropod arrived, as in the case of Lark Quarry in Australia (Thulborn and Wade, 1979). Perhaps exceptions to normal behavior occurred as a result of environmental conditions when these dinosaurs passed by; water depth may have varied, or the animal's stride length may have been abnormal. The calculation speed would be wrong if true H did not comply with the formulas proposed by Alexander (1976) and Thulborn (1990). The limb anatomy of the dinosaurs that left the quick, small theropod footprints in 1ST is special.

According to Pérez-Lorente (1996a), as the value of l (footprint length) increases, the trackmaker speed slows (208 trackways studied). The opposite is not necessarily true because not all small theropod footprints are on fast trackways. It is likely, therefore, that the theory of Farlow et al. (2000) regarding the injury threshold is one of the limitations to high speed in these animals. Only small animals could run safely.

The rocks are well conserved and seem able to resist the passage of time. Colonization of plants in the footprint shafts and natural cracks on the site has been detected in 2ST.

History

This site was found in 1994 by a worker who was working on an access road to the village of Treguajantes. That same year, the site was cleared, the footprints were studied, a map was published, and results were obtained. In 2004, a paper was presented at a conference citing some of the footprints here as being the largest for a theropod in the Cameros Basin. These include a cast found in the neighboring province of Soria.

In 2006, it was discovered that unknown persons had cleared part of the surrounding area and had discovered two more footprints located behind the start of the trackway, thereby increasing the original six steps on the site to eight.

Camino a Treguajantes

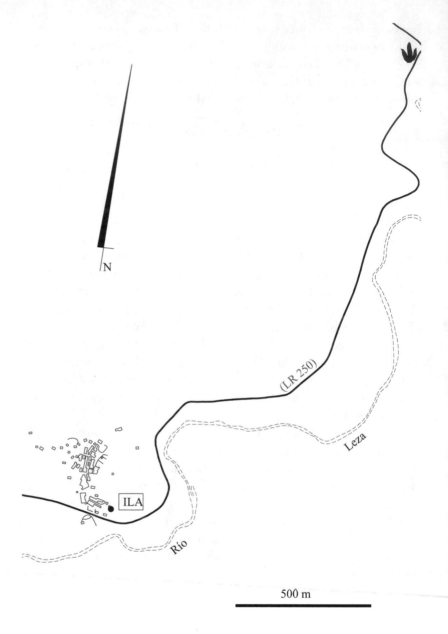

3.211. Location of site of Camino Treguajantes (TR) and neighboring sites La Ilaga (ILA) and an undescribed outcrop indicated by footprint symbol. Local road is labeled LR 250.

Description of Location

Camino a Treguajantes (TR) is a prominent area of clean, small rock in the middle of loose fragments. The outcrop is not continuous. On one side is the TR1 trackway with five footprints, and on the other side are the three remaining footprints. The large footprints are on the top of a limestone layer, while the two small ones are at a lower level of the same layer that surfaced as a result of erosion of the upper layer. Access is easy;

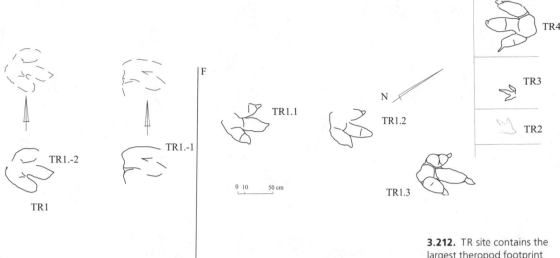

TR1.1

TR1.2

TR1.-1

TR1.-2

TR1

N

F

TR1.3

TR4

TR3

TR2

0 10   50 cm

3.212. TR site contains the largest theropod footprint stamps in Spain. Dashed outlines show inferred positions of footprints TR1.–1 and TR1.–2 before motion along fracture line, F. Three footprints in same trackway are shown in their actual position to right of fracture line. Footprints TR2 and TR3 are smaller prints at a lower level than main trackway. Print TR4 is another big print at the western end of the tracksite, but it seems to be unrelated (in a different orientation and with no continuity) to trackway TR1.

the site is next to the road where the worker who found the footprints was working.

The two closest sites (Fig. 3.211) are La Ilaga, 600 m away (Hernández Medrano and Pérez-Lorente, 2002, 2003), and one as yet unnamed and unexplored site located 2600 m to the north. About 3600 m to the south is another site with unstudied footprints in the vicinity of San Roman. Just under 4000 m to the north is Soto 3 (3ST) and a site with dinosaur bones not yet studied. The status of the site with bones is currently uncertain, but all the rest of the sites are in the Oncala Group.

Lithology

The footprints are on two superimposed levels of limestone, separated by a probable discontinuous layer of shale. The lamina is seen only at those points where the upper level is missing. On the top of the lower level there are mud cracks, while the upper layer is smooth with no appreciable structures. Algal laminites are seen in almost all layers of the limestone outcrop.

The sedimentary environment attributed to these rocks is a protected shallow area on the edge of a lake.

Footprint Mud Structures

The footprints from both the upper and lower layers sank little into the mud. Despite their size, the prints in the upper level are impressed barely 4 cm deep. Both sets of footprints have relatively well-preserved outlines and crease structures between the toe pads.

The mud cracks in the lower level are deformed by small footprints in it. Deformation at lower level (undertracks) induced by the footprints

3.213. Right footprint TR1.3, in which characters of ichnite are well represented: size, well-developed claws, digital pads, spread toes, prominent "heel."

at higher levels was not noticed. In both cases, the tracking surface is a hardened surface with a thin, plastic layer underneath.

Footprint Features (Ichnotypes)

All the footprints on the site are theropod. Two types are distinguished by their size, the stratigraphic level at which they are located, and their shape.

SMALL THEROPOD FOOTPRINTS

On the lower level (Fig. 3.212) there are two small isolated footprints (TR2, TR3). Their size is at the limit between large and small footprints of Thulborn (1990). Both have thin, well-separated, tapered digits. The digits probably have several pads because there are side constrictions on their outlines. The back part of TR2 is prominent. There are no skid marks to indicate that this form is due to slippage of the foot during the T phase. They would be classified as small theropod footprints.

LARGE THEROPOD FOOTPRINTS

The large theropod footprints are separated into two sets. One of these (TR1) comprises a sequence of five footprints, and the other (TR4) consists of a single footprint. When the site was studied (Casanovas et al., 1995f)

TR1 only had three footprints. Now there are five, the result of two more footprints (TR1.-1 and TR1.-2) being found after clearing debris at the beginning of the trackway. Between TR1.-1 and TR1.-2 there is a fracture that displaces the first two footprints. It has made the reconstruction of the original trackway possible (Fig. 3.212).

The six footprints in trackways TR1 and TR4 are long tridactyl footprints (Fig. 3.213). The average length of 72 cm makes them giant footprints—among the largest tridactyls in the world (Boutakiout et al., 2008; Boutakiout, Hadri et al., 2009). The variation in length in relation to the width ([l − a]/a = 0.15) places them near to wide tracks (Pérez-Lorente, 2001a).

The digits are long, relatively thick, and well separated. According to Farlow (2001), large theropods have relatively thick digits, as these footprints do, but big ornithopods have rather short, stout toes (Farlow et al., 2006). The claw mark is narrow and distinct at the tip of each digit. The length of these footprints includes the distal tip of the claw mark. The pads are well impressed, but not sufficiently to determine their number. There are two pads on digit II; none is distinguishable on digit III; and digit IV has at least three, including the heel. The most posterior of the pads has an elongated tip aligned with the long axis of digit IV. This causes a lateral posterior asymmetric projection in the posterior outline of the heel. The heel is wide and asymmetric. The most backward part is toward the side of digit IV, with a small indentation at its intersection with the axis of the foot. The divarication of the digits is low, as in most theropod footprints.

The acetabular height is 306 cm according to the allometric formula of Thulborn (1990).

The trackway is narrow (Ar/a less than 0.5), but the trackmaker made slow progress. This is consistent with the general trend of large dinosaurs in La Rioja, both ornithopods and theropods (Casanovas et al., 1995a; Pérez-Lorente, 1996a). The is z/l ratio is 4, which is typical of dinosaurs with thick limbs (Table 3.19).

COMPARISON WITH OTHER LARGE THEROPOD FOOTPRINTS IN LA RIOJA

Several authors have studied giant theropod footprints and include those in La Rioja (Barco et al., 2004; Boutakiout, Hadri, et al., 2009; Farlow et al., 2006). An objective in each of these studies was to compare them with others in other parts of the world.

Theropod trackways from La Rioja whose average length is 50 cm or more are shown in Fig. 3.214. Semiplantigrade and very irregular footprints have been excluded. Among those drawn, several have deformed stamps (LVC1, LVC5, LVC16, VA4, LN6, LN23, LN25, LN27). Others can be assumed to be undertracks (BAC1 [Barranco de Acrijos], LN1).

However, removing some undertracks (LN1) allows features from the rest to be grouped together (Table 3.19). Considering the morphological

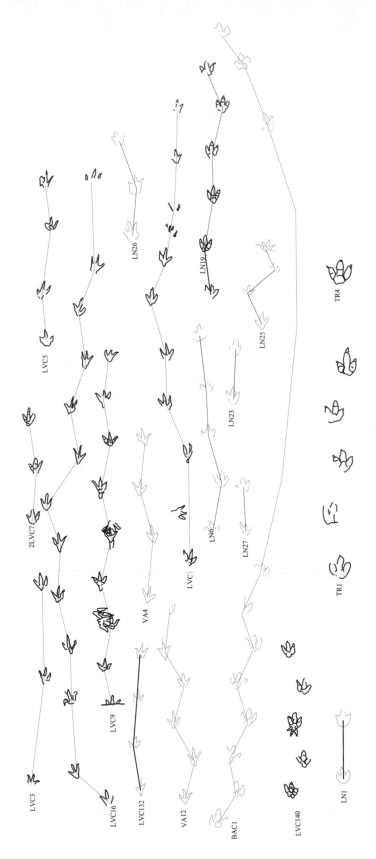

3.214. Trackways in La Rioja with footprints of average length greater than 50 cm.

Table 3.19. Largest theropod footprints in La Rioja.

Trackway	l	a	Ap	z	H	v	(l – a)/a	Ar/a	II^IV	z/l
LVC1	50	42	167	310	239	6	0.16	0.2	75	6
LN19	52	38	159	265	247	5	0.4	0.3	39	5
BAC1	54	50	140	215	255	3	0.08	0.4	40	4
LVC140	50	42	145	216	239	4	0.16	0.4	60	4
TR1/4	70	63	155	286	318	3.5	0.15	0.4	55	4

Note: Abbreviations as in Table 2.1. Measurements are the average dimensions of all footprints for each trackway.

features, the following types are seen: footprints with thin digit marks (LVC1, LVC3, LVC5, LVC16, LVC132, 2LVC7, VA4, VA12), and footprints with a clear claw mark for digit II. This is seen well at the site of Las Navillas (LN6, LN7, LN19, LN25); footprints of trackway BAC1; footprints of trackway LVC140; and footprint TR4 and from trackway TR1.

The features in the above groups are also found in other theropod footprints in La Rioja, although these are smaller than 50 cm. Trackway BAC1 (Fig. 3.215) is from the Barranco de Acrijos site (not described in this book), but the length of its footprints places them among the largest.

The features for separating ichnotypes in the La Rioja trackways mentioned are still under study. They have not yet been expressed numerically but are visible in the drawings, as in Fig. 3.215.

Type 1 footprints have long, slender digits; the heel is prominent and pointed, and the heel pad is about the same width as the medial and distal pads of digit IV; the proximal pad for digit IV is set backward with respect to digit II; the separation of the proximal pads of digits II and IV from the axis of the foot is relatively small. Examples of this type of footprint, but smaller, are found at many sites of La Rioja. It is assigned to *Megalosauripus* in some sites (Fig. 3.215).

Type 2 footprints have wider digits than the previous ones; the heel projects less toward the rear and is wider and less pointed; the proximal pad of digit II is less forwardly advanced compared with digit IV; the separation of proximal pads of II and IV from the axis of the foot is relatively small; there is a claw in digit II that stands out from the other two in all the footprints. Similar theropod footprints with the digit II claw prominent are cited in several points in La Rioja. Footprints in trackway LN19 are a good example.

Type 3 footprints have wider digits than the preceding ones; the heel is broad and rounded; the heel pad is wider than the digit pads; and the position of the heel pad is symmetrical with respect to the footprint axis. It is pointless discussing the position of the proximal pads of digits II and IV with respect to the axis of the foot because the footprints in trackway BAC1, the exemplars of this footprint type, are probably undertracks.

Type 4 footprints have wide digits, although narrower than those of the preceding print types; the heel is prominent but not pointed; the heel pad is approximately the same width as those at the digits; the proximal

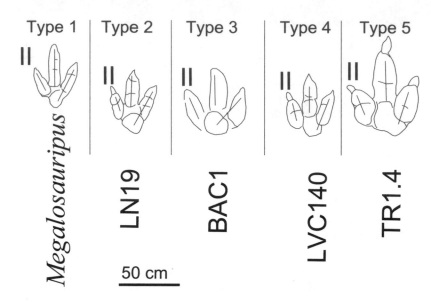

3.215. Five possible ichnotypes for large theropod footprints in La Rioja.

Type 1 | Type 2 | Type 3 | Type 4 | Type 5

Megalosauripus | LN19 | BAC1 | LVC140 | TR1.4

50 cm

part of the first pad of digit II is probably not very forwardly advanced compared with the same part of digit IV; and the separation of the proximal pads of digits II and IV of the axis of the foot is small. The claw mark of digit II is probably larger than that of digit IV. These footprints are only seen in La Rioja in trackway LVC140.

Type 5 footprints have thick digits; the heel protrudes very little and is wide; the heel pad is broad and asymmetrical, and slightly elongated as a continuation of the back part of digit IV; the proximal part of the first pad in digit II is slightly advanced with respect to that in digit IV; toe II is the most separated from the long axis all the above footprints (except BAC1?); and there are strong claw marks in all three digits.

The age of the footprints is different, depending on which groups they are found in, as follows.

TYPE 1 (at the base of the Enciso Group [upper Barremian–Aptian]) Those of the Virgen del Campo and Barranco de Valdecevillo sites in siliceous packages. Footprints with the same features, but smaller, are found in all groups.

TYPE 2 (in the limestone units of the Enciso Group [Aptian]) Those of the LN19 type. This same type of footprint but less than 50 cm in size is cited in other parts of La Rioja, the stratigraphic location of which is not stated.

TYPE 3 (in the lower part of the Urbión Group [upper Hauterivian–Barremian]) BAC1.

TYPE 4 The LVC140 trackway is also in the site of La Virgen del Campo, like those in the first section (upper Barremian–Aptian).

TYPE 5 TR1–4 have been placed in the Oncala Group (Berriasian).

Barco et al. (2004) comment that the described large theropod ichnotaxa of the Lower Cretaceous are *Megalosauripus, Bueckeburgichnus, Irenesauripus,* and *Abelichnus.* The ichnogenus *Eubrontes* is reserved for footprints with long digits and clear pads, especially in the Triassic and Lower Jurassic. The same authors compared the large footprints of Treguajantes with those cited and decided that they cannot be attributed to *Megalosauripus* Lockley, Meyer, and Santos, 1998. One of the diagnostic features of *Megalosauripus* (Lockley, Meyer, and Santos, 1998) is its elongated heel in relation to the length of digit III. In addition, *Irenesauripus* and *Abelichnus* have the thinnest and most separated digits. These authors mention the problems of *Bueckeburgichnus,* with hallux impressed and a wide and prominent rear part (metatarsal impression?). They therefore conclude that it is a new ichnotaxon.

According to Thulborn (2001) there are many problems with the ichnotaxon *Bueckeburgichnus.* Regardless of the doubts of Thulborn (1990) regarding whether *Bueckeburgichnus* in the site of Los Cayos (Moratalla and Sanz, 1997) actually represents the ichnogenus, the footprints on this site (Los Cayos) are not direct structures but undertracks. Thulborn (2001) reviews the names and states that *Megalosauripus* Lessertisseur, 1955, is appropriate and that *Bueckeburgichnus* Kuhn, 1958, is a later synonym for the same ichnogenus.

Type 1 footprints could be assigned to *Megalosauripus,* but the remaining large theropod footprint types do not belong to any described ichnogenus. Until the outcome of a research project, like others before it, repeating the examination of autopod bones to try to find applicable anatomic synapomorphies to the footprints, there are insufficient data to choose among these types of footprints.

The largest digitigrade theropod footprints without metatarsal marks in the Iberian Peninsula are at this site. There are other places with large theropod footprints, but they either have metatarsal marks or show sliding of the foot. TR1-4 has digital pad marks, which suggests that they are stamps. In Portugal, there were two giant theropod footprints, *Megalosauripus pombali* Lapparent and Zbyszewski, 1957, and *Megalosauripus* sp. (Lockley, Meyer, and Santos, 1998). The first is a deformed footprint with insufficient data. The second disappeared after the layer containing it collapsed. The age of the Portuguese sites is Upper Jurassic. In the Jurassic Museum of Asturias (García-Ramos, Piñuela, and Lires, 2007) there is also a large footprint (82 cm long). This footprint is not comparable with those of Treguajantes because it is a cast and it has sliding grooves parallel to the longitudinal axis of the foot.

General Information

The track density is low (0.3 footprints/m^2), and it does not seem that any further digging is likely to increase it. The site is on a slope without vegetation and with many fragments of loose rock and clay. More of the surface may present itself for digging, but the result is not predictable because the layers are displaced by small faults. It is not known if the layer with the large footprints is below ground at the sides of the site or was higher and has been eroded.

The dinosaur's estimated walking speed is between 4.5 and 4.7 km/h, according to which formula used in the derivation, which is not fast considering the size of the footprints.

The footprints are in good condition but exposed to the atmosphere. We do not know the rate of degradation of the rock, although the worst problem seems to be the dissolution of the limestone. No peeling or fractures are seen, and for now, the roots of plants are not too aggressive, as there is no brush covering the layer with footprints.

The geometry of the proximal pads (perhaps related to the phalangeal relative position) and the development of claw II are the principal characters for the different great theropod footprints (type 1 to type 5). It is difficult to apply the metric measures of phalangeal toes (Farlow et al., 2013) on the discrimination of the footprints of La Rioja.

Valdemayor

History

The Valdemayor (CBC) site was discovered in 1989 by a team from the Universidad Autónoma de Madrid and the energy company Iberdrola. At the time of its discovery, it was the first Spanish trackway of a quadrupedal ornithopod. Until 1994, it was the only site known in the area where the discovery occurred (there were none in the village of Cabezón de Cameros or in neighboring villages). Today this is not true mainly for two reasons: first, people know about dinosaur footprints and report the outcrops they see; and second, there are prospecting campaigns around the reported sites. Since 1989, five new sites have been found in the vicinity of Cabezón de Cameros, and a study of them was published in 2007 (Díaz-Martínez et al., 2007).

Description of Location

The site is in a layer located on the side of a ravine. It is not reachable by car. There are two paths from Cabezón de Cameros, the nearest village, that pass near the site. Although the distance between Valdemayor and the village is only about 700 m as the crow flies, the path is difficult.

The nearest places with footprints are the sites of Los Chopos at 1900 m, Cabezón de Cameros (CCM) 1 at 1300 m away, 2CCM at 930 m away, and 3CCM at 760 m away (Fig. 3.216). The area has not been searched for fossils, and there are no references to skeletal material in places nearby.

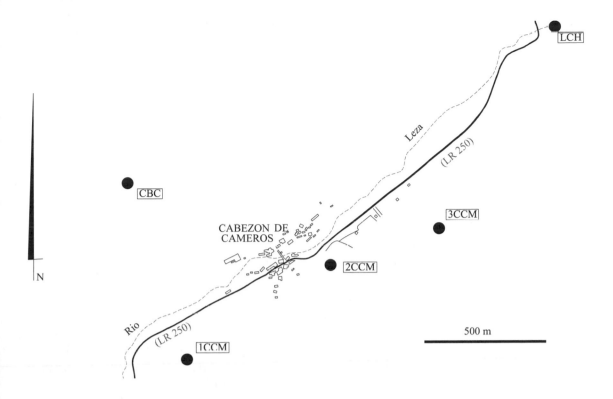

Lithology

3.216. Location of site of Valdemayor (CBC) and nearby localities with footprints: Los Chopos (LCH) and Cabezón de Cameros 1, 2, and 3 (1CCM, 2CCM, 3CCM). Local road is labeled LR 250.

The rock containing the footprints is a fine-grained shale sandstone about 30 cm thick. According to Moratalla (1993) it is toward the top of a paleochannel of fluvial origin.

It is assumed that the stratigraphic unit containing it (Ramírez et al., 1990) is in the lower part of the Enciso Group. It is therefore likely to be dated as Barremian in age (Doublet, 2004).

Footprint Mud Structures

The footprints are of variable depth. Trackway CBC-R1 is deep (hind foot prints up to 11 cm), which means that the mud viscosity was low. The extrusion rims surrounding the footprints are not very pronounced. The depth of the forefoot marks is less than that of the hind feet—about 3.5 cm on average (Moratalla, 1993).

Footprint Features (Ichnotypes)

Moratalla described the best-preserved tracks in Valdemayor. These have well-pronounced forefoot and hind foot marks, especially in trackway CBC-R1. The hind foot prints are somewhat longer (58 cm) than they are

3.217. Map of trackway CBC-R1. Modified from Moratalla (1993).

Table 3.20. Average numerical data of the CBC–R1 trackway.

Foot	l	a	P	z	Ap	O	H	Ar	Lr	v_1	v_2	(l − a)/a	Ar/a	z/l
Hind	58	55	124	224	152	−19	239	15	76	2.2	3.2	0.06	0.3	3.8
Fore	15	27	127	215	120	45		43	105			−0.42	1.6	

Note: Abbreviations as in Table 2.1.

wide (55 m), with a broad and symmetrical heel. The digits are thick and sturdy. The forefoot prints are wider (26.5 cm) than they are long (15.5 cm) and bilobed. The medial lobe is smaller than the lateral one.

The footprints are attributed to the ichnotaxon *Iguanodonipus cuadrupedae* Moratalla, 1993, which has the following features: footprints of medium and large size, usually between 40 and 60 cm in length, but maybe greater; wide digit marks, short and robust with rounded distal ends; similar development of digits II and IV; hypex symmetrical; wide plantar area; and rounded proximal heel contour. There are two notches on the sides of the heel (medial and lateral), occasionally very well marked. The ratio of digit III length to (foot length − digit III length) ranges between 0.6 and 1.3. It is a relatively wide biped trackway, occasionally quadrupedal with the front imprints transversely elongated and bilobed. Negative orientation of the hind feet and positive orientation for the forefeet are evident. The forefoot prints are in front of the hind foot prints or even further away from the midline. The heteropody index is high.

According to Díaz-Martínez et al. (2009), the ichnogenus is defined taking into account the foot mark features: size, shape of digits, and shape of heel. It includes quadruped and biped footprints from the Upper Jurassic to the Lower Cretaceous.

The trackway is of a quadruped (Fig. 3.217), and the forefeet and hind feet are considered separately in the trackway study. Some data (Table 3.20) preserved here are similar to those provided by Moratalla et al. (1992), while other data differ significantly. Some of this may be due to the difference between taking measurements directly on the ground and data plotted in a diagram. However, other data are different because of the method used for determining the values. For example, the height of the acetabulum (392 cm), calculated in accordance with the allometric formulas of Thulborn (1990), is different from that obtained by previous authors (239 cm). Because Thulborn's (1990) formulas are those that have been used in the calculation of all the tracksites described in this book, I also apply them in this case. By using this height for the acetabulum, the speed (v_1) according to the formulas of Alexander (1976) or (v_2) Demathieu (1986) is less than that provided by the above authors (4 km/h) and is consistent with the low speed obtained in large ornithopod dinosaur trackways.

According to the ratios of the footprint and trackway features, the foot is broad (a = 55 cm) but relatively narrow ([l − a]/a = 0.06), although almost circular, with thick hind limbs (z/l = 3.8), and the trackway is relatively (Ar/a = 0.3) narrow. The forefoot trackway width (Lr) is 105 cm.

The glenoacetabular distance according to Moratalla et al. (1992) is 190 cm (H = 239 cm), calculated for the alternating trot-type gait. Apparently this type of walk is characteristic of animals whose stylopods are not in the parasagittal plane (Demathieu, 1987b). In this type of motion, the spine participates in the mechanics of walking with its rotation. It seems more likely that quadruped dinosaurs used an alternating or pace type of walk (glenoacetabular distance of 230 cm and 290 cm) (Demathieu, 1987b).

General Information

The trackway is too short for general conclusions to be drawn from it, but its data are important because it is one of five Spanish ornithopod trackways with forefoot marks, and one of the few cited in the world.

The manus prints have a positive orientation, unlike those of the hind feet. This also occurs in the rest of the few known cases (see Pérez-Lorente et al., 1997:fig. 8, and Lim, Lockley, and Kong, 2012:fig. 4). This is normal and happens when the marks of the forefeet are long or are placed parallel to the midline or the orientation is strongly positive. Similarly, it shows that the form of forefoot prints is not uniform among ornithopod trackways. According to Moratalla et al. (1992), this trackway supports the theory that the shape of the forefoot marks may depend on both anatomic parameters and physical properties of the substrate and methods of locomotion. This is one of the examples used to support the theory of quadrupedal walking in ornithopod dinosaurs. This is not supported in La Rioja because it is the only quadrupedal ornithopod trackway known from a total of 125 ornithopod trackways (Pérez-Lorente, 2003b). In this site, as in many others, the quadrupedal ornithopod trackway may be a part of a mixed trackway (bipedal/quadrupedal). There are many long ornithopod trackways recognized in the world containing the two types of gait—that is, sectors with and sectors without the manus print (cf. Pérez-Lorente et al., 1997; Torcida et al., 2003).

In reconstructions of ornithopod dinosaur skeletons (see Norman, 2004), the glenoacetabular distance is usually similar or slightly less than the height of the acetabulum. The glenoacetabular distance calculated according to Moratalla et al. (1992) is much smaller than the height of the limb, which is not consistent with reconstructions of ornithopod skeletons. If the glenoacetabular distance is calculated for other types of walking (trot or pace), this significantly increases the distance mentioned. The interpretation of these estimates is problematic because the inclination of the femur in ornithopod dinosaurs is not known; nor is it known whether the shoulder girdle participates in the movement of the forelimbs.

The site track density is very low (0.4), equivalent to one track every 2.5 m². There may have been more footprints in the area. The rock is highly fractured, and intense weathering and erosion may have destroyed them.

The site has a number of areas with broken and slippery rocks. They could possibly contain footprints in large numbers, which of course would be in as many pieces as the fractured strata containing them, and therefore difficult to see or recognize.

Main Features of Site

This is the only quadrupedal ornithopod dinosaur trackway in La Rioja. In the Cameros Basin there are two other sites, but these are in the province of Burgos, one with two trackways studied by Bengoechea et al. (1993) and Moratalla, Sanz, and Jiménez (1994), and the other, adjacent to the previous, with an isolated trackway, studied by Torcida et al. (2003). In the Iberian Peninsula, there was until recently only one other known trackway (Pérez-Lorente et al., 1997), at the site of Las Cerradicas (Galve); but another similar and parallel trackway has emerged after recent work. One of the three trackways in Burgos also has the mark of a tail. They were all assigned to the Lower Cretaceous, although recently the site at Las Cerradicas was changed to Upper Jurassic. The length of the footprints ranges between 47 and 63 cm for the Cretaceous localities and 23 cm for the Jurassic occurrence.

Although the observation conditions at Valdemayor are not clear, it can be seen that there are some ornithopod hind foot tracks associated with forefoot tracks and others that are not. This is true for all Spanish sites with quadrupedal ornithopod footprints (except Las Cerradicas).

Conservation of the Tracksites

THE SITES WITH DINOSAUR FOOTPRINTS IN LA RIOJA HAVE BEEN OF interest for many years. Concern for their conservation has spread throughout the population of this region. They are an asset that must remain in the countryside, and the inhabitants of the villages with the tracks view them as part of their heritage. If it were not so, dealers, collectors, and even some of the local inhabitants would have removed the best of them. Sites of interest must have some value that ideally can be quantified. Objects that have value are treated as assets, prompting states and other governmental entities to declare them part of the heritage of the land via legislation or other official regulations.

From the beginning, the teams studying the footprints have been interested in their conservation and protection as well as their publication (Pérez-Lorente, 2001b). Relevant to this is political action, which began in 1980 with the first proposals to include outcrops as places covered by laws protecting natural areas. Thanks to the government of La Rioja and its institutions, there have been grants for almost all fieldwork or research projects. The Instituto de Estudios Riojanos has funded research projects on dinosaur tracks in La Rioja and has published about a hundred articles in its journal, *Zubía*, relating to footprints, their conservation, and their relationship with the geological heritage. Colleges and universities have lent their staff and their experience. For example, the University of La Rioja currently offers summer courses on dinosaur footprints and conservation of sites. Support for these institutions has led to other such initiatives supporting their publicity and preservation, which can be classified as legal protection, physical protection, and maintenance protection.

LEGAL PROTECTION The laws, decrees, and legislative processes affecting sites with footprints. Included in this section are declarations or proposals for places of geological interest.

PHYSICAL PROTECTION Fences and covers so that trace fossils are not directly exposed to the elements or walked over by people or livestock. Signs and notice boards have been placed in almost all La Rioja sites to disseminate information and for visitors to learn of their importance.

MAINTENANCE PROTECTION Direct action on rocks exposed to the open air to prevent their destruction and ensure longer life for the footprints. Actions involved here are the use of products to protect and conserve the rocks.

Legislation and Legal Denominations

For many years, inhabitants of the villages in La Rioja have considered that sites with dinosaur footprints are a part of the heritage of the region. The same has happened at higher political levels: the government of the autonomous region of La Rioja and the government of Spain have made an effort for them to be declared world heritage sites. As cultural heritage sites, mechanisms must be in place for their legal protection and the management of field activities.

There have been several laws and decrees by way of protecting and conserving the sites and the ring of land around them (Montes, 2006; Pérez-Lorente, 2000a, 2000b). The first effective legislation was the Artistic and Historic Heritage Law (law 16/1985, June 25) regulating exploration and controlling the pieces removed. Removal or trade in Spanish fossils is not allowed, which of course includes those from La Rioja. A resolution of the government of La Rioja was passed on December 5, 1994, approving PEPIR (a special project of dinosaur ichnite protection in La Rioja). This plan regulates land use in areas containing footprints in accordance with scientific criteria. On the basis of cultural property law, the government of La Rioja prepared a decree (decree 34/2000, June 20) declaring 40 areas, encompassing all dinosaur footprints in La Rioja, as historic sites. The name according to the decree is Bienes de Interés Cultural (Properties of Cultural Interest), or BIC, in accordance with the Spanish Artistic Heritage law. Following this was an environmental impact assessment law (law 6/2001, May 8), making it a legal requirement to perform a study of natural elements that may be modified by the human works. The cultural, historical, and artistic heritage law of La Rioja (law 7/2004, October 18) assumes the protection of all dinosaur footprint sites, which are considered places of cultural interest, within the category of historical sites and paleontological areas. Law 42/2007, regarding natural heritage and biodiversity, provides for the protection of Spanish natural heritage sites, and within this, sites of geological interest, known as LIG. Most of the sites with dinosaur footprints in La Rioja are on the inventory of LIGs. As a result, dinosaur footprint sites in La Rioja are legally protected under three different modalities: cultural (BIC; historical sites), environmental (PEPIR), and natural (LIG).

In addition to the nomenclature above, the same sites proposed as LIGS are also proposed as GEOSITES (IUGS Global Geosites Working Group). The GEOSITES project defines and selects the most important geological frameworks at an international level for each country and locations of geological interest (framework: fossils and trace fossils from the Lower Cretaceous; geosites: dinosaur tracks in La Rioja). The law therefore protects places so that they cannot be acted upon without permission and appropriate controls. A nature protection police service, SEPRONA, monitors and controls activities, and requires the necessary permits for those carrying out any activity in locations with tracks. The government of La Rioja has created a research and information center, the Centro Paleontológico de Enciso. In addition, for its assessment and for greater ease of operation, it has created the Fundación Patrimonio Paleontológico de La Rioja, which deals with all matters under the government of La Rioja related to footprint sites.

Evaluation Systems

The aim of site assessment is to provide quantitative (numerical) information to establish comparative scales. A trace fossil is the remains of the vital activity of an organism, and trace fossils are the most important remains for investigating the behavior of extinct animals. The formation of dinosaur footprints involves three main components: the animal that produced it, the state of the ground walked upon, and ultimately the behavior of the dinosaur. The footprints make an impression on the ground of the behavior and of soft tissue that rarely fossilizes with the bones; the footprints also reflect the function of limb joints during locomotion. The problems to be solved in the study of the most important fossil footprints of dinosaurs are the identification of ichnotaxa; indirect analysis of the dynamic functions of limb anatomy; determination of behavior; different types of dinosaur trackmakers; and determination of habitat.

Sites with footprints are also concerned with spreading scientific information about research topics and for development of tourism. The scales are different depending on the intended aims, such as education, entertainment, and heritage conservation. Cendrero (1996) and Elízaga, Gallego, and García Cortés (1991) established criteria and scales referring to scientific interest in terms of fossil accuracy, scientific interest, and scarcity or rarity (Pérez-Lorente, 1999c). The Caro and Pérez-Lorente (1997a, 1997b, 1998) criteria for evaluating sites (Table 4.1) establish a rating value for a site between 1 and 4. A site having a new ichnogenus (4) and a trackway with more than 10 steps (3) does not receive a rating of (4 + 3 =) 7, but the higher rating of 2 (4). Almost all the sites described in this book score 4 on this scale.

Physical Protection

Fossil footprint sites are places that cannot be moved or degraded by removal of part of their content. All remains have to be kept where they are found and thus are exposed to the weather. Restoration must be a work in progress; covering them is expensive, and some systems can provide an inappropriate environment. In several sites in La Rioja, the main processes that cause site deterioration over time have been established. Basic conservation indicators are based on macroscopic and microscopic forms of imperfections shown by the rocks. Ordaz and Esbert (1988) published a glossary of terms relating to the deterioration of stone outcrops. The terms are clear and descriptive; they are often used by researchers and professionals when referring primarily to the visible condition of the rocks and macroscopic changes in them. The different indicators of the imperfections and destruction processes used by the Fundación Patrimonio Paleontológico de La Rioja are provided in Table 4.2.

Table 4.1. Tracksite valuation criteria.

Criteria	Scale
Property accuracy: Faithful reproduction of the property structures. Regarding the fossil: · Evaluation of content. · Best place and unique site criteria.	4 = Identification of a new dinosaur footprint. 3 = Well-marked tracks with features that are different from those defined in published papers. 2 = Type of footprints on site can be classified. 1 = Fossilized footprints present.
Data of scientific interest obtained from the property: dinosaur activity deduced from footprints. Regarding the trackway: · Evaluation of content and its environment (individual behavior). · Original place and unique place criteria.	4 = At least 1 sequence of steps leading to a new finding regarding locomotion. 3 = Sequence of steps greater than 10. 2 = At least 1 trackway with sequence of steps between 5 and 10. 1 = Trackways have fewer than 5 steps.
Data of scientific interest obtained from property, its environment, and their relationship: dinosaur activity/habitat deduced from structures (behavior/environment). Regarding individual behavior: · Data of scientific interest to evaluate from property (trackway), its environment (behavior) and their relationship (paleoecology). · Original place and unique place criteria.	4 = Contains sequence leading to deduction of rare individual behavior. 3 = Associations of footprints similar to those described elsewhere. 2 = Top of outcropping strata is totally covered with footprints (dinoturbation). 1 = Only traces of footprints seen.
Data of scientific interest obtained from environment: footprint and trackway layout and features. Regarding the behavior of groups: · Data of scientific interest to evaluate from property (tracks), its environment (behavior) and relationship between property and environment (paleoecology). · Original place and unique place criteria.	4 = Contains combination of footprints, trackways, or both, leading to deduction of particular individual or group behavior. 3 = Footprints are noted for their size (either very large or very small).

Source: After Caro et al. (1997).

Covers, Sheds, Fences, and Posters

The first covering (Fig. 3.68) installed was in 1989 on the site of Los Cayos by Iberdrola, an energy company. The structure consisted of an assembly of beams and iron columns, with a roof of corrugated plates and a fence surrounding the masonry and metal. Its advantages are evident today. Apart from some weak plants, no grass or shrubbery grow on the inside, and the covering has prevented erosion by water and fragmentation of the rock due to freezing.

The government of La Rioja started building covers in 1998 in Peñaportillo (Fig. 3.165). The design came from a group of architects using different materials. Four cast iron columns supported translucent roof panels, and the area was fenced off with steel cable. The materials do not release oxides or components that attack the rock. Two more shelters made of iron bars and a metal plate roof were raised in the village of Enciso to protect Icnitas 3 and trackway 4VA at the bottom of the Barranco de Valdecevillo. The first fence was installed by Iberduero (now Iberdrola)

Table 4.2. Deterioration of stone outcrops: indicators of imperfections and processes of destruction.

Imperfections	Conservation/destruction indicators	Destruction processes
Breaking of the stone in sheets parallel to the rock face	Plates and scales. Sheets. Parallel to one another and independent of the structure of the rock.	Deplating and scaling.
	Sheets. Parallel to the structural planes of rock.	Exfoliation.
	Sheets. Parallel to slaty cleavage planes.	Fissility.
Small-scale globular cavities	Microkarst.	Dissolution.
	Pitting.	Differential erosion.
	Alveoli.	
Grooved cavities	Grooves. Striation and vermiculation.	
Crevices, broken rock (stable or displaced)	Fracture.	Traction or shearing stress.
	Fissure.	Preexisting as a result of mechanical stress or cyclic, alternating stress.
	Crack.	Opening of discontinuity.
Smothering of surface, weathering	Abrasion.	Erosion due to friction or impact.
Cracked or broken surface with missing pieces	Vegetation, tourism, chipping, fragments.	Plant erosion, human erosion, saline erosion, hydration, expansion–contraction, freezing–thawing.
Incisions, mutilations	Excoriation.	Human destruction.
Coloring of rock, granular disaggregation	Chromatic alteration.	Oxidation: change in stone color due to chemical alteration.
	Patina stain.	Oxidation: aging due to exposure to the elements.
	Decohesion and granular disaggregation.	Oxidation: increase in resulting oxides.
Loss of stone mass, crumbling	Disaggregation and granular disaggregation.	Dissolving of cement or matrix.

in 1976 to encircle the same trackway 4VA. Later, another fence was installed along the whole of Barranco de Valdecevillo, preventing visitors from walking on the site but allowing them to view the footprints from an elevated point. Other fences (Fig. 3.126) have been installed recently in La Virgen del Campo and La Senoba.

The sheds seem the best system for protecting the rock. They virtually eliminate peeling, plaque formation, and rupture of the blocks. Although some snow does penetrate in the winter and the wind blows rainwater under the roofing, the amount that enters is much reduced. Also, because the structures are open, the water soon evaporates, so they approximate an environment free of water. The sun's rays do not impinge on the rock, so the temperature never goes any higher than the outdoor air temperature. Rocks without this protection become hotter than the air during the summer (almost all are dark), so during the months of July and August the daytime temperature variation is about 25 to 30°C.

Conservation of Tracksites and Footprints

It is assumed that the aim is to maintain sites with dinosaur footprints and to protect them as much as possible from degradation and destruction (Pérez-Lorente, 1992b; Pérez-Lorente and Romero-Molina, 2001a). The restoration of the rock must be done in the most minimally invasive way possible, and with maximum concealment. The site must appear exactly the way it was. No attempt must be made to fool anyone. The components must not be substituted, and the modifications must be clearly recognizable. Finally, maximum durability and faithfulness to the original must be sought at all times. It is recommended to replace any worn material with the most resistant copies available.

In La Rioja, we are particularly careful to be minimally invasive. Very open fractures are sealed with mortar; they are not filled to the top but left a few centimeters short, to be completed in the future. The tracks are part of our heritage and are to be preserved in their current state, so no replacing of components is performed.

Rehabilitation and Conservation of Rock

The process of developing a conservation program begins with a preliminary analysis of the site (Caro, Pavía, and Pérez-Lorente, 2002, 2003; Caro, Pavía, and Requeta, 2006; Caro, Pérez-Lorente, and Requeta, 2002; Caro et al., 2000), including an examination of the forms of macroscopic changes; a comparison of the weathered rock, the surface, and deep rock (with no apparent changes); and an assessment of the status of any fracturing or erosion due to abrasion, runoff, and so on. Factors causing deterioration at the site are determined, including potentially aggressive organic or inorganic compounds. This is done after an analysis of the mineralogy and features related to water absorption capacity, capillary response, quantification of the response to physical cycles (dryness, humidity; freezing, thawing), the determination of porosity (total and effective), and the existing microcracking in the rock.

After the above, the main deterioration factors can be deduced and the most suitable compounds to apply in the field can be sought. Trials for sealing cracks and pores are done, as are trials for the total protective coating, to see the result of the rocks with the products applied. Because most degraded sites are outdoors, these are given priority over the most resistant sites and those protected by shelters or covers.

The treatments applied include the following: clearing the outcrops by elimination of colonizing plants, sealing cracks and other fractures; and adhering loose fragments. Also, shelters or covers have been proposed to protect the rocks from direct sunlight and water (rain, snow, frost), and fences have been installed to prevent people and livestock from walking over the tracks. The strength of the rock improves when cracks and fissures are sealed. This prevents water, the main agent for chemical and physical weathering, from destroying the rocks. Resin has been used to cover cracks and paste on fragments, and silicone and cement are used to fill the cracks. The effectiveness of cement applied over 15 years ago is confirmed by witnesses in the area.

Recovery of the site and application of the recommended products is done with the help of assistants in the field. If not, the budget for the operation would be too expensive. The total cost of using 25 people in a month, working 70 km from the nearest city and housed in a village, has not been calculated. These 25 people are trained and work as specialists. On the site there are also between one and three experts, depending on how much advice is required and how much care is needed in applying repair and maintenance products, which in turn depends on the conservation status of the rock.

The first operation undertaken is clearing. The rock is hardly ever continuous, without any cracks or loose fragments. Fissures (joints, faults) from separated walls, even if they are empty and clean, almost always contain plant detritus and roots. The problems and the method are complex and depend on the density of fractures, the separation of the walls from the fractures, and the depth to which the detritus reaches, as well as the density and strength of the roots. Sometimes slabs of rock are lifted, cleaned, and put back like a jigsaw puzzle. Sometimes it is impossible to remove the clay between the fragments without risk of destroying the rock. In this case, resins are applied that completely fill the altered fragment.

The second operation is to apply products to stick loose fragments and fill fractures. In operations carried out to date, the following have been used: limestone cement, siliceous sand, silicone, and epoxy resin. Each of these products is used according to the extent of the crack or hole to clean up. Currently a black coloring (carbon black or graphite) is mixed with cement and sand to change the color of the mixture.

When applying mortar in cracks, a gap of about 1 cm is left at the very top. This is filled later to ensure the top of the layer with tracks is treated carefully. Silicone is applied with a commercial spray gun. After cleaning the cracks, they are given an epoxy resin treatment to soak into the fragments and fissures that cannot be reached. Applying silicone is the most complicated step because it is not easy to completely fill these cracks. Silicone has been used because it has the advantage of staying soft for several years, so it is easy to remove and reinject into the same spot. So as not to see the silicone immediately after application, dirt is thrown over the outcrop. This is then lightly pressed in with the fingers so that it adheres to the silicone. Resin is used in smaller cracks (joints, plates, flakes, etc.), and injected with syringes.

Systematic work on dinosaur footprints in La Rioja began in 1980. At that time, only the sites published by Brancas, Martínez, and Blaschke (1979) were known, so the first work must have been an adventure. One of these scientists, Blaschke, helped form the original team and continue the adventure. Every year since 1980, there have been dinosaur footprint student-based activities in the field.

The first few years were spent digging and exploring, but in view of the destruction caused by weathering, the cultural department of the government of La Rioja decided to stop new digs if they had no plans to conserve what they had unearthed. In the beginning, people came for the fieldwork; they were interested in nature, in living with others, and in the dinosaurs. Gradually the approach became more structured, and now there are five work camps of 15 days each. The total number of students involved is 125 per year. Three of these camps involve intensive activity, with 8 hours of work per day. They include a summer course on paleoichnology of dinosaurs and site conservation. The other two camps involve more free time, with students working only 4 hours a day on the sites.

The students learn to work with footprints and apply the conservation techniques used in La Rioja. Thanks to them, large areas that would otherwise have been destroyed are now preserved. They have worked on the sites of Las Navillas, Era del Peladillo, Barranco de Valdecevillo, La Senoba, Las Losas, Peñaportillo, and La Pellejera. One of the aims of the summer courses and series of conferences is to spread information about the sites in La Rioja and to teach our study and work techniques. Students who attend the intensive courses must be attending a university. Their subject is immaterial, but the attendees must be receptive to what we are trying to convey. The intensive courses are official La Rioja University courses, with the recognition of a course of 110 hours.

There is also a series of conferences in which many well-known paleoichnologists have participated. The issues addressed are all directly or indirectly related to the dinosaur tracks in La Rioja. They may be related to dinosaur paleontology, environmental or natural heritage matters, or educational ones applicable to La Rioja or, of course, dinosaur ichnology. Speakers share their knowledge with the attendees; speakers and attendees can also view the tracks on the ground and learn the methods we use to describe, identify, and classify them. The outside specialists provide us with an external control system for our work.

A

LAS MORTAJERAS (LM)

1m

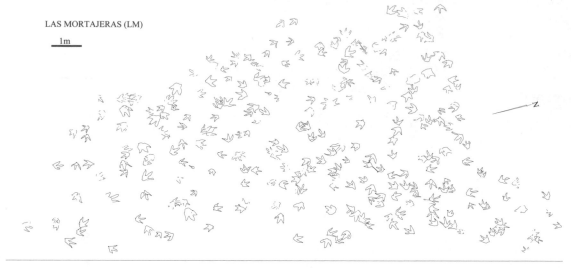

B

C

Summary

5

AS IN OTHER PARTS OF THE WORLD, FOOTPRINTS OF THE SAME TYPE (theropod, sauropod, or ornithopod) dominate in small outcrops. They were probably made because in some cases, dinosaurs of similar taxa are dominant or exclusive in areas of definite habitats. Formerly it was said that the different dinosaurs were separated in space and time, I suppose because large outcrops were unknown – so much so that sometimes the same ichnotaxon was applied to all footprints found in the same place. It was assumed, therefore, that the habitat, the age of rocks, and the habits of dinosaurs were sufficient conditions for the dinosaurs to be in widely dispersed groups. In La Rioja we can see that as the size of the site increases, the ichnogroups present also increase (Casanovas et al., 1999).

Because theropod footprints are the most abundant, so are outcrops with theropod footprints. If all theropod footprints are of carnivorous dinosaurs, in La Rioja there is a problem of interpretation. Were there more carnivorous than plant-eating dinosaurs? Several hypotheses have been put forward on this topic. In the Igea Paleontology Center, the deposited dinosaur bones are identified as *Baryonyx* (Viera and Torres, 1995a). In that same center, the authors offer the theory that many of the theropod dinosaurs in La Rioja may have been piscivores (*Baryonyx*). This would justify, they say, the presence of large numbers of theropod footprints and the scarcity of herbivorous dinosaurs. However, the overabundance of theropod footprints can also be produced by the greater activity levels on the part of carnivores than herbivores (Coombs, 1980; Leonardi, 1984; Farlow, 1987).

In La Rioja there are many large outcrops with footprints of different ichnogroups (Table 5.1). In these outcrops the conclusion is reached that dinosaurs of different ichnotaxa occupied different sectors. For example, in the same bed and location at Era del Peladillo, more than five sets of footprints have been found separated by sectors and ichnogroups. The physical conditions of Era del Peladillo also changed over time, where the stratum was able to make molds of footprints: in 1PL there are shallow theropod footprints in the lower part, together with footprints of ornithopods walking on ground of different resistance; in 3–4PL, there was soft sediment, where the feet of the ornithopods and sauropods sank deeply into the mud. In 3PL, after the passage of the sauropods, there were synsedimentary movements making parts of the site slip, erasing the tracks they contained. After this, small theropod dinosaurs passed over, leaving footprints with collapse structures as the feet sank into the mud (incisions and grooves on the inside of the digit marks). In 5PL, sauropod

Concentration of Footprints

5.1. Las Mortajeras (LM) site. (A) Chaotic distribution of footprints. (B) Arrangement of paces. (C) More orderly arrangement of trackways (midlines).

Table 5.1. Number and percentage of La Rioja site 156 tracksites containing various trackways.

Trackway	n (%)
Theropod	126 (81)
Ornithopod	71 (45)
Sauropod	32 (20)
Unidentified	38

footprints are also deep with high extrusion rims, while ornithopod and theropod footprints are shallow. It is not known exactly when different dinosaurs passed by, nor is it known what type of different dinosaurs passed in each stage (time interval); apparently different dinosaur groups did not coincide in either space or time.

To separate different trackmaker groups, the study surfaces are analyzed. In the same outcrop, the study surface can be the tracking surface for some of the footprints, but not all (stamps and undertracks). The tracking surface conditions may be different according to the different environmental conditions prevailing during the times of passage. Dinosaurs passed over the same surface at different times when the physical conditions of the mud were different (shallow or deep footprints; La Pellejera) or where the water depth was greater or less (traces of swimming and running on the same layer; El Villar-Poyales). Sometimes there are sedimentary movements, induced by changes in the slope of the basin or by earthquakes (Era del Peladillo, La Virgen del Campo), between the time when one group of dinosaurs passed and another passed by on the same tracking surface. It is not strange to see trackways of large theropod footprints crossing those of ornithopods, without any consequences (VA11/VA12: ornithopod/theropod).

The tracking surface conditions may be different according to the particular location of the surface (e.g., relative to the shoreline of the lake) in the same period of passage. Certain types of footprints are probably found more in certain states of mud. For example, many sauropod footprints are on sites with deep marks (Las Navillas, 2–6PL, 2ST center), but there are many other places where the footprints are shallow (7PL, Barranco de Valdecevillo, 2ST side). Nevertheless, it seems that where there is a concentration of sauropod footprints, the footprints are usually deep. Something similar can be said for ornithopod footprints.

Deep theropod footprints may be closed by the collapse of mud walls, such that many of these dinosaur footprints may have deep structures masked by falling mud and thus have gone unnoticed. It cannot be said that shallow theropod footprints outnumber deep ones in La Rioja. Deep theropod footprints are found at La Virgen del Prado, 2–3PL, Los Cayos, El Villar-Poyales, La Senoba, La Virgen del Campo, Las Losas, Las Mortajeras, La Barguilla, Santisol, and probably at other sites. Shallow theropod footprints are found in Mina Victoria, Las Navillas, 1PL, 5–7PL, 1TT, Navalsaz, EVP–Icnitas 3, Barranco de Valdecevillo, East Valdecevillo, La Senoba, La Virgen del Campo, Peñaportillo, Munilla, Hornillos, La Pellejera, 1ST, 3ST, and San Vicente de Robres. This proportion does not favor the cautious dinosaur hypothesis, which suggests that theropods are disadvantaged in soft mud.

Three tendencies can be distinguished: (1) groups of sauropods and ornithopods with many individuals leave deep footprints; (2) no shallow groups of footprints have been seen for these types; and (3) in heterogeneous (size) groups of ornithopods, and in isolated sauropod and ornithopod trackways, in general, the footprints are shallow. Moreover, theropod footprints do not show any definite trend.

It might be thought that the mud layers harden over time, but this is not the case. The situation is more complex. It can be seen in many outcrops that undertracks and subtrack prints are deeper than stamps, i.e., footprints made on the tracking surface (stamps) are shallower than subtracks, undertracks, and axial downfolds made several levels above. This is observed at several outcrops. For example, in the same area of 1LVC there are shallow stamps (LVC140, less than 3 cm deep) with pad and nail marks along with undertracks of about 15 cm depth appearing beneath. A trackway (LVC140) with large footprints has been chosen for comparison with deep undertracks. The undertracks, of a similar size to the footprints in LVC140, were left by footfalls made on layers 20 cm above them. The dimensions of the footprints do not determine the depth because there are large shallow footprints (stamps) and large deep footprints (undertracks). The stamps and undertracks are from theropod footprints.

The same occurs at La Pellejera, where stamps on the study surface sink less than 5 cm, while undertracks and subtracks sink up to 20 cm. Appearing in the recesses of the subtracks are up to four sedimentary levels (two of sandstone and two of sandy shale) squashed under pressure in the sediment. The behavior of the levels varies. Upper sandstone levels are fragile and break, while the sandstone level below bends. The viscosity of the layers and sedimentary levels, which should seemingly increase over time, can also decrease. There may be components that harden (dried algal mats? other agents?) or make the ground surface more resistant while in contact with water, which then disappear and the sediment loses its firmness.

The oldest footprints are no deeper than the undertracks that appear subsequently in the same layer in several of the outcrops of La Rioja. We cannot know what type of dinosaur exerts more pressure on the ground under its feet, which makes the previous effect paradoxical: more recent tracks should be less deep. One would also expect that the thicker the layer of sediments on the study surface, the shallower the undertracks. Stamps of the same size as undertracks can be chosen from all ichnotypes for comparison, and on the same study surface there can be undertracks that are much deeper than stamps.

Isolated trackways are the exception at large sites, which usually have several trackways and/or several groups of footprints of the same ichnotype. Trackways of the same ichnotype almost always follow parallel paths. In small outcrops, it is easy to see that there is only one trackway with one type of footprint (Treguajantes, Valdenocerillo, La Cuesta de Andorra, Barranco de la Sierra del Palo). In some long sites, only one

Mud Condition Over Time in the Sites

Isolated Individuals, Groups, and Herds

trackway of an ichnotype may also be seen (VA13). However, in the same site, there could be parallel trackways hidden under the adjacent rock layers or missing as a result of erosion.

In the La Rioja sites there is only one long trackway that we can safely say was made by an individual of a particular size and that has no other similar trackways made by other dinosaurs parallel to it. This is 1LP17, which is 30 m long and would contain 40 footprints if there were no intermediate ones missing. The other La Pellejera footprints of similar caliber, features, and tracking surface have no apparent relationship to it.

One example of the precautions that have to be taken is demonstrated by the case of the three parallel theropod trackways 1PL8, 1PL9, and 1PL10. The distance between them ranges from 3 to 7 m. Any small outcrop (6 to 14 m in width) containing part of one of these trackways would suggest that the area was traversed only by a single theropod dinosaur.

Groups of only a few individuals have been found that are moving in the same direction and have common features. In the case cited above (1PL8, 1PL9, 1PL10) the average length of the footprint is between 26 and 30 cm and the width between 24 and 25 cm, with clear pad and claw marks. To the east, beside 1PL10, the tracking surface has no outcrop, so it is possible that more dinosaurs left parallel tracks. The footprints are not deep, on the order of a centimeter, and are similar for all footprints. The simplest interpretation would be that three dinosaurs passed by at the same time and they were part of a group of a few individuals. The dinosaurs are large (the average height calculated for the acetabulum is 175 cm) and have been interpreted as a family group or a hunting group (Pérez-Lorente et al., 2001; García-Ortiz and Díaz-Martínez, 2008). The presence of both trackways 1LP6 and 1LP7, with similar but somewhat larger footprints (l = 34 to 36 cm) and an oblique direction, introduces uncertainty to the previous interpretation. It is not known whether the five trackways are related, and if they were, what reason there might be. Such long, parallel trackways with so few individuals have not been detected before in La Rioja.

Another example of a few parallel trackways going in one direction and with footprints of the same type is the group of three ornithopod dinosaurs in the Barranco de Valdecevillo, VA9, VA10, and VA11. These are two relatively large adjacent ornithopod trackways with a smaller central one (l = 33.7, 22.5, and 32.5 cm) that are parallel and cross the stratum. There are no similar footprints throughout the rest of the tracksite. They have been interpreted as a family group of two adults and a baby dinosaur.

There is a pairing of two large theropod dinosaurs PP144 and PP145 (l = 30 and 42 cm) and three smaller ones accompanying them, PP147, PP148, and PP149 (l = 13 to 16 cm). Here the outcrop is small; a little more space to the west is needed to see if these five trackways are accompanied by any others. Peñaportillo was uncovered in 1991–1992, but the rock was so hard that it was difficult to continue with the excavation. Pérez-Lorente et al. (2001) suggested that it may be a family group of theropods.

It is seen that in all the sites, the trackways are pointing in certain preferred directions (Martín Escorza, 1986, 1988, 2001; Moratalla and Hernán, 2010). If the trackways are grouped by orientation intervals or if diagrams are drawn of their paths, then there are always some directions taken more than others. The more trackways there are, the more peaks are found and the more difficult it is to find individualized predominant directions. If some trackways change their direction on the site (e.g., at Las Losas), one or both trajectories (initial and final part) will coincide with one of the existing ones already taken. It may be, therefore, that the number of coincident paths the trackways take is dependent on their number on the site. Seemingly chaotic sites seem much more orderly when the trackways are examined (Fig. 5.1). The matching directions of the trackways may simply be an apparent effect, especially in certain outcrops.

Only if tracks have equivalent features and are very different from the rest can we talk about herd behavior (first case) or of conduct regulated by some external agent or natural barrier (second case). This is the cases for the Barranco de Valdebrajés sites and the small theropod footprints in Era del Peladillo 3 (3PL).

In the first case, eight or nine small dinosaurs left their footprints on a small outcrop. There may have been more footprints on the sides of the outcrop. Whatever type of dinosaur left them, there are many individuals with similar features unlike any others found in La Rioja. The concentration and the parallelism of the footprints are an important part of making the site exclusive. Several important questions remain unanswered regarding the Barranco de Valdebrajés site. It is not clear whether the trackways are of theropod or ornithopod dinosaurs, or even if they are of adults or their young. They are too large to be hatchlings. In this case, the herd behavior proposed by Aguirrezabala, Torres, and Viera (1985) seems appropriate. It is not clear whether it is a herd or a brood as a result of the size of the footprints, which suggest the acetabulum was 39 to 48 cm in height. It must be assumed that these animals formed part of a gregarious group of small dinosaurs. The footprints do not give any indication of whether the dinosaurs were young or small mature animals. There are no other similar footprints to compare with them in La Rioja.

In the second case, eight trackways (3PL11 to 3PL18) have the same orientation but go in opposite directions in a narrow strip of land. The footprints in the trackways are completely different from others in the environment. The strip of land is a kind of channel that occupies a depression in the ground formed by a slip in the top of the layer. This slip took place after the passage of the graviportal dinosaurs (sauropods and ornithopods) that left deep footprints. The eight trackways were probably made by the same theropod dinosaurs that walked in the channel, first in one direction and then the other.

There are sites in which trackways of the same ichnotype show no preferred direction. More than three trackways without preferential orientation and isolated footprints of the same ichnotype can be distinguished in them. There are many theropod footprints and relatively few ornithopod ones. In groups of theropod and ornithopod footprints, it is not uncommon to find footprints of varying size (theropod: 3I; ornithopod: 5PL, Navalsaz) if they are the same ichnotype. Sauropod groups are often in large outcrops.

There are outcrops where footprints are superimposed, leaving a chaotic mixture of ichnites. In such places, it is difficult to distinguish trackways. In La Rioja there are four possible chaotic groups of sauropod footprints (2–6PL, 2ST, 1SM, 2TVR) and one of ornithopod footprints (1PL). In the two large sauropod groups (2–6PL, 2TVR) the direction of movement of the dinosaurs is not known, while in the two small outcrops it is a south–southeast direction. In the ornithopod group (1PL), the trackmakers traveled in a southerly direction. This group is flanked by two parallel trackways of two separate individuals of the same type: one at 5 m from the herd (1PL12) and the other at 20 m (2PL173). All the four groups are interpreted as the traces of four herds of dinosaurs.

No data can be provided on the behavior of the sauropod dinosaurs in other outcrops. There are 11 more sites with sauropod footprints in isolated trackways and one, Las Navillas, with separate, seemingly unrelated individuals. In these places, the surface with footprints is so small that assumptions about the behavior of sauropods in the region cannot easily be made. The 11 tracksites with a sauropod trackway or a linear accumulation of sauropod footprints are La Virgen del Prado, 7PL, LCS, Barranco de Valdecevillo, El Sobaquillo (previously mentioned) and Camino de Soria (Casanovas et al., 1995b), La Cela A and La Cela C (Casanovas et al., 1995g), San Vicente de Robres (Casanovas et al., 1990b; Díaz, Pina, and Ponce, 1990), Valdemurillo (Moratalla, 2009a; Moratalla, Sanz, and Jiménez, 1993), and La Moga (Caro, Pérez-Lorente et al., 2002).

In the first studies performed at La Rioja, it was assumed that many of the rounded undertracks were of sauropod dinosaurs, the result of their abundance and their chaotic patterns. This interpretation is incorrect because there are many theropod and ornithopod footprints with rounded undertracks.

Interdigital Webs

In 1PL (*Hadrosaurichnoides igeensis*) and EVP (*Theroplantigrada encisensis* and EVP18) footprints with interdigital webs have been described. However, they cannot be interpreted as deformations induced by the digits, even bearing in mind the conclusions of the works of Manning (2004) and Falkingham et al. (2009) on the deformation of the interdigital spaces of the footprints. The first location has more than 200 ornithopod footprints (digits with a pad; heel also with a pad that is broad and rounded; relatively wide feet). Almost all the footprints have the front part of the contour line – that is, the part running along the front of the digits – forming an unbroken arc in the interdigital spaces. The edges of the shafts are abrupt. The depth of the base of the interdigital spaces in the footprint is less than for the rest of the footprint; however, it is still below the ground level surrounding the footprints. The mud sinks into the sole of the pads and the interdigital spaces. Fine wrinkles run lengthwise along these spaces. Several different reasons have been offered for this phenomenon. If the digits did in fact have an interdigital web, the mud sinks between them, dragged by the webs, leaving wrinkles as a result of the skin of the web itself.

If the digits did not have interdigital webs, the lateral sides of the digits join together at least while stepping, probably as a result of the expansion of the pads. The wrinkles may come from the same skin or may be the folded sedimentary laminites formed by the movement to expel the mud as a result of the pressure of the digits on the ground. The mud is not flexible because the edges of the footprints are so steep. This interpretation conflicts with observations of the continuous nature of the edge of the footprint at the tips of the digits—that is, there are no indentations along the boundary line between the toes. The digits have wide claws that mark the base near the distal ends of the footprints.

The second tracksite, El Villar-Poyales, has a wide range of footprint types. There are shallow, well-marked footprints showing interdigital pads. There are claw marks of at least one dinosaur floating in the water that intersect with the above trackways, and finally footprints with metatarsal marks and possible interdigital webs. The footprints with these marks are semiplantigrade, with the impression of four digits and the metatarsus. Between the four digits, there is a depression linking them. This depression does not link the hallux with the rear end of the metatarsus, as the metatarsus—digit I distance is less than that for digit I—digit II.

It has been proposed that an interdigital web linked the four digits because the lack of a link between digit I and the rear metatarsal part is not demonstrated, despite the strangeness of the proposal. There are no criteria for suggesting that the physical conditions of the sediment (flexibility) or a vegetation layer may be the cause of the structure. Nor is it a structure induced during the footstep; there are no indications of forward movement in the footprints. For these forms to have been produced by a thixotropic flow of mudflow between the four digits, the movement of the foot had to have been such that this fluidization affected only the front part of the footprint.

EVP18 is an isolated footprint. No generalizations can be made about it.

Semiplantigrade Footprints

The first semiplantigrade footprints in La Rioja were discovered at Barranco de la Sierra del Palo tracksite (Brancas, Martínez, and Blaschke, 1979), followed by those in El Villar-Poyales. In the intervening period, footprints with metatarsal marks were described in the neighboring province of Soria (Aguirrezabala and Viera, 1980). W. A. S. Sarjeant convinced us to use the term *semiplantigrade* when the dinosaurs did not use a supported tarsus when walking.

At El Villar-Poyales there are two types of footprints with metatarsal marks. EVP1 has a rounded rear end, and the boundary line enclosing it is approximately parallel to the axis of the metatarsal mark, while EVP4 has a sharp rear end with a V-shaped outline. In the first case, Pérez-Lorente (2001a) proposed that the dinosaur supported the metatarsus flat on the ground, while in the second the metatarsus was inclined and penetrated the mud. I attribute the sharp end to the mark of the flexor tendon of the digits.

At the same site, the angle between the axis of the metatarsus and the acropod is noteworthy. This angle is a recurring phenomenon of semiplantigrade footprints that warrants further study. The implications for the form of walking are important because the pigeon-toed walk (or negative orientation of the feet) in dinosaurs may be related to the metatarsal joint movement.

The nonparasagittal position of the metatarsus with respect to the foot is also responsible for the structures of some footprints. If the dinosaur's foot sinks into the mud after sliding forward in the T phase, the metatarsus does not follow the axis of the acropod but moves parallel to the midline and toward the outer digit (digit IV). However, in the next phase, the W phase, the animal lifts the metatarsus when advancing, and this moves toward the axis of the acropod so it is positioned vertical to the first phalanx of digit III.

The speed of dinosaurs when walking semiplantigrade has been estimated, and as Kuban (1989) says, the result is abnormally high. The calculations were performed using two alternatives: with the theoretical height of the limb and with this number subtracting the metatarsal length. Because the speed depends on the length of the stride, the explanation must be that the dinosaur lengthens its stride when adopting this form of locomotion.

At Las Losas there are many semiplantigrade trackways as well as some trackways that show both semiplantigrade and digitigrade ways of walking. This site showed that at least some dinosaurs had the option of using either method. However, the length of the trackways is too short to see if this change of walking mode changes the speed. At Las Losas semiplantigrade trackways coincide more with areas of irregular footprints, the result of the softer mud in these sectors, than in the rest of the site. Here it seems that this method of semiplantigrade walking is related to the accumulation of softer mud.

At the Las Losas site there are many footprints where the hallux points forward. Thulborn (1990) said that the hallux mark of ornithopod dinosaurs points forward and that of theropods points backward. At Las Losas, the large theropod footprints, with digits apart, sharp claw marks, and several digit pads, also have the mark of the hallux forward.

La Rioja Bonanza

Bonanza is a mining term used when finding a rich vein of ore. The term has also been used by Lockley to refer to places with an abundance of

5.2. Two semiplantigrade trackways discovered during construction of dam near Enciso. Left of trackways is represented by three fragmentary footprints. Note hole in position of hallux mark in more complete right trackway.

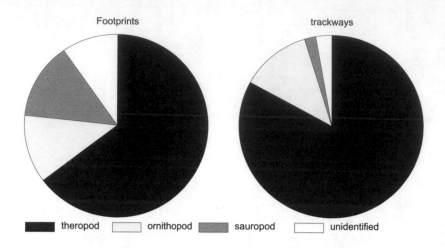

5.3. Average of La Rioja footprints and trackways (2003 data). Black, theropods; light gray, ornithopods; dark gray, sauropods; white, unidentified.

Footprints   trackways

theropod   ornithopod   sauropod   unidentified

footprints, and I think it could also be applied to La Rioja. The bonanza occurs because there are many levels with tracks, because the surfaces of the strata stand up well to erosion, because of the alternation of easily eroded and resistant beds, and because the strike and dip of the strata are appropriate.

There are countless levels with footprints in the Enciso Group. Almost everywhere where there is a surface outcrop with footprints and the stratigraphic profile is appropriate, various superimposed levels are recognized. Similar superimposed levels with footprints are also found in the excavations for public works or mining, which involve mass removal of rock (Caro, 1998; Melero and Pérez-Lorente, 2011). Many footprints are destroyed in these operations because the excavation process is done on many fronts with fine fragmentation of the rock. Examples of superimposition include El Villar-Poyales with seven levels, La Virgen del Campo with four or five levels, Los Cayos (LCA, LCB, LCC), where footprint levels are also recognized within the strata, the Navajún mine, and surfaces with footprints that have emerged in the construction of the Enciso dam wall (Fig. 5.2).

The different resistance of the layers to erosion and their behavior during deformation have led to beds with ichnites being well preserved. In the Enciso Group resistance to erosion between shale and sand/limestone levels is different. Many ravines and small valleys are oriented in the same way as the layers as a result of shale eroding more rapidly than sandstone and limestone. In the Urbión Group, river courses are not so marked in that direction because there is almost no limestone, and all other rocks are weathered with the same intensity. Concerning deformation, the rocks in the Urbión Group have more cracks than those of the Enciso Group. It is likely that in the Enciso Group shale is better at absorbing the deformation like plastic levels, and therefore the limestone and sandstone do not fracture so much. The fact is that outcropping stratification surfaces are better conserved. Finally, the geometry of the synclinal structure is such that dip of the layers is close to the slope of the hillside.

Table 5.2. Percentage of Cretaceous theropod, ornithopod, sauropod and unidentified footprints in La Rioja.

Type of footprint	Average of footprints and trackways according to:						
	Moratalla, and Hernán (2010)	Casanovas et al. (1999)		La Rioja (Pérez-Lorente, 2003b)		Spain (Pérez-Lorente, 2003b)	
	Footprints %	Footprints %	Trackways %	Footprints%	Trackways, %	Footprints, %	Trackways, %
Theropod	75	37	65	64	83	54	79
Ornithopod	16	25	19	12	12	11	14
Sauropod	8	25	4	14	2	27	5
Unidentified		13	12	10	3	8	2

Abundance

In 2003 a count was made of Cretaceous footprints in Spain (Pérez-Lorente, 2003b). As expected, in La Rioja the number of theropod dinosaur tracks is much greater than all other footprints (Fig. 5.3A), and the same goes for the trackways (Table 5.2). On the absolute or relative abundance of footprints, Farlow (1987) says that the abundance of a type of dinosaur footprint does not necessarily reflect the relative abundance of the trackmakers. Theoretically, abundance depends on the habitat and secondly on the animals' activity (sedentary animals leave fewer footprints).

The number of footprints cannot be used to determine the number of each type of dinosaur. If it is tried with trackways (Fig. 5.3B), it can be seen that there is no correlation with the number of footprints for each ichnotype. The footprints of theropods and ornithopods do not usually overlap as much as those of sauropods, such that in herds or groups of sauropods trackways cannot be separated; there are many sauropod footprints and few trackways. This would confirm the gregarious nature of these dinosaurs were it not for the number of sites with isolated sauropod dinosaur ichnites that exist.

One possibility for analyzing the frequency of distribution is to find out the number of sites occupied by footprints of each type. Table 5.1 separates surfaces with different footprints, including those within the same sites. A site can have various tracking surfaces (the various tops of strata with footprints), which were considered independent when making this table. Footprints on the same tracking surface are not considered independent even if extensive.

Theropod footprints (Tables 5.1, 5.2) are present at almost all sites, so the habitat of La Rioja was favorable for this type of animal. The least represented group are sauropods; sauropod footprints are present at 20% of sites, although the number of individuals may have been similar to ornithopods. The lack of sauropod trackways and low occupancy suggests either herd behavior or more restricted habitats for these dinosaurs.

Size

To date there are no dinosaur footprints of less than 8 cm that have been discovered in La Rioja. The largest for biped dinosaurs is 80 cm (ornithopods), and for quadrupeds it is 76 cm (sauropods) (Casanovas et al., 1993c, 1997a; García-Ortiz et al., 2009). Biomass studies in relation to the habitat area of the dinosaurs cannot be done (Farlow, 1993), at least for large theropods (TR1–4), because they are in the Oncala Group, which is poorly represented in La Rioja. For calculations in the Enciso Group, those dinosaurs that are piscivorous and those that are carnivorous would need to be considered, as would the production of biomass, taking into account the periods when water covered everything and when water salinity restricted living conditions in the lakes.

Theropod footprints show an almost continuous sequence of sizes. There are only two discontinuities: for small footprints (Pérez-Lorente, 1996a) and for the largest. The largest theropod footprints are smaller than 55 cm, except those in the Camino a Treguajantes and in Las Navillas. There are many sites in La Rioja with theropod footprints between 50 and 55 cm. There are no digitigrade theropod footprints or those without extramorphological deformation measuring between 55 and 70 cm in length. In all the outcrops of La Rioja there is only one point with footprints of one or two giant individuals (Camino a Treguajantes). These footprints also have relatively wide digits, unlike most of the large theropod footprints, the digits of which are narrower. A foot with these features should mean that the animal has less ability to run, and therefore it indicates that it probably had different eating habits.

Large ornithopod footprints do not show such a numerical jump (55 to 70 cm). Although there is a footprint measuring 75 cm in Navalsaz, the average length obtained from all footprints in the trackway is 63 cm (Casanovas et al., 1993c). There are also two large casts, one at the site of the Sol de la Pita (Moratalla, Sanz, and Jiménez, 1997b) and the other, 63 cm long, deposited in the Centro Paleontológico de Enciso (Pérez-Lorente, 2000c). As with theropod footprints, the length of ornithopod footprints is grouped into three modal sizes separated by two minima. There are two gaps in the frequency distribution of footprint lengths: between the large and very large, and among the small. There is a marked decrease in the number of large ornithopod trackways with a length of more than 54 cm and in smaller ones with length of less than 28 cm.

The largest sauropod footprints are those of El Sobaquillo, whose average length is 76 cm. It is estimated that the acetabulum of this dinosaur would have been 320 cm above the ground. The smaller ones are in San Martín 1 (1SM) measuring between 24 and 34 cm in length. These two sites have shallow footprints. The length of the feet is probably close to that of the footprints. No statistical or comparative study has yet been done for the remaining sites because the amount of information concerning the length of sauropod stamps and sole feet marks is low.

Giant biped dinosaurs seem to be either isolated individuals or in small groups (Camino a Treguajantes, Navalsaz). It must be remembered that these tracks are in small sites, so the observations are partial. However, it is clear that the stamps of giant theropod dinosaurs are only found at one site, which is significant for the estimation of the number of animals of this type in the Cameros Basin. Also, these footprints are not found in the Aptian or Albian areas but in the lowermost Cretaceous (Berriasian), distinct from most sites in La Rioja (Enciso Group). Something similar happens with the largest ornithopod footprints. One of the footprints (Sol de la Pita) is of Berriasian? age (Oncala Group?), separate from the sites of the Barremian–lower Albian, which are rich in ornithopod footprints. Large ornithopod footprints have also been found in the latter, but only at the site of Navalsaz are there footprints comparable in size to the footprint found in Sol de la Pita.

The top of the beds containing footprints is highly sensitive to the processes that occur in it. Footprints may be recorded only in the upper part of a sediment that is being formed, or they may pass through and penetrate below it. At the same time rims, are formed as a result of the extrusion of mud, which in Era del Peladillo exceed 50 cm above surface level. The sedimentation conditions prevailing during the stages (before the formation of the first footprints; those that occurred between the passing of two distinct groups of dinosaurs; and later ones) are clearly seen in many tracksites. The geological processes sometimes include erosion of already deposited sediments, from partial footprint erosion to the total disappearance of some layers.

The difference in depth between some footprints and others, the regularity and irregularity of the prints, and the difference in gait (plantigrade or digitigrade) along the same trackway have helped identify the flooded and drier parts at Las Losas (Romero-Molina et al., 2003). In the outcrop of La Virgen del Campo, one can also distinguish the variation of sediment strength during the passage of the dinosaurs. In La Rioja the continuity and persistence of the conditions can be deduced because there are strata (extensive and narrow) with footprints. The enormous extension of shallow water and calm conditions are deduced from the lateral continuity of carbonate levels (several of them checked over more than 15 km of outcrops with a constant thickness and no lateral change).

On the top of the layers with footprints in the Cameros Basin there are also indications of tectonic activity, mainly synsedimentary faults, slump structures, and mudflow, the relative age and bathymetry of which can be checked with great accuracy as a result of the passing of dinosaurs beforehand (Pérez-Lorente, Fernández, and Uruñuela, 1986), and even the passing of dinosaurs before and after the activity (Meléndez and Pérez-Lorente, 1996).

In order to carry out certain studies, there must be a large number of footprints and large outcrops in addition to the footprints being well conserved. The reliability of results depends on the amount of data available to consider, which is directly related to the conservation status, the number and variety of footprints, and the extension of the outcrops. As new footprints are found and identified, the fossil record improves and our understanding of the evolution of a group of animals that became extinct at the end of the Cretaceous advances. One of the problems with the fossil population is the number of animals in different regions of the planet and the proportion of different types. Perhaps the most important future contribution to the history of life that can be obtained from the La Rioja dinosaur footprints is the behavior revealed by their trackmakers (Moratalla, Sanz, and Jiménez, 1990; Pérez-Lorente et al., 2001) or, in general, the behavioral patterns of an important number of terrestrial vertebrates during the Cretaceous.

References

R

Abel, O. 1935. Vorzeitliche Lebenspuren. Gustav Fischer, Jena, Germany.

Aguirre, J., E. Aramburu, E. Campos, M. Gallego, A. Garritz, I. Gómez Cubillo, R. Gómez Díez, A. González Díaz, C. González Ferrero, P. González Martín, C. Herreros, M. Izco, I. Juaniz, S. López, C. Marca, M. Marín, J. Monreal, M. Ochoa, E. Osés, A. Ramos, A. Rodríguez, F. Sánchez, C. Suescun, L. Vega, and I. Viteri. 2001. Las anomalias de una rastrillada terópoda (6bLT16). Afloramiento de La Torre 6, Igea (La Rioja, España). Zubía 19:97–113.

Aguirrezabala, L. R., and L. J. Viera. 1980. Icnitas de dinosaurios en Bretún (Soria). Munibe 3–4:257–279.

Aguirrezabala, L. R., J. A. Torres, and L. J. Viera. 1985. El weald de Igea (Cameros–La Rioja). Sedimentologia, bioestratigrafia y paleoicnología de grandes reptiles (Dinosaurios). Munibe 37:111–118.

Alexander, R. M. 1976. Estimates of speed of dinosaurs. Nature 261:129–130.

Alexander, R. M. 1989. Dynamics of Dinosaurs and Other Extinct Giants. Columbia University Press, New York, New York.

Allen, J. R. L. 1997. Subfossil mammalian tracks (Flandrian) in the Severn Estuary, SW Britain: mechanics of formation, preservations and distribution. Philosophical Transactions of the Royal Society of London B 352:481–518.

Alonso, A., and R. Mas. 1993. Control tectónico e influencia del eustatismo en la sedimentación del Cretácico Inferior de la Cuenca de los Cameros, España. Cuadernos de Geología Ibérica 17:285–310.

Alonso, A., A. Meléndez, and R. Mas. 1986–1987. Los arrecifes coralinos del Malm en la Sierra de los Cameros (La Rioja, España). Acta Geológica Hispánica 21–22:296–306.

Alvarez, S., E. Aragón, S. Azofra, J. Barrau, M. P. Benzal, S. J. Berger, J. M. Camarón, I. Catalán, A. Coda, L. Clome, E. Gallego, A. Garrido, M. Gonzalo, A. Hernáez, P, Del Hoyo, S. López, N. Lorente, F. J. Madariga, M. E. Marín, D. Martínez, J. Menéndez, A. Moneo, J. Pérez, A. Rodríguez, V. Sáenz, M. E. Sáenz, M. Sáenz, I. Torrontegui, M. Yanguas, and C. Zorzano. 2000. El yacimiento de huellas de dinosaurios de Hornillos de Cameros (La Rioja, España). Zubía 18:73–95.

Ansorena, P., I. Díaz-Martínez, and F. Pérez-Lorente. 2007–2008. Mina Victoria (Navajún) y Valdeperillo (Cornago). Nuevos yacimientos de icnitas de dinosaurio en el Grupo de Urbión (Cuenca de Cameros, La Rioja, España). Zubía 25 26.

Antunes, M. T. 1976. Dinosaurios eocretácicos de Lagosteiros. Ciencias da Terra 1:1 35.

Avanzini, M. 1998. Anatomy of a footprint: bioturbation as a key to understanding dinosaur walk dynamics. Ichnos 6:129–139.

Barale, G., and L. I. Viera. 1989. *Tempskia riojana* nov. sp. fougere arborescente du Cretace Inferieur du nord-ouest de l'Espagne. Palaeontographica B 212:103–122.

Barale, G., and L. I. Viera. 1991. Description d'une nouvelle paleoflore dans le Cretace Inferieur du Nord de l'Espagne. Munibe 43:21–35.

Barco, J. L., J. I. Canudo, J. I. Ruiz-Omeñaca, F. Pérez-Lorente, and J. L. Rubio. 2004. Ichnological evidence of the presence of gigantic theropods in the Berriasian (Lower Cretaceous) of Spain; pp. 18–19 in Abstracts of the First International Congress on Ichnology (Ichnia 2004), Trelew, Argentina.

Bengoechea, A., L. A. Izquierdo, J. M. Martínez, J. L. Molinero, D. Montero, F. Torcida, and V. Urién. 1993. Icnitas de dinosaurio en el sureste de la provincia de Burgos. Boletín Geológico y Minero 140:243–258.

Bermúdez-Rochas, D. D., G. Delvene, and J. Hernán. 2008. Estudio preliminar del contenido paleontologico del Grupo de Urbión (Cretácico Inferior, Cuenca de Cameros, España): restos ictiologicos y malacologicos. Boletín Geológico y Minero 117:531–536.

Bird, R. T. 1941. A dinosaur walks into the museum. Natural History 47:75–81.

Bird, R. T. 1944. Did *Brontosaurus* ever walk on land? Natural History 53:60–67.

Blanco, M. I., S. Caro, F. Pérez-Lorente, E. Requeta, and M. Romero-Molina. 1999b. Paleo-ichnological sites in La Rioja: some examples in Enciso, Igea and Munilla; pp. 51–74 in The Geological and Paleontological Heritage of Central and Eastern Iberia (Iberian Range, Spain). Seminario de Paleontologia de Zaragoza, Zaragoza, Spain.

Blanco, M. I., S. Caro, F. Pérez-Lorente, E. Requeta, and M. Romero-Molina. 2000. El yacimiento de icnitas de dinosaurio del Cretácico Inferior de Las Losas (Enciso, La Rioja, España). Zubía 18:97–138.

Blanco, M. I., S. Caro, A. López, F. Pérez-Lorente, E. Requeta, and M. Romero-Molina. 1999a. The Las Losas paleoichnological site; p. 26 in Programme and Abstracts of the IV European Workshop of Vertebrate Paleontology, Albarracín (Teruel, Spain).

Boutakiout, M., L. Ladel, I. Díaz-Martínez, and F. Pérez-Lorente. 2009. Prospecciónes paleoicnológicas en el sinclinal de Iouaridène (Alto Atlas, Marruecos). 2: Parte oriental. Geogaceta 47:33–36.

Boutakiout, M., M. Hadri, J. Nouri, S. Caro, and F. Pérez-Lorente. 2006. The syngenetic structure suite of dinosaur footprints in finely laminates sandstones: site no. 1 of Bin el Ouidane (1BO; Central Atlas, Morocco). Ichnos 13:69–79.

Boutakiout, M., H. Hadri, J. Nouri, I. Díaz-Martínez, and F. Pérez-Lorente. 2008. Icnitas terópodas gigantes. Sinclinal de Iouaridene. Jurásico Superior. Marruecos; pp. 21–22 in XXIV Jornadas de la Sociedad española de paleontología. Libro de resúmenes, Oviedo, Spain.

Boutakiout, M., H. Hadri, J. Nouri, I. Díaz-Martínez, and F. Pérez-Lorente. 2009. Rastrilladas de icnitas terópodas gigantes del Jurásico Superior (Sinclinal de Iouaridene, Marruecos). Revista Española de Paleontología 24:31–46.

Brancas, R., J. Martínez, and J. Blaschke. 1979. Huellas de dinosaurio en Enciso. Gonzalo de Berceo, Logroño, Spain.

Brenner, P., and J. Wiedmann. 1974. Nuevas aportaciones al conocimiento del "Weald" celtibérico septentrional y sus relaciones paleogeográficas; pp. 125–134 in 1er symposium sobre el Cretácico de la Cordillera Ibérica, Cuenca, Spain.

Brown, T. 1999. The Science and Art of Tracking. Berkley Books, New York, New York.

Brusate, S. L. 2012. Dinosaur Paleobiology. John Wiley and Sons, Chichester.

Calzada, S. 1977. Un yacimiento Barremiense en Cameros (Logroño). Boletín de la Real Sociedad Española de Historia Natural Geología 75:35–38.

Cámara, A., and O. Durantez. 1982. Mapa geológico de España. Scale 1:50,000. Villoslada de Cameros. Instituto Geológico y Minero de España, Madrid, Spain.

Canudo, J. I., R. Royo-Torres, and G. Cuenca-Bescós. 2008. A new sauropod: Tastavinsaurus sanzi gen. et sp. nov. from the Early Cretaceous (Aptian) of Spain. Journal of Vertebrate Paleontology 28:712–731.

Canudo, J. I., J. L. Barco, X. Pereda-Suberbiola, J. I. Ruiz-Omeñaca, L. Salgado, F. Torcida, and J. M. Gasulla. 2009. What Iberian dinosaurs reveal about the bridge said to exist between Gondwana and Laurasia in the Early Cretaceous. Bulletin de la Société Géologique de France 180:1, 5–11.

Canudo, J. I., J. M. Gasca, M. Aurell, A. Badiola, H.-A. Blain, P. Cruzado-Caballero,D. Gómez-Fernández, M. Moreno-Azanza, J. Parrilla, R. Rabal-Garcés, and J. I. Ruiz-Omeñaca. 2010. La Cantalera: an exceptional window onto thevertebrate biodiversity of the Hauterivian–Barremian transition in the Iberian Peninsula. Journal of Iberian Geology 36:295–224.

Carvalho, I. S. 2001. Pegadas de dinosauros em depositos estuarinos (Cenomaniano) da Bacia de Sao Luis (MA), Brasil; pp. 245–264 in D. F. Rossetti, A. M. Goés and W. Truckenbrodt (eds.), O Cretáceo na Bacia de SãoLuís-Grajaú. Museu Paraense Emílio Goeldi, Coleção Friedrich Katzer, Belém, Pará, Brazil.

Caro, S. 1998. Supervisión paleoicnológica de los trabajos en la obra de ensanche y mejora de la carretera LR-356 de la LR-115 a la LR-123 por Rincón de Olivedo. Tramo Enciso-Navalsaz (La Rioja). Estrato 9:105–110.

Caro, S., and S. Pavía. 1998. Alteración y conservación de los yacimientos de huellas de dinosaurio de La Rioja. "La Virgen del Campo" (Enciso) y la "Era del Peladillo" (Igea). Zubía 16:199–232.

Caro, S., and F. Pérez-Lorente. 1997a. Concepto y valoración del patrimonio paleoicnológico de La Rioja. Zubía 15:35–38.

Caro, S., and F. Pérez-Lorente. 1997b. Definición de concepto y propuesta de valoración del patrimonio paleoicnológico (pisadas de dinosaurio) de La Rioja, España. Zubía 15:39–43.

Caro, S., and F. Pérez-Lorente. 1998. El patrimonio paleoicnológico de La Rioja. Propuesta de restauración; pp. 224–226 in IV Congreso Internacional de rehabilitación del Patrimonio Arquitectónico y Edificación, Havana, Cuba.

Caro, S., S. Pavía, and F. Pérez-Lorente. 1997. Los yacimientos con huellas de dinosaurio de La Rioja; pp. 15–18 in L. Pallí y J. Carreras (eds.), Patrimonio Geológico de España. Universidad de Gerona, Gerona, Spain.

Caro, S., S. Pavía, and F. Pérez-Lorente. 2002. La intervención en la conservación de las huellas de dinosaurio de La Rioja (España); p. 14 in F. Pérez-Lorente (ed.), Congreso internacional sobre dinosaurios y otros reptiles mesozoicos en España. Resúmenes, Logroño, Spain.

Caro, S., S. Pavía, and F. Pérez-Lorente. 2003. Intervenciones en la conservación de las huellas de dinosaurio de La Rioja (España); pp. 225–250 in F. Pérez-Lorente (ed.), Dinosaurios y otros reptiles mesozoicos de España. Ciencias de la Tierra 26. Instituto de Estudios Riojanos, Logroño, Spain.

Caro, S., S. Pavía, and L. E. Requeta. 2006. Alteración de la roca con huellas de dinosaurio y evaluacion de los productos para su recuperacion y conservación; pp. 217–242 in F. Torcida (ed.), Huellas que perduran. Fundación Patrimonio Histórico de Castilla y León, Valladolid, Spain.

Caro, S., F. Pérez-Lorente, and E. Requeta. 2002. Restauración de yacimientos paleoicnológicos en Villoslada de Cameros (La Rioja, España). Geogaceta 30:27–31.

Caro, S., C. Fuentes, M. Meijide, and F. Pérez-Lorente. 1995. Una rastrillada nueva de ornitópodo encontrada en el Barranco de la Muga (Soria–La Rioja, España); pp. 25–26 in F. Pérez-Lorente (ed.), Huellas fósiles de dinosaurios de La Rioja. Nuevos yacimientos. Ciencias Tierra 18. Instituto de Estudios Riojanos, Logroño, Spain.

Caro, S., S. Pavía, F. Pérez-Lorente, and M. Romero-Molina. 2000. Restauración y conservación de yacimientos de huellas de dinosaurio de La Rioja (España): método de trabajo; pp. 219–224 in Abstracts of the First International Congress on Quarry–Laboratory–Monument, Pavía, Italy.

Casamiquela, R. M., G. R. Demathieu, H. Haubold, G. Leonardi and W. A. S. Sarjeant. 1987. Glossary and Manual of Tetrapod Footprint Palaeoichnology. Departamento Nacional da Produçao Mineral, Brasilia, Brazil.

Casanovas, M. L., and J. V. Santafé. 1971. Icnitas de reptiles mesozoicos en la provincia de Logroño. Acta Geológica Hispánica 5:139–142.

Casanovas, M. L., and J. V. Santafé. 1974. Dos nuevos yacimientos de icnitas reptiles mesozoicos en la región de Arnedo. Acta Geológica Hispánica 3:88–91.

Casanovas, M. L., and J. V. Santafé. 1995. Icnitas de La Rioja. Investigación y Ciencia 6–12.

Casanovas, M. L., F. Pérez-Lorente, and J. V. Santafé. 1989. Huellas de dinosaurio en Valdenocerillo (Cornago, La Rioja, España). Zubía 7:29–35.

Casanovas, M. L., A. Fernández, F. Pérez-Lorente, and J. V. Santafé. 1989. Huellas de dinosaurio de La Rioja. Yacimientos de la Virgen del Campo, La Senoba y Valdecevillo, 1–190. Ciencias Tierra 12. Instituto de Estudios Riojanos, Logroño, Spain

Casanovas, M. L., A. Fernández, F. Pérez-Lorente, and J. V. Santafé. 1992. Icnitas de dinosaurios en Valdevajes (La Rioja). Nota de contrarréplica. Revista Española de Paleontología 7:97–99.

Casanovas, M. L., A. Fernández, F. Pérez-Lorente, and J. V. Santafé. 1995f. Un terópodo carnosaurio en el camino a Treguajantes (La Rioja, España); pp. 13–14 in F. Pérez-Lorente (ed.), Huellas fósiles de dinosaurios de La Rioja. Nuevos yacimientos. Ciencias Tierra 18. Instituto de Estudios Riojanos, Logroño, Spain.

Casanovas, M. L., A. Fernández, F. Pérez-Lorente, and J. V. Santafé. 1995g. Icnitas terópodas y saurópodas en La Cela, Muro en Cameros (La Rioja, España); pp. 17–24 in F. Pérez-Lorente (ed.), Huellas fósiles de dinosaurios de La Rioja. Nuevos yacimientos. Ciencias Tierra 18. Instituto de Estudios Riojanos, Logroño, Spain.

Casanovas, M. L., A. Fernández, F. Pérez-Lorente, and J. V. Santafé. 1995h. Icnitas de terópodos y saurópodos del yacimiento de Las Navillas (La Rioja, España); pp. 33–44 in in F. Pérez-Lorente (ed.), Huellas fósiles de dinosaurios de La Rioja. Nuevos yacimientos. Ciencias Tierra 18. Instituto de Estudios Riojanos, Logroño, Spain.

Casanovas, M. L., A. Fernández, F. Pérez-Lorente, and J. V. Santafé. 1995i. Dinosaurios terópodos del yacimiento de Munilla (La Rioja, España); pp. 53–57 in F. Pérez-Lorente (ed.), Huellas fósiles de dinosaurios de La Rioja. Nuevos yacimientos. Ciencias Tierra 18. Instituto de Estudios Riojanos, Logroño, Spain.

Casanovas, M. L., A. Fernández, F. Pérez-Lorente, and J. V. Santafé. 1996. Itinerario por pisadas de dinosaurio en los alrededores de Enciso; pp. 89–100 in F. Pérez-Lorente (ed.), Excursiones Geológicas por La Rioja. Ciencias Tierra 19. Instituto de Estudios Riojanos, Logroño, Spain.

Casanovas, M. L., A. Fernández, F. Pérez-Lorente, and J. V. Santafé. 1997a. Sauropod trackways from site El Sobaquillo (Munilla, La Rioja, Spain) indicate amble walking. Ichnos 5:101–107.

Casanovas, M. L., A. Fernández, F. Pérez-Lorente, and J. V. Santafé. 1998. Ocho nuevos yacimientos de huellas de dinosaurio. Zubía 16:117–152.

Casanovas, M. L., F. Pérez-Lorente, J. V. Santafé, and A. Fernández. 1985. Nuevos datos icnológicos del Cretácico Inferior de la Sierra de Cameros (La Rioja). Paleontología y Evolución 19:3–18

Casanovas, M. L., R. Ezquerra, A. Fernández, F. Pérez-Lorente, and J. V. Santafé. 1990a. Huellas de dinosaurio en Soto de Cameros. La Rioja (España). Zubía 8:49–71.

Casanovas, M. L., R. Ezquerra, A. Fernández, F. Pérez-Lorente, and J. V. Santafé. 1990b. Huellas de dinosaurio en San Vicente de Robres. La Rioja (España). Zubía 8:33–47.

Casanovas, M. L., R. Ezquerra, A. Fernández, F. Pérez Lorente, and J. V. Santafé. 1991a. Dinosaurios coelúridos gregarios en el yacimiento de Valdebajes (La Rioja, España). Revista Española de Paleontología 6:177–189.

Casanovas, M. L., R. Ezquerra, A. Fernández, F. Pérez-Lorente, and J. V. Santafé. 1992a. Revisión del yacimiento "Icnitas 3" de huellas de dinosaurio (Enciso, La Rioja, España). Zubía 10:31–44

Casanovas, M. L., A. Fernández, M. M. Romero-Molina, and J. V. Santafé. 1999. Empreintes de dinosaures dans la Rioja: dinosaurs in the Mediterranean. Revue Editée par la Cité des Sciences á Tunis, Almadar 11:109–132.

Casanovas, M. L., A. Fernández, F. Pérez-Lorente, J. V. Santafé, and F. Torcida. 1995j. La Era del Peladillo 4 (La Rioja.España); pp. 45–52 in F. Pérez-Lorente (ed.), Huellas fósiles de dinosaurios de La Rioja. Nuevos yacimientos. Ciencias Tierra 18. Instituto de Estudios Riojanos, Logroño, Spain.

Casanovas, M. L., A. Fernández, F. Pérez-Lorente, J. V. Santafé, and F. Torcida. 1997b. Pisadas de ornitópodos, terópodos y saurópodos en la Era del Peladillo, 5 (La Rioja, España). Zubía 15:229–246.

Casanovas, M. L., F. Pérez-Lorente, M. Ruiz, J. V. Santafé, and F. Torcida. 1991c. Terópodos carnosaurios en la Virgen del Campo II. Enciso (La Rioja, España). Zubía 9:113–126.

Casanovas, M. L., R. Ezquerra, A. Fernández, F. Pérez-Lorente, J. V. Santafé, and F. Torcida. 1991b. Huellas de dinosaurio en el camino de Igea a Valdebrajes (La Rioja, España). Zubía 9:89–111.

Casanovas, M. L., R. Ezquerra, A. Fernández, F. Pérez-Lorente, J. V. Santafé, and F. Torcida. 1992b. Un grupo de saurópodos en el yacimiento Soto 2. La Rioja (España). Zubía 10:45–52.

Casanovas, M. L., R. Ezquerra, A. Fernández, F. Pérez-Lorente, J. V. Santafé, and F. Torcida. 1993a. Huellas de dinosaurios palmeados y de terópodos en la Era del Peladillo, Igea (La Rioja). Zubía 11:11–53.

Casanovas, M. L., R. Ezquerra, A. Fernández, F. Pérez-Lorente, J. V. Santafé, and F. Torcida. 1993b. Tracks of a herd of webbed ornithopods and other footprints found in the same site. Revue de Paleobiologie 7, vol. sp.: 29–36.

Casanovas, M. L., R. Ezquerra, A. Fernández, F. Pérez-Lorente. 1993c. Icnitas de dinosaurios. Yacimientos de Navalsaz, Las Mortajeras, Peñaportillo, Malvaciervo y la Era del Peladillo 2. Zubía monográfico 5:9–133.

Casanovas, M. L., R. Ezquerra, A. Fernández, F. Pérez-Lorente, J. V. Santafé, and F. Torcida. 1993d. Icnitas digitígradas y plantígradas de dinosaurios en el afloramiento de El Villar–Poyales (La Rioja, España). Zubía monográfico 5:135–163.

Casanovas, M. L., R. Ezquerra, A. Fernández, F. Pérez-Lorente, J. V. Santafé, and F. Torcida. 1995b. Dos nuevos yacimientos de icnitas de dinosaurios en La Rioja y en la provincia de Soria. Coloquios de Paleontología 47:9–23.

Casanovas, M. L., R. Ezquerra, A. Fernández, F. Pérez-Lorente, J. V. Santafé, and F. Torcida. 1995c. Huellas de dinosaurio en la Era del Peladillo 3. Primera nota. Zubía 13:83–101.

Casanovas, M. L., R. Ezquerra, A. Fernández, F. Pérez-Lorente, J. V. Santafé, and F. Torcida. 1995d. Huellas de dinosaurio en el yacimiento Soto 3. La Rioja (España); pp. 27–32 in F. Pérez-Lorente (ed.), Huellas fósiles de dinosaurios de La Rioja. Nuevos yacimientos. Ciencias Tierra 18. Instituto de Estudios Riojanos, Logroño, Spain.

Casanovas, M. L., A. Fernández, M. C. Ondiviela, F. Pérez-Lorente, J. V. Santafé, and M. R. Serrano. 1995e. El rastro del barranco de Acrijos (Cornago, La Rioja, España); pp. 15–16 in F. Pérez-Lorente (ed.), Huellas fósiles de dinosaurios de La Rioja. Nuevos yacimientos. Ciencias Tierra 18. Instituto de Estudios Riojanos, Logroño, Spain.

Casanovas, M. L., R. Ezquerra, A. Fernández, D. Montero, F. Pérez-Lorente, J. V. Santafé, F. Torcida, and L. V. Viera. 1995a. El yacimiento de La Canal (Munilla, La Rioja, España). La variación de velocidad en función del tamaño del pie de los ornitópodos. Zubía 13:55–81.

Casas, A., Castillo, E., Gil, J., Mata, M. P., Muñoz, A., and F. Pérez-Lorente. 1996. Itinerario geológico por el valle del Cidacos; pp. 59–87 in F. Pérez-Lorente (ed.), Excursiones Geológicas por La Rioja. Ciencias Tierra 19. Instituto de Estudios Riojanos, Logroño, Spain.

Cendrero, A. 1996. Propuesta sobre criterios para la clasificación y catalogación del patrimonio geológico. El patrimonio geológico. pp. 29–38. Ministerio de Obras Públicas, Madrid, Spain.

Colbert, E. H., and D. Merrilees. 1967. Cretaceous dinosaur footprints from Western Australia. Journal of the Royal Society Western Australia 50:21–25.

Coombs, W. P. 1980. Swimming ability of carnivorous dinosaurs. Science 207:1198–1200.

Costeur, L., and R. Ezquerra. 2008. Early Cretaceous fish trails from La Rioja, Spain.

Currie, P. J. 1983. Hadrosaur trackways from the Lower Cretaceous of Canada. Acta Palaeontologica Polonica 28:63–73.

Currie, P. J. 1995. Ornithopod trackways from the Lower Cretaceous of Canadá; pp. 431–443 in W. A. S. Sarjeant (ed.), Vertebrate Fossils and the Evolution of Scientific Concepts. Gordon and Breach, Reading, U.K.

Dalla Vecchia, F., A. Tarlao, G. Tunis, and S. Venturini. 2000. New dinosaur track sites in the Albian (Early Cretaceous) of the Istria Peninsula (Croatia). Memorie di Scienze Geologiche 57:192–193.

Day, J. J., D. B. Norman, A.-S. Gale, P. Upchurch, and H. P. Powell. 2004. A middle Jurassic dinosaur trackway site from Oxfordshire, U.K. Paleontology 47:319–348.

Delair, J. B., and A. B. Lander. 1973. A short history of the discovery of reptilian footprints in the Purbeck beds of Dorset, with notes on their stratigraphical distribution. Proceedings of the Dorset natural History and archaeological Society 94:17–20.

Delvene, G., and M. Munt. 2011. New Trigonioidoidea (Bivalvia; Unionoida) from the Early Cretaceous of Spain. Palaeontology 54:631–638.

Demathieu, G. 1984. Utilisation des lois de la mécanique pour l'estimation de la vitesse de locomotion des vertébrés tétrapodes du passé. Geobios 17:439–446.

Demathieu, G. 1986. Nouvelles recherches sur la vitesse des vertébrés, auteurs de traces fossiles. Geobios 19:327–333.

Demathieu, G. R. 1987a. Use of statistical methods in palaeoichnology. Apparent limbs. Thickness of the footprints reliefs and its significance: research on the distribution of the weigths upon the autopodia; pp. 52–62 in G. Leonardi (ed.), Glossary and Manual of Tetrapod Footprint

Palaeoichnology. Departamento Nacional da Produçao Mineral, Brasilia, Brazil.

Demathieu, G. R. 1987b. Plate 8, C–F; p. 92 in G. Leonardi (ed.), Glossary and Manual of Tetrapod Footprint Palaeoichnology. Departamento Nacional da Produçao Mineral, Brasilia, Brazil.

Demathieu, G. R. 1990. Problems in discrimination of tridactyl dinosaur footprints, exemplifired by the Hettangian trackways, the Causes, France. Ichnos 1:97–110.

Demathieu, G. R., G. Gand, J. Sciau, and P. Freytet. 2002. Les traces de pas de dinosaures et autres archosaures du Lias inferieur des Grands Causses, Sud de la France. Palaeovertebrata 31:1–143.

Díaz, E., C. M. Pina, and P. Ponce. 1990. Estudio de unas trazas icnológicas en el Cretácico inferior de San Vicente de Robres (La Rioja). Geogaceta 7:78–81.

Díaz-Martínez, I., F. Pérez-Lorente, J. I. Canudo, and X. Pereda-Suberbiola. 2008. Causas de la variabilidad en icnitas de dinosaurio y su aplicación en icnotaxonomía. IV Jornadas internacionales sobre paleontología de dinosaurios y su entorno, 207–220. Actas, Salas de los Infantes, Spain.

Díaz-Martínez, I., F. Pérez-Lorente, X. Pereda-Suberbiola, and J. I. Canudo. 2009. Iguanodon-like footprints from the Enciso Group (Aptian, Lower Cretaceous) of La Rioja (Cameros Basin, Spain). Darwin Bernisart meeting, 9–13. Bruselas.

Díaz-Martínez, I., E. Garcia-Ortiz de Landaluce, R. Ibisate, and F. Pérez-Lorente. 2007. Nuevas aportaciones al registro paleoicnológico en Cabezón de Cameros (La Rioja, España). Geogaceta 42:87–90.

Doublet, S. 2004. Controles tectonique et climatique de l'enregistrement stratigraphique dans un basin continental de rift: le bassin de Cameros. These doctoratl. Université de Dijon (Francia).

Doublet, S., and J. P. Garcia. 2004. The significance of dropstones in a tropical lacustrine setting, eastern Cameros Basin (Late Jurassic-Early Cretaceous, Spain). Sedimentary Geology 163:293–309.

Dutuit, J. M., and A. Ouazzou. 1980. Decouverte d'une piste de dinosaure sauropode sur le site d'empreintes de Demnat (Haut Atlas marocain). Memoires de la Societe Geologique de France 139:95–102.

Elízaga, E., E. Gallego, and A. García Cortés. 1991. Inventaire national des sites d'interet geologique en Espagne: methodologie et deroulement. 1er Symposium international sur la protection du patrimoine géologique. Comunications, Digne.

Ellenberger, P. 1974. Contribution a la classification des Pistes de Vertebres du Trias: Les types du Stormberg d'Afrique du Sud (I^{eme}Partie: Le Stromberg superieur. I Le biome de la zone B/$_1$ au niveau de Moyeni: ses biocenoses). Palaeovertebrata. Memoire Extraordinaire, Montpelier.

Extremiana, I., and V. Lanchares. 1995. Análisis de la relación I con v$_1$ y v$_2$ según los datos de la tabla 9. Zubía 13:77–81.

Ezquerra, R., and L. Costeur. 2009. Fish trails from the early Cretaceous of La Rioja, Spain.

Ezquerra, R., and F. Pérez-Lorente. 2002. Reptiles nadadores en el sector oeste del yacimiento 4LVC. La Virgen del Campo (Enciso, La Rioja, España); p. 16 in F. Pérez-Lorente (ed.), Congreso internacional sobre dinosaurios y otros reptiles mesozoicos en España. Resúmenes, Logroño, Spain.

Ezquerra, R., and F. Pérez-Lorente. 2003. Reptiles nadadores en el sector oeste del yacimiento de la Virgen del Campo (4LVC, Enciso, La Rioja, España); pp. 215–224 in F. Pérez-Lorente (ed.), Dinosaurios y otros reptiles mesozoicos de España. Ciencias de la Tierra 26. Instituto de Estudios Riojanos, Logroño, Spain.

Ezquerra, R., L. Costeur, and F. Pérez-Lorente. 2010. Los dinosaurios también nadaban. Investigación y ciencia diciembre: 2–8.

Ezquerra, R., L. Costeur, S. Doublet, P. M. Galton, and F. Pérez-Lorente. 2004. Lower Cretaceous swimming theropod trackways from "La Virgen del Campo" (La Rioja, Spain); p. 150 in Abstracts of the Paleontological Association Annual Meeting, Lille, France.

Ezquerra, R., S. Doublet, L. Costeur, P. M. Galton, and F. Pérez-Lorente. 2007. Were non-avian Theropod dinosaurs able to swim? Supportive evidence from an Early Cretaceous trackway, Cameros Basin (La Rioja, Spain). Geology 35:507–510.

Falkingham, P. L., L. Margrits, I. Smith, M., and P. L. Manning. 2009. Reinterpretation of palmate and semi-palmate (webbed) fossil tracks; insights from finite element modelling.

Palaeogeography, Palaeoclimatology, Palaeoecology 271:69–76.

Farlow, J. O. 1981. Estimates of dinosaur speeds from a new trackway site in Texas. Nature 294:747–748.

Farlow, J. O. 1987. Lower Cretaceous dinosaur tracks, Paluxy River Valley, Texas. South Central Section, Geological Society of America, Baylor University, Waco, Texas.

Farlow, J. O. 1992. Sauropod tracks and trackmakers: integrating the ichnological and skeletal records. Zubía 10:89–138.

Farlow, J. O. 1993. On the rareness of big, fierce animals: speculations about the bodie sizes, population densities, and geographic ranges of predatory mammals and large carnivorous dinosaur. American Journal of Science 231:167–199.

Farlow, J. O. 2001. *Acrocanthosaurus* and the maker of Comanchean large-theropod footprints; pp. 408–427 in D. H. Tanke and K. Carpenter (eds.), Mesozoic Vertebrate Life. Indiana University Press, Bloomington, Indiana.

Farlow, J. O., and R. M. Elsey. 2010, Footprints and trackways of the American alligator, Rockefeller Wildlife Refuge, Montana; pp. 31–439 in M. J. Milàn, S. G. Lucas, M. G. Lockley, and J. A. Spielmann (eds.), Crocodile Tracks and Traces. Bulletin 51. New Mexico Museum of Natural History and Science, Alburquerque, New Mexico.

Farlow, J. O., and J. M. Hawthorne. 1989. Stop 4. Comanchean dinosaur footprints; pp. 23–30 in A. Winkler, P. A. Murry, and L. L. Jacobs (eds.), Field Guide to the Vertebrate Paleontology of the Trinity Group, Lower Cretaceous of Central Texas. Institute for the Study of the Earth and Man, Dallas Southern Methodist University, Dallas, Texas.

Farlow, J. O., J. G. Pittman, and J. M. Hawthorne. 1989. *Brontopodus birdi,* Lower Cretaceous sauropod footprints from the U.S. Gulf coastal plain; pp. 371–394 in D. D. Gillette and M. G. Lockley (eds.), Dinosaur Tracks and Traces. Cambridge University Press, Cambridge, U.K.

Farlow, J. O., M. B. Smith, and J. M. Robinson. 1995. Body mass, bone "strength indicator" and cursorial potential of *Tyrannosaurus rex.* Journal of Vertebrate Paleontology 15:713–725.

Farlow, J. O., R. E. Chapman, B. Breithaupt, and N. Matthews. 2012. The scientific study of dinosaur footprints; pp. 713–759 in M. K. Brett-Surman, T. R. Holtz Jr., and J. O. Farlow (eds.), The Complete Dinosaur. Indiana University Press, Bloomington, Indiana.

Farlow, J. O. T. R. Holtz Jr., T. H. Worthy, and R. E. Chapman. 2013. Feet of the fierce (and not so fierce) pedal proportions in large theropods, other non-avian dinosaurs, and large ground birds; pp. 88–132 in J. M. Parrish, R. E. Molnar, P. J. Currie, and E. B. Koppelhus (eds.), Tyrannosaurid Paleobiology. Indiana University Press, Bloomington, Indiana.

Farlow, J. O., S. M. Gatesy, T. R. Holtz, J. R. Hutchinson, and J. M. Robinson. 2000. Theropod locomotion. American Zoologist 40:640–663.

Farlow, J. O., W. Langston, E. Everett, R. Solis, W. Ward, B. L. Kirkland, S. Hovorka, T. L. Reece, and J. Whitcraft. 2006. Texas giants: dinosaurs of the Heritage Museum of the Texas Hill Country. Heritage Museum, Texas Hill Country, Canyon Lake, Texas.

Fornós, J. J., R. G. Bromley, L. B. Clemensen, and A. Rodríguez-Pérez. 2002. Tracks and trackways of *Myotragus balearicus* Bate (Artiodactyla, Caprinae) in Pleistocene Aeolianites from Mallorca (Balearic Islands, Western Mediterranean). Palaeogeography, Palaeoclimatology, Palaeoecology 180:277–313.

Galton, P. M. 1974. The ornitischian dinosaur Hypsilophodon from the Wealden of the Isle of Wight. Bulletin of the British Museum (Natural History) Geology 25:1–152.

Galton, P. M., and J. O. Farlow. 2003. Dinosaur State Park, U.S.A. History, footprints, trackways, exhibit. Zubía 21:129–173.

Ganse, B., A. Stahn, S. Stoinski, T. Suthau, and H.-C. Gunga. 2011. Body mass stimation, thermoregulation, and cardiovascular physiology of large sauropods; pp. 105–118 in N. Klein, K. Remes, C. T. Gee, and P. M. Sander (eds.), Biology of the Sauropod Dinosaurs: Understanding the Life of Giants. Indiana University Press, Bloomington, Indiana.

García-Ortiz, E., and I. Díaz-Martínez. 2008. Aportaciones de algunos yacimientos representativos de La Rioja al estudio del comportamiento de los dinosaurios; pp. 207–219 in J. Esteve and G. Meléndez (eds.), Palaeontologica Nova (VI EJIP). Publicaciones del Seminario de Paleontología (SEPAZ) 8.

García-Ortiz, E., J. M. Ortega-Girela, A. Hurtado-Reyes, and I. Díaz-Martínez. 2009. Revisión de las huellas terópodas y saurópodas de mayor y menor tamaño de La Rioja (España) y su comparación con el registro mundial: los Guinness world record. Paleolusitania 1:201–209.

Garcia Raguel, M., I. Cuevas, I. Díaz-Martínez, and F. Pérez-Lorente. 2009. Fragmentos de roca con huellas de ave en el terciario de Alcanadre (La Rioja). Descripción, estructuras y problemas de identificación. Zubía 27:81–158.

García-Ramos, J. C., L. Piñuela, and J. Lires. 2002a. Terópodos precavidos y refugios para saurópodos. Hipótesis basadas en icnitas de dinosaurios del Jurásico de Asturias; p. 24 in F. Pérez-Lorente (ed.), Congreso internacional sobre dinosaurios y otros reptiles mesozoicos en España. Resúmenes, Logroño, Spain.

García-Ramos, J. C., L. Piñuela, and J. Lires. 2002b. Icnitas de dinosaurios, tipos de sedimento y consistencia del sustrato; pp. 25–26 in F. Pérez-Lorente (ed.), Congreso internacional sobre dinosaurios y otros reptiles mesozoicos en España. Resúmenes Logroño, Spain.

García-Ramos, J. C., L. Piñuela, and J. Lires. 2007. Atlas del Jurásico de Asturias. Nobel Press, Oviedo, Spain.

Gatesy, S. M. 2003. Direct and indirect tracks features: what sediment did a dinosaur touch? Ichnos 10:91–98.

Gatesy, S. M., K. M. Middleton, F. A. J. Jenkins, and N. H. Shubin. 1999. Three-dimensional preservation of foot movements in Triassic theropod dinosaurs. Nature 399:141–144.

Gierlinski, G., and A. Potemska. 1987. Lower Jurassic dinosaur footprints from Gliniany Las, northern slope of the Holy Cross Mountains, Poland. Neues Jahrbuch für Geologie und Paläontologie Abhandlungen 175:107–120.

Gierlinski, G., G. Niedzwiedzki, and P. Nowacki. 2009. Small theropod and ornithopod footprints in the Late Jurassic of Poland. Acta Geológica Polonica 59:221–234.

Gomes, J. P. 1915–1916. Descoberta de restos de saurios gigantescos no Jurássico do Cabo Mondego. Comunicações Comissão Serviços Geológicos de Portugal 9:132–134.

Hadri, M., X. Pereda, M. Boutakiout, and F. Pérez-Lorente. 2007. Icnitas de posibles dinosaurios tireóforos del Jurásico Inferior (Alto Atlas, Goulmima,

Marruecos). Revista Española de Paleontología 22:147–156.

Haubold, H. 1971. Ichnia amphibiorum et reptiliorum fossilium; pp. 1–124 in O. Kuhn (ed.), Handbuch der Paläoherpetologie, Volume 18. Gustav Fischer Verlag, Stuttgart, Germany.

Henderson, D. M. 2003. Footprints, trackways, and hip heights of bipedal dinosaurs–testing hip height predictions with computer models. Ichnos 10:109–114.

Hernández, A., J. I. Ramírez, A. Olivé, A. Alvaro, J. Ramírez del Pozo, M. J. Aguilar, and A. Meléndez. 1990. Mapa geológico de España. Scale 1:50,000. Hoja nº 242 Munilla. Instituto Geológico y Minero de España, Madrid, Spain.

Hernández Medrano, N., and F. Pérez-Lorente. 2002. Un nuevo yacimiento de icnitas. La Ilaga (Terroba, La Rioja, España). Suelo flexible y rastrilladas paralelas; p. 27 in F. Pérez-Lorente (ed.), Congreso internacional sobre dinosaurios y otros reptiles mesozoicos en España. Resúmenes, Logroño, Spain.

Hernández Medrano, N., and F. Pérez-Lorente. 2003. Un nuevo yacimiento de icnitas. La Ilaga (Terroba, La Rioja, España). Suelo flexible y rastrilladas paralelas; pp. 195–214 in F. Pérez-Lorente (ed.), Dinosaurios y otros reptiles mesozoicos de España. Ciencias de la Tierra 26. Instituto de Estudios Riojanos, Logroño, Spain.

Hernández Medrano, N., E. Requeta, and F. Pérez-Lorente. 2006. La Pellejera, ejemplo de nuevos yacimientos icníticos en Cameros (La Rioja–Soria, España); pp. 235–252 in Actas III Jornadas internacionales. Paleontología de dinosaurios y su entorno (2004), Salas de los Infantes, Spain.

Hitchcock, E. 1848. An attempt to discriminate and describe the animals that made the fossil footmarks of the United States, and especially of New England. Memoirs of the American Academy of Arts and Sciences, n.s., 3:129–256.

Hitchcock, E. 1858. Ichnology of New England. A report on the sandstone of the Connecticut Valley, specially its fossil footmarks. William White, Boston, Massachusetts.

Huerta, P., F. Torcida, J. Farlow, and D. Montero. 2012. Exceptional preservation processes of 3D dinosaur footprint casts in Costalomo (Lower Cretaceous, Cameros Basin, Spain). Terra Nova 24:136–141.

Hunt, A. P., and S. G. Lucas. 2004. Multiple parallel dinosaur tail drags from the Early Cretaceous of New Mexico. Journal of Vertebrate Paleontology 24:73A.

Irby, G. V. 1996. Paleoichnological evidence for running dinosaurs worldwide; pp. 109–112 in M. Morales (ed.), The Continental Jurassic. Bulletin 60. Museum of Northern Arizona, Flagstaff, Arizona.

Ishigaki, S. 1986. Dinosaur footprints from Atlas Mountains (1). Nature Study 31:5–8

Ishigaki, S. 1989. Footprints of swimming sauropods from Morocco; pp. 83–86 in D. D. Gillette and M. G. Lockley (eds.), Dinosaur Tracks and Traces. Cambridge University Press, Cambridge, U.K.

Ishigaki, S., and M. G. Lockley. 2010. Didactyl, tridactyl and tetradactyl theropod trackways from thye Lower Jurassic of Morocco: evidence of limping, labouring and other irregular gaits. Historical Biology 24:100–108.

Ishigaki, S., and Y. Matsumoto. 2009. "Off-tracking"-like pohenomenon observed in the turning sauropod trackway from the Upper Jurassic of Morocco. Memoir of the Fukui Prefectural Dinosaur Museum 8:1–10.

Isles, T. E. 2009. The socio-sexual behaviour of extant archosaurs: implications for undestanding dinosaur behaviour. Ichnos 21:139–214.

Jenny, J., and J. A. Jossen. 1982. Découverte d'empreintes de pas de Dinosauriens dans le Jurassique inférieur (Pliensbachien) du Haut Atlas central (Maroc). Comptes Rendues des Séances de l'Academie des Sciences, Paris, II, 294:223–226.

Jiménez, S. 1978. Los dinosaurios. Iberduero, Bilbao, Spain.

Jiménez Vela, A., and F. Pérez-Lorente. 2005. A new great dinosaur footprint site: Barranco de Valdegutiérrez (La Rioja, Spain); p. 37 in Abstracts of the International Symposium on Dinosaurs and Other Vertebrates' Palaeoichnology, Fumanya, Barcelona.

Jiménez Vela, A., and F. Pérez-Lorente. 2006–2007. El Corral del Totico. Dos nuevos yacimientos con pistas singulares. (La Rioja, España). Zubía monográfico 18–19:115–144.

Kuban, G. J. 1989. Elongate dinosaur tracks; pp. 57–72 in D. D. Gillette and M. G. Lockley (eds.), Dinosaur Tracks and Traces. Cambridge University Press, Cambridge, U.K.

Kuhn, O. 1958. Die Fahrten der vorzeitlichen Amphibien und Reptilien. Verlagshaus Meisenbach, Bamberg.

Kumagai, C. J., and J. O. Farlow. 2010. Observations on traces of the American crocodile (Crocodylus acutus) from northwestern Costa Rica; pp. 41–49 in M. J. Milàn, S. G. Lucas, M. G. Lockley, and J. A. Spielmann (eds.), Crocodile Tracks and Traces. Bulletin 51. New Mexico Museum of Natural History and Science, Alburquerque, New Mexico.

Langston, W. 1960. A hadrosaurian ichnite. Natural History Paper, National Museum of Canada 4:1–19.

Lapparent, A. F. de. 1965. Une trace de pas d'un petit dinosaure dans le Kimmeridgien d'Arroyo Cerezo (Valencia). Boletín de la Real Sociedad Española de Historia Natural (Geología) 63:225–230.

Leakey, M. D. 1987. Animal print and tails; pp. 451–489 in M. D. Leakey and J. M. Harris (eds.), Laetoli: A Pliocene Site in Southern Tanzania. Clarendon Press, Oxford, U.K.

Leonardi, G. 1979. Nota preliminar sobre seis pistas de dinossauros Ornithischia da Bacia do rio do Peixe, em Sousa, Paraiba, Brasil. Anais Academia Brasileira das Ciências 51:501–516.

Leonardi, G. 1984. Le impronte fossili di dinosauri; pp 165–166 in J. F. Bonaparte, E. H. Colbert, P. J. Currie, A. de Ricqles, Z. Kielan-Jaworowska, G. Leonardi, N. Morello, and P. Taquet (eds.), Sulle orme dei dinosauri. Erizzo Editrice, Venice, Italy.

Leonardi, G. (ed.). 1987. Glossary and Manual of Tetrapod Footprint Palaeoichnology. Departamento Nacional da Producao Mineral, Brasilia, Brazil.

Lim, J.-D., M. G. Lockley, and D.-Y. Kong. 2012. The trackway of a quadrupedal ornithopod from the Jindong Formation (Cretaceous) of Korea. Ichnos 19:101–104.

Lockley, M. G. 1991. Tracking Dinosaurs. Cambridge University Press, Cambridge, U.K.

Lockley, M. G. 2000. An amended description of the theropod footprint Bueckeburgichnus maximus Kuhn 1958, and its bearing on the Megalosaur tracks debate. Ichnos 7:169–181.

Lockley, M. G. 2002. A Guide to the Fossil Footprints of the World. University of Colorado, Dinosaur Trackers and the Friends of Dinosaur Ridge, Morrison, Colorado.

Lockley, M. G. 2009. New perspectives on morphological variation in tridactyl footprints: clues to widespread convergence in developmental dynamics. Geological Quarterly 53:415–432.

Lockley, M. G., and C. Meyer. 1999. Dinosaur tracks and other fossil footprints of Europe. Columbia University Press, New York, New York.

Lockley, M. G., and A. Rice. 1990. Did *"Brontosaurus"* ever swim out to sea? Evidence from brontosaur and other dinosaur footprints. Ichnos 1:81–90.

Lockley, M. G., and J. L. Wright. 2001. Trackways of large quadrupedal ornithopods from the Cretaceous: a review; pp. 428–442 in D. H. Tanke and K. Carpenter (eds.), Mesozoic Vertebrate Life. Indiana University Press, Bloomington, Indiana.

Lockley, M. G., J. O. Farlow, and C. A. Meyer. 1994. *Brontopodus* and *Parabrontopodus* ichnogen. nov. and the significance of wide- and narrow-gauge sauropod trackways. Gaia 10:135–145.

Lockley, M. G., K. J. Houck, and N. K. Prince. 1986. North America's largest dinosaur trackway site: implications for Morrison paleoecology. Bulletin of the Geological Society of America 97.1163–1176.

Lockley, M. G., A. P. Hunt, J. J. Moratalla, and M. Matsukawa. 1994. Limping dinosaurs? Trackway evidence of anormal gait. Ichnos 3:193–202.

Lockley, M. G., C. A. Meyer, and J. J. Moratalla. 1998. *Therangospodus:* trackway evidence for the widespread distribution of a Late Jurassic theropod with well-padded feet. Gaia 15:339–353.

Lockley, M. G., C. A. Meyer, and V. F. Santos. 1998. *Megalosauripus,* and the problematic concept of *Megalosaur* footprints. Gaia 15:317–327.

Lockley, M. G., B. H. Young, and K. Carpenter. 1983. Hadrosaur locomotion and herding behavior: evidence from footprints in the Mesaverde Formation, Grand Mesa Coal Field, Colorado. Mountain Geologist 20:5–14.

Lockley, M. G., V. F. Santos, C. Meyer, and A. Hunt. 1998. A new dinosaur tracksite in the Morrison Formation, Boundary Butte, Southeastern Utah; pp. 317–330 in K. Carpenter, D. Chure and J. Kirkland (eds.), Upper Jurassic Morrison Formation: An Interdisciplinary Study. Modern Geology 23.

Loope, D. B. 1986. Recognizing and utilizing vertebrate tracks In cross-section: Cenozoic hoofprints from Nebraska. Palaios 1:141–151.

Manning, P. L. 2004. A new approach to analysis and interpretacion of tracks: examples from the Dinosauria; pp. 93–123 in D. McIlroy (ed.), The Application of Ichnology to Palaeoenvironmental and Stratigraphic Analysis. Special Publication 228. Geological Society, London, U.K.

Martín Escorza, C. 1986. Las icnolineaciones de dinosaurios wealdenses en Enciso-Navalsaz. Zubía 4:33–43.

Martín Escorza, C. 1988. Orientación de las icnitas en el valle del Cidacos (La Rioja). Boletín de la Real Sociedad Española de Historia Natural, Actas, 84:32–34.

Martín Escorza, C. 1992. Gregarismo y dinosaurios. Cartas al editor. Revista Española de Paleontología 7:97.

Martín Escorza, C. 2001. Orientación de las icnitas de dinosaurios en la Sierra de Cameros. Zubía 19:139–163.

Marty, D. 2008. Sedimentology, taphonomy, and ichnology of Late Jurassic dinosaur tracks from the Jura carbonate platform (Chevenez–Combe Ronde tracksite, NW Switzerland): insights into the tidal-flat palaeoenvironment and dinosaur diversity, locomotion and palaeoecology. Ph.D. thesis, University of Fribourg, Fribourg, Switzerland.

Mas, R., M. I. Benito, J. Arribas, A. Serrano, J. Guimera, A. Alonso, and J. Alonso-Azcarate. 2002. La Cuenca de Cameros: desde la extensión finijurasica–eocretacica a la inversión terciaria–implicaciones en la exploración de hidrocarburos. Zubía monográfico 14:9–64.

Meléndez, A., and F. Pérez-Lorente. 1996. Comportamiento gregario aparente de dinosaurios condicionado por una deformación sinsedimentaria (Igea, La Rioja, España). Estudios Geológicos 52:77–82.

Melero, M., and F. Pérez-Lorente. 2011. Huellas en las obras. Reconocimiento y estudio de huellas fósiles de dinosaurio en las obras de la presa de Enciso (La Rioja, España). Zubía 29:31–60.

Milan, J., L. B. Clemensen, and N. Bonde. 2004. Vertical sections through dinosaur tracks (Late Triassic lake deposits, East Greenland)–undertacks and other subsurface deformation structures revealed. Lethaia 37:285–296.

Montes, P. 2006. El ejemplo de La Rioja. Los yacimientos de icnitas de dinosaurio; pp. 141–170 in F. Torcida (ed.), Huellas que perduran. Fundación Patrimonio Histórico de Castilla y León, Valladolid, Spain.

Moratalla, J. J. 1993. Restos indirectos de dinosaurios del registro español: paleoicnología de la Cuenca de Cameros (Jurásico superior–Cretácico inferior) y paleoología del Cretácico superior. Ph.D. thesis, Universidad Autónoma, Madrid, Spain.

Moratalla, J. J. 2002. Cameros Basin megasequence (Spain): an overview on body and ichnological biodiversity from the European Cretaceous. Journal of Vertebrate Paleontology 22:90A.

Moratalla, J. J. 2008. The story of the Cameros Basin dinosaurs (Lower Cretaceous, Spain) written in their tracks. Geophysical Research Abstracts 10.

Moratalla, J. J. 2009a. Sauropod tracks of the Cameros Basin (Spain): identification, trackway patterns and changes over the Jurassic–Cretaceous. Geobios 42:797–811.

Moratalla, J. J. 2009b. The story of Iberian theropod dinosaurs inferred from the Cretaceous ichnological record of the Cameros Basin (Spain): discrimination, identification, distribution and evolution; pp. 85–86 in A. Buscalioni and M. Fregenal (eds.), Program and Abstracts of the 10th Symposium on Mesozoic Terrestrial Ecosystems and Biota, Teruel, Spain.

Moratalla, J. J., and J. Hernán. 2005. Field trip guide to the La Rioja Fossil tracksites; pp. 1–31 in International Symposium on Dinosaurs and Other Vertebrates' Paleoichnology. Libro de excursiones, Fumanya, Barcelona.

Moratalla, J. J., and J. Hernán. 2007. Dinosaur icnocenosis and the Cameros Basin as an oblige pass area during the Lower Cretaceous of the Iberian Plate. Journal of Vertebrate Paleontology 27:119A–120A.

Moratalla, J. J., and J. Hernán. 2008. Los Cayos S y D: dos afloramientos con icnitas de saurópodos, terópodos y ornitópodos en el Cretácico inferior del área de Los Cayos (Cornago, La Rioja, España). Estudios Geológicos 64:161–173.

Moratalla, J. J., and J. Hernán. 2009. Turtle and pterosaur tracks from the Los Cayos dinosaur tracksite, Cameros basin (Cornago, La Rioja, Spain): tracking the Lower Cretaceous bio-diversity. Revista Española de Paleontología 24:59–77.

Moratalla, J. J., and J. Hernán. 2010. Probable palaeogeographical influences of the Lower Cretaceous Iberian rifting phase in the eastern Cameros Basin (Spain) on dinosaur trackway orientations. Palaeogeography, Palaeoclimatology, Palaeoecology 295:116–130.

Moratalla, J. J., and J. L. Sanz. 1992. Icnitas aviformes en el yacimiento del Cretácico Inferior de Los Cayos (Cornago, La Rioja, España). Zubía 10:153–160.

Moratalla, J. J., and J. L. Sanz. 1997. Cameros basin megatracksite; pp. 87–89 in P. J. Currie and K. Padian (eds.), Encyclopedia of Dinosaurs. Academic Press, New York, New York.

Moratalla, J. J., J. Hernán, and S. Jiménez. 2003. Los Cayos dinosaur tracksite. An overview of the Lower Cretaceous ichno-diversity of the Cameros Basin (Cornago, La Rioja Province, Spain). Ichnos 10:229–240.

Moratalla, J. J., J. L. Sanz, and S. Jiménez. 1988. Nueva evidencia icnológica de dinosaurios en el Cretácico Inferior de La Rioja (España). Estudios Geológicos 44:119–131

Moratalla, J. J., J. L. Sanz, and S. Jiménez. 1990. Naturaleza y significado de las icnitas de dinosaurios. Estrato 2:25–30.

Moratalla, J. J., J. L. Sanz, and S. Jiménez. 1992. Hallazgo de nuevos tipos de huellas en La Rioja. Estrato 4:63–66.

Moratalla, J. J., J. L. Sanz, and S. Jiménez. 1993. Nuevos hallazgos de icnitas de dinosaurios en Préjano e Inestrillas. La Rioja. Estrato 5:75–76.

Moratalla, J. J., J. L. Sanz, and S. Jiménez. 1994. Dinosaur tracks from the Lower Cretaceous of Regumiel de la Sierra (province of Burgos, Spain): inferences on a new quadrupedal ornithopod trackway. Ichnos 3:89–97.

Moratalla, J. J., J. L. Sanz, and S. Jiménez. 1996. Nuevos yacimientos en Aldeanueva de Cameros y en Trevijano. Estrato 7:111–113.

Moratalla, J. J., J. L. Sanz, and S. Jiménez. 1997a. Acondicionamiento de los yacimientos paleoicnológicos riojanos Estrato 8:98–102.

Moratalla, J. J., J. L. Sanz, and S. Jiménez. 1997b. Dinosaurios en La Rioja. Guía de yacimientos paleoicnológicos. Gobierno de La Rioja and Iberdrola, Logroño, Spain.

Moratalla, J. J., J. L. Sanz, and S. Jiménez. 1999. Nuevos hallazgos de dinosaurios y pterosaurios en el Cretácico inferior de La Rioja. Estrato 10:91–96.

Moratalla, J. J., J. L. Sanz, and S. Jiménez. 2000a. Yacimiento de icnitas de La Virgen del Prado (Iniestrillas, Aguilar del Rio Alhama) y excavaciones en el Sol de la Pita 2 (Préjano) y La Llana (Aguilar del Rio Alhama). Estrato 11:92–97.

Moratalla, J. J., J. L. Sanz, and S. Jiménez. 2000b. El yacimiento de icnitas de dinosaurios "Vuelta de los manzanos." Estrato 12:130–133.

Moratalla, J. J., J. L. Sanz, and S. Jiménez. 2001. El área de Los Cayos: nuevos afloramientos, diversidad y perspectiva. Estrato 13:144–149.

Moratalla, J. J., J. L. Sanz, and S. Jiménez. 2004. El area de Los Cayos (Cornago, La Rioja): un ejemplo de biodiversidad paleoicnológica en el CretácicoInferior español; pp. 336–346 in E. Baquedano and E. Rubio (eds.), Paleontología: Miscelánea en homenaje a Emiliano Aguirre. Museo Arqueológico Nacional, Madrid, Spain.

Moratalla, J. J., J. L. Sanz, S. Jiménez, and J. Hernán. 2002. Icnitas de dinosaurios titanosauridos en el Cretácico Inferior de Los Cayos (Cornago, La Rioja, España); p. 36 in F. Pérez-Lorente (ed.), Congreso internacional sobre dinosaurios y otros reptiles mesozoicos en España. Resúmenes, Logroño, Spain.

Moratalla, J. J., J. L. Sanz, S. Jiménez, and M. G. Lockley. 1992. A quadrupedal ornithopod trackway from the lower cretaceous of La Rioja (Spain): inferences on gait and hand structure. Journal of Vertebrate Paleontology 12:150–157.

Moratalla, J. J., J. L. Sanz, S. Jiménez, and F. Ortega. 1994. Restos de un cocodrilo e icnitas de pterosaurios en el Cretácico Inferior de la Cuenca de Cameros. Estrato 6:90–93.

Moratalla, J., J. Sanz, I. Melero, and S. Jiménez. 1988. Yacimientos Paleoicnológicos de La Rioja (Huellas de dinosaurios). Gobierno de La Rioja–Iberduero, Logroño, Spain.

Moratalla, J. J., J. Garcia-Mondejar, V. F. Santos, M. G. Lockley, J. L. Sanz, and S. Jiménez. 1994. Sauropod trackway from the Lower Cretaceous of Spain. Gaia 10:75–83.

Mulas, E., S. Jiménez, S. Martín, and E. Jiménez. 1988. Columna estratigrafica del yacimiento de icnitas de dinosaurios del Cretácico Inferior de Cornago (La Rioja). Studia Geológica Salmanticiensa 25:161–168.

Norman, D. B. 2004. Basal Inguanodontia; pp. 413–437 in D. B. Weishampel, P. Dodson, and H. Osmólska (eds.), The Dinosauria (2nd edition). University of California Press, Berkeley, California.

Nouri, J. 2007. La paleoichnologie des empreintes de pas des dinosauriens imprimees dans les couches du Jurassique du Haut-Atlas Central. Ph.D. thesis, Mohammed V University, Rabat, Morocco.

Ordaz, J., and R. M. Esbert. 1988. Glosario de términos relacionados con el deterioro de las piedras de construcción. Materiales de construcción 38:39–45.

Ortega, F., F. Escaso, and J. L. Sanz. 2010. A bizarre, humped Carcharodontosauria (Theropoda) from the Lower Cretaceous of Spain. Nature 467:203–206.

Ostrom, J. 1972. Were some dinosaurs gregarious? Palaeogeography, Palaeoclimatology, Palaeoecology 11:287–301.

Paik, I. S., H. J. Kim, and Y. I. Lee. 2001. Dinosaur tracks bearing deposits in the Cretaceous Jindong Formation, Korea: occurrence, palaeoenvironments and preservation. Cretaceous Research 22:79–92.

Palacios, P., and R. Sánchez Lozano. 1885. La formación Wealdense en las provincias de Soria y Logroño. Boletín de la Comisión del Mapa Geológico de España 9:109–140.

Pereda-Suberbiola, X., and P. M. Galton. 2001. Thyreophoran ornithischian dinosaurs from the Iberian Peninsula; pp. 147–161 in Actas de las I Jornadas internacionales sobre Paleontología de Dinosaurios y su entorno, Salas de los Infantes, Spain.

Pereda-Suberbiola, X., C. Fuentes, M. Meijide, F. Meijide-Fuentes, and, M. Meijide-Fuentes Jr. 2007. New remains of the ankylosaurian dinosaur Polacanthus from the Lower Cretaceous of Soria, Spain. Cretaceous Research 28:583–586.

Pérez-Lorente, F. 1988. Huellas de dinosaurio en el Wealdiense del Grupo de Enciso; pp. 309–314 in III Coloquio de Estratigrafia y Paleogeografia del Jurásico de España. Ciencias de la Tierra 11. Instituto de Estudios Riojanos, Logroño, Spain.

Pérez-Lorente, F. 1990. Excavaciones sobre icnitas de dinosaurio en Enciso e Igea (La Rioja). Estrato 2:47–50.

Pérez-Lorente, F. 1991. Excavaciones en Enciso, Igea y Munilla. Estrato 3:9–11.

Pérez-Lorente, F. 1992a. Trabajos sobre icnitas en Igea, Munilla y Enciso. Estrato 4:59–62.

Pérez-Lorente, F. 1992b. ¿Icnitas de dinosaurio? Tierra y tecnología 11:29–33.

Pérez-Lorente, F. 1993a. Excavaciones en Enciso, Igea y Munilla. Estrato 5:77–79.

Pérez-Lorente, F. 1993b. Dinosaurios plantígrados de La Rioja. Zubía monográfico 5:181–228.

Pérez-Lorente, F. 1993c. Huesos de dinosaurio en Prejano. Estrato 5:80.

Pérez-Lorente, F. 1994. Icnitas en Enciso, Igea y Munilla. Estrato 6:94–95.

Pérez-Lorente, F. 1996a. Pistas terópodas en cifras. Zubía 14:37–55.

Pérez-Lorente, F. 1996b. Determinación y análisis de icnitas de dinosaurio. Enseñanza de las Ciencias de la Tierra 4:100–106.

Pérez-Lorente, F. 1996c. Huellas de dinosaurio de La Rioja. Actividades del verano 1995. Estrato 7:119–122.

Pérez-Lorente, F. 1997. Huellas de dinosaurio de La Rioja. Restauración de yacimientos. Estrato 8:103–106.

Pérez-Lorente, F. 1998. Restauración de yacimientos con huellas fósiles de dinosaurio. Estrato 8:100–103.

Pérez-Lorente, F. 1999a. Trabajos de restauración, mantenimiento y limpieza en los yacimientos de la Era del Peladillo 5, La Virgen del Campo y Las Losas. Estrato 9:85–90.

Pérez-Lorente, F. 1999b. Il Cretácico di La Rioja; pp. 161–165 in G. Pinna (ed.), Storia naturale d'Europa. Jaca Books, Rome, Italy.

Pérez-Lorente, F. 1999c. Patrimonio y valoración de yacimientos paleontológicos; pp. 5–18 in IV Sesión científica de la Sociedad para la defensa de Patrimonio Geológico y Minero, Escuela de Minas, Bélmez, Spain.

Pérez-Lorente, F. 2000a. Experiencias de Geoconservación en La Rioja; pp. 179–196 in D. Barettino, W. A. P. Wimbledon, and E. Gallego (eds.), Patrimonio Geológico y Gestión. Instituto Tecnológico y Geominero de España, Madrid, Spain.

Pérez-Lorente, F. 2000b. Restauración, mantenimiento en los yacimientos de la Era del Peladillo 1 y 5 (Igea) La Virgen del Campo y Las Losas (Enciso) y Peñaportillo (Munilla). Prospección en la Escurquilla y en el camino de Hornillos a Larriba. Limpieza y estudio del yacimiento del Contadero (Torremuña). Estrato 11:98–102.

Pérez-Lorente, F. 2000c. Restauración y catalogación de nuevos yacimientos durante la campaña del año 2000. Estrato 12:125–129.

Pérez-Lorente, F. 2001a. Paleoicnología. Los dinosaurios y sus huellas en La Rioja. Apuntes para los cursos y campos de trabajo de verano. Cultural joven, Gobierno de La Rioja and Fundación Caja Rioja, Logroño, Spain.

Pérez-Lorente, F. 2001b. Buscando la protección de los yacimientos de huellas de dinosaurio. Proyecto Icnitas de dinosaurio de la Península Ibérica (IdPI). Tierra y tecnología 22:25–30.

Pérez-Lorente, F. 2001c. Campaña 2001: actividades sobre yacimientos de icnitas. Estrato 13:134–138.

Pérez-Lorente, F. 2002a. La distribución de yacimientos y de tipos de huellas de dinosaurio en La Cuenca de Cameros (La Rioja, Burgos, Soria, España). Zubía monográfico 14:191–210.

Pérez-Lorente, F. 2002b. Aportaciones de los yacimientos de La Barguilla, Santisol y Santa Juliana (Hornillos de Cameros, La Rioja, España); pp. 43. Congreso internacional sobre dinosaurios y otros reptiles mesozoicos en España. Resúmenes Logroño, Spain.

Pérez-Lorente, F. (ed.). 2003a. Dinosaurios y otros reptiles mesozoicos de España. Ciencias de la Tierra 26. Instituto de Estudios Riojanos, Logroño, Spain.

Pérez-Lorente, F. 2003b. Icnitas de dinosaurio del Cretácico de España; pp. 49–108 in F. Pérez-Lorente (ed.), Dinosaurios y otros reptiles mesozoicos de España. Ciencias de la Tierra 26. Instituto de Estudios Riojanos, Logroño, Spain.

Pérez-Lorente, F. 2003c. Aportaciones de los yacimientos de La Barguilla, Santisol y Santa Juliana (Hornillos de Cameros, La Rioja, España); pp. 161–194 in F. Pérez-Lorente (ed.), Dinosaurios y otros reptiles mesozoicos de España. Ciencias de la Tierra 26. Instituto de Estudios Riojanos, Logroño, Spain.

Pérez-Lorente, F. 2005. El trabajo con las huellas de dinosaurio en La Rioja. Tierra y Tecnología. 28:64–67.

Pérez-Lorente, F. 2006. La candidatura idpi a patrimonio de la humanidad; pp. 365–390 in F. Torcida (ed.), Huellas que perduran. Fundación Patrimonio Histórico de Castilla y León, Valladolid, Spain.

Pérez-Lorente, F. 2007. Estudio de huellas de dinosaurio en La Rioja; pp. 145–158 in J. Arnaez y J. M. Garcia Ruiz (eds.), Espacios naturales y paisajes en La Rioja. Ciencias de la Tierra 27. Instituto de Estudios Riojanos, Logroño, Spain.

Pérez-Lorente, F., and J. Herrero. 2007. El movimiento de un dinosaurio deducido de una rastrillada terópoda con estructuras de inmersión de los pies en el barro y de arrastre de cola (Formación Villar del Arzobispo, Galve, Teruel, España). Revista Española de Paleontología 22:157–174.

Pérez-Lorente, F., and A. Jiménez Vela. 2006–2007. El Barranco de Valdegutierrez. Un nuevo gran yacimiento con huellas de dinosaurio en La Rioja (España). Zubía monográfico 18–19:9–20.

Pérez-Lorente, F., and M. Romero-Molina. 2001a. Problemática de una zona con yacimientos de interés; pp. 163–174 in Las huellas de dinosaurio de La Rioja (España). I Jornadas Internacionales sobre paleontología de Dinosaurios y su entorno, Salas de los Infantes, Spain.

Pérez-Lorente, F., and M. M. Romero-Molina. 2001b. Icnitas terópodas del Cretácico Inferior de La Rioja (España). Zubía 19:115–138.

Pérez-Lorente, F., A. Fernández, and L. Uruñuela. 1986. Pisadas fósiles de dinosaurio. Algunos ejemplos de Enciso. Gobierno de La Rioja, Logroño, Spain.

Pérez-Lorente, F., M. M. Romero, and F. Torcida. 2002. Spanish dinosaur footprints. DinoPress 7:106–115.

Pérez-Lorente, F., M. Romero-Molina, and J. C. Pereda. 2000. Icnitas ornitópodos de El Contadero (Torremuña, La Rioja, España); pp. 29–41 in Tomo homenaje a J. L. Fernández Sevilla y M. Balmaseda Aróspide. Instituto de Estudios Riojanos, Logroño, Spain.

Pérez-Lorente, F., I. Díaz-Martínez, N. Hernández Medrano, and L. E. Requeta. 2009. Heterocronía de huellas de dinosaurio. La Pellejera (Cretácico Inferior, Cuenca de Cameros, La Rioja, España). Reunión anual de comunicaciones de la Asociación Paleontológica Argentina. pp. 63–64. Buenos Aires.

Pérez-Lorente, F., M. Romero-Molina, E. Requeta, M. Blanco, and S. Caro. 2001. Dinosaurios. Introducción y análisis de algunos yacimientos de sus huellas en La Rioja. Ciencias de la Tierra 24. Instituto de Estudios Riojanos, Logroño, Spain.

Pérez-Lorente, F., G. Cuenca-Bescos, M. Aurell, J. I. Canudo, A. R. Soria, and

J. I. Ruiz-Omeñaca. 1997. Las Cerradicas tracksite (Berriasian, Galve, Spain): growing evidence for quadrupedal ornithopods. Ichnos 5:109–120.

Pérez-Moreno, B., J. L. Sanz, A. D. Buscalioni, J. J. Moratalla, F. Ortega, F., and D. Rasskin-Gutman. 1994. A unique multitoothed ornithomimosaur dinosaur from the Lower Cretaceous of Spain. Nature 370:363–367.

Peterson, W. 1924. Dinosaur tracks and the roofs of coal mines. Natural History 24:388–391.

Pienkowski, G., and G. Gierlinski. 1987. New finds of dinosaur footprints in Liassic of the Holly Cross Mountains and its palaeoenvironmental background. Prezeglad Geologiczny 35:199–205.

Ramírez, J. I., A. Olivé, A. Hernández, and M. Alvaro. 1990. Mapa geológico de España. Escala 1:50,000. Hoja nº 241 Anguiano. Instituto Geológico y Minero de España, Madrid, Spain.

Requeta, E. 1999. Obtención y análisis de imágenes en tres dimensiones de huellas de dinosaurio. Temas Geológico Mineros 26:481–483

Requeta, L. E., N. Hernández Medrano, and F. Pérez-Lorente. 2006–2007. La Pellejera: descripción y aportaciones. Heterocronía y variabilidad de un yacimiento con huellas de dinosaurio de La Rioja (España). Zubía monográfico 18–19:21–114.

Richter, G. 1930. Die Iberischen Ketten zwischen Jalón und Demanda. Beiträge zur Geologie der westlichen Mediterrangebiete Mathematisch-Physikalische Klasse, (5) 16: 3. Abhandlungen der Gesellschaft der Wiessenshaften zu Göttingen, Berlin, Germany.

Romano, M., M. A. Whyte, and S. J. Jackson. 2007. Trackway ratio: a new look at trackway gauge in the analysis of quadrupedal dinosaur trackways and its implications for ichnotaxonomy. Ichnos 14:257–270.

Romero-Molina, M., F. Pérez-Lorente, and P. Rivas. 2001. Estructuras asociadas con huellas de dinosaurio en La Rioja (España). Zubía 19:61–96.

Romero-Molina, M. M., F. Pérez-Lorente, and P. Rivas. 2002. Análisis de la parataxonomia utilizada con las huellas de dinosaurio; pp. 52–53 in F. Pérez-Lorente (ed.), Congreso internacional sobre dinosaurios y otros reptiles mesozoicos en España. Resúmenes, Logroño, Spain.

Romero-Molina, M. M., F. Pérez-Lorente, and P. Rivas. 2003. Análisis de la parataxonomia utilizada con las huellas de dinosaurio; pp. 13–32 in F. Pérez-Lorente (ed.), Dinosaurios y otros reptiles mesozoicos de España. Ciencias de la Tierra 26. Instituto de Estudios Riojanos, Logroño, Spain.

Romero-Molina, M. M., W. A. S. Sarjeant, F. Pérez-Lorente, A. López, and E. Requeta. 2003. Orientation and characteristics of theropod trackways from the Las Losas Palaeoichnological site (La Rioja, Spain). Ichnos 10:241–254.

Ruiz-Omeñaca, J. I. 2011. *Delapparentia turolensis* nov. gen et sp., un nuevo dinosaurio iguanodontoideo (Ornithischia: Ornithopoda) en el Cretácico Inferior de Galve. Estudios Geológicos 67:83–110.

Ruiz-Omeñaca, J. I., J. I. Canudo, M. Aurell, B. Badenas, G. Cuenca-Bescós, and J. Ipas. 2004. Estado de las investigaciones sobre los vertebrados del Jurásico Superior y el Cretácico Inferior de Galve (Teruel). Estudios Geológicos 60:179–202.

Sacchi, E. 2004. Metodi, potencialità e problematiche dell'ichnologia di dinosauromorfi. Lo sviluppo di un nuovo approccio metodologico. Ph.D. thesis, Sapienza University of Rome, Rome, Italy.

Sanz, J. L., J. J. Moratalla, and M. L. Casanovas. 1985. Traza icnologica de un dinosaurio iguanodontido en el Cretácico Inferior de Cornago (La Rioja, España). Estudios Geológicos 41:85–91.

Sanz, J. L., A. D. Buscalioni, M. L. Casanovas, and J. V. Santafé. 1987. Dinosaurios del Cretácico Inferior de Galve (Teruel, España). Estudios Geológicos, vol. extr.: 45–64.

Sarjeant, W. A. S. 1989. "Ten paleoichnological commandments": a standardized procedure for the description of fossil vertebrate footprints; pp. 369–370 in D. D. Gillette and M. G. Lockley (eds.), Dinosaur Tracks and Traces. Cambridge University Press, Cambridge, U.K.

Sarjeant, W. A. S., J. B. Delair, and M. G. Lockley. 1998. The footprints of *Iguanodon:* a history and taxonomic study. Ichnos 6:183–202.

Sarjeant, W. A. S., M. M. Romero-Molina, A. López, F. Pérez-Lorente, and E. Requeta. 2002. Algunas implicaciones de la orientación y características de las huellas terópodas digitigradas, semiplantígradas irregulares y con marcas de uña del yacimiento de Las Losas (La Rioja, España); p. 63 in F. Pérez-Lorente (ed.), Congreso internacional sobre dinosaurios y otros reptiles mesozoicos en España. Resúmenes, Logroño, Spain.

Sowerby, J. 1816. The mineral conchology of Great Britain. B. Meredith, London.

Sternberg, C. M. 1926. Dinosaur tracks from the Edmonton Formation of Alberta. Canada Department of Mines Geological Survey Bulletin (Geological Series) 44 (46): 85–87.

Sternberg, C. M. 1932. Dinosaur tracks from Peace River, British Columbia. National Museum of Canada Annual Report (1930), 59–85.

Suárez-González, P., I. E. Quijada, J. R. Mas, and M. I. Benito. 2011. Nuevas aportaciones sobre la influencia marina y la edad de los carbonatos de la Fm Leza en el sector de Préjano (SE de La Rioja). Cretácico Inferior, Cuenca de Cameros. Geogaceta 49:7–10.

Thulborn, R. A. 1990. Dinosaur Tracks. Chapman and Hall, London.

Thulborn, T. 2001. History and nomenclatures of the theropod tracks *Bueckeburgichnus* and *Megalosauripus.* Ichnos 8:207–222.

Thulborn, R. A., and M. Wade. 1979. Dinosaur stampede in the Cretaceous of Queensland. Lethaia 12:275–279.

Thulborn, R. A., and M. Wade. 1989. A footprint as a history of movement; pp. 51–56 in D. D. Gillette and M. G. Lockley (eds.), Dinosaurs Tracks and Traces. Cambridge University Press, Cambridge, U.K.

Tischer, G. 1966. Uber die Wealden–Ablagerung und die Tektonil der ostlichen Sierra de los Cameros in den nordwestlichen Iberischen Ketten (Spanien). Geologisches Jahrbuch 44:123–164.

Torcida, F. 1996. Actividad didáctica de geología de campo. La Era del Peladillo (Igea); pp. 43–57 in F. Pérez-Lorente (ed.), Excursiones Geológicas por La Rioja. Ciencias Tierra 19. Instituto de Estudios Riojanos, Logroño, Spain.

Torcida, F. 2003. Didáctica sobre dinosaurios en Museos y centros educativos: experiencias desarrolladas en España; pp. 423–432 in F. Pérez-Lorente (ed.), Dinosaurios y otros reptiles mesozoicos de España. Ciencias de la Tierra 26. Instituto de Estudios Riojanos, Logroño, Spain.

Torcida, F., J. I. Canudo, P. Huerta, D. Montero, X. Pereda-Suberbiola, and L. Salgado. 2011. *Demandasaurus darwini,* a new rebbachisaurid sauropod from the Early Cretaceous of the

Iberian Peninsula. Acta Palaeontologica Polonica 56:535–552.

Torcida, E., L. A. Izquierdo, P. Huerta, D. Montero, G. Pérez Martínez, and V. Urién. 2006. El yacimiento de icnitas de dinosaurio de Costalomo (Slalas de los Infantes, Burgos, España): un nuevo escenario; pp. 313–348 in F. Torcida and P. Huerta (eds.), Actas III jornadas de paleontología de dinosaurios y su entorno. Colectivo Arqueológico–Paleontológico de Salas C.A.S., Salas de los Infantes, Spain.

Torcida, F., D. Montero, P. Huerta, L. A. Izquierdo, G. Pérez, F. Pérez-Lorente, and F. V. Urien. 2003. Rastro ornitópodo de andar cuadrúpedo con marca de cola. Cretácico Inferior (Burgos, España); pp. 109–118 in F. Pérez-Lorente (ed.), Dinosaurios y otros reptiles mesozoicos de España. Ciencias de la Tierra 26. Instituto de Estudios Riojanos, Logroño, Spain.

Torres, J. A., and L. I. Viera. 1994. *Hypsilophodon foxii* (Reptilia, Ornithischia) en el Cretácico Inferior de Igea (La Rioja, España). Munibe 46:3–41.

Torres, J. A., and L. I. Viera. 1997. Nuevo dinosaurio para el registro paleontológico de La Rioja. Estrato 8:94–97.

Valle, J. M. 1993. Metodología y resultado del levantamiento por fotogrametría terrestre del yacimiento de huellas de dinosaurio de Valdecevillo en Enciso (La Rioja). Zubía monográfico 5:165–186.

Viera, L. I. 1991. *Condfusiscala mirambelensis* (Gastropoda) en el Neocomiense de Igea (La Rioja). Consecuencias. Munibe 43:3–7.

Viera, L. I., and L. M. Aguirrezabala. 1982. El Weald de Munilla (La Rioja) y sus icnitas de dinosaurios. I. Munibe 34:245–270.

Viera, L. I., and J. A. Torres. 1979. El Wealdico de la zona de Enciso (Sierra de los Cameros) y su fauna de grandes reptiles. Munibe 31:141–157.

Viera, L. I., and J. A. Torres. 1992. Sobre "Dinosaurios coelúridos gregarios en el yacimiento de Valdevajes (La Rioja, España)." Nota de réplica y crítica. Revista Española de Paleontología 7:93–96.

Viera, L. I., and J. A. Torres. 1995a. Presencia de *Baryonyx walkeri* (Saurischia, Theropoda) en el Weald de La Rioja (España). Nota previa. Munibe 47:57–61.

Viera, L. I., and J. A. Torres. 1995b. Análisis comparativo sobre dos rastros de dinosaurios theropodos. Forma de marcha y velocidad. Munibe 47:53–56.

Viera, L. I., and J. A. Torres. 1996. Nuevos datos paleontológicos en el area de Hornillos de Cameros. Estrato 7:114–118.

Viera, L. I., J. A. Torres, and L. M. Aguirrezabala. 1984. El Weald de Munilla (La Rioja) y sus icnitas de dinosaurios. II. Munibe 36:3–22.

Wade, M. 1989. The stance of dinosaurs and the Cossak Dancer syndrome; pp. 73–82 in D. D. Gillette and M. G. Lockley (eds.), Dinosaur Tracks and Traces. Cambridge University Press, Cambridge, U.K.

Weems, R. E. 1992. A re-evalution of the taxonomy of extensive statistical data from a recently exposed tracksite near Culpeper, Virginia; pp. 113–127 in P. C. Sweet (ed.), Proceedings of the 26th Forum on the Geology of Industrial Minerals. Publication 119. Virginia Division of Mineral Resources, Charlotte, Virginia.

Weishampel, D. B., P. Dodson, and H. Osmólska (eds.). 1990. The Dinosauria. University of California Press, Berkeley, California.

Whyte, M. A., M. Romano, and D. J. Elvidge. 2007. Reconstruction of Middle Jurassic dinosaur-dominated communities from the vertebrate ichnofauna of the Cleveland Basin of Yorkshire, U.K. Ichnos 14:117–129.

Wilson, J. A. 2005. Integrating ichnofossil and body fossils record to estimate locomotor posture and spatiotemporal distribution of early sauropod dinosaurs: a stratocladistic approach. Paleobiology 31:400–423.

Wilson, J. A., and M. T. Carrano. 1999. Titanosaurs and the origin of "wide-gauge" trackways: a biomechanical and systematic perspective on sauropod locomotions. Paleobiology 25:252–267.

Wilson, J., and D. Fisher. 2003. Are manus-only trackways evidence of swimming, sinking or wading? Journal of Vertebrate Paleontology 23:111A.

Wright, J. L. 2005. Steps in understanding sauropod biology; pp. 252–284 in K. A. Curry, and J. A. Wilson (eds.), The Sauropods: Evolution and Paleobiology. University of California Press, Berkeley, California.

Index

FÉLIX PÉREZ-LORENTE is a retired professor of geology at the Universidad de La Rioja with interests in the stratigraphy, petrology, and tectonic activity of the Iberian Hercynian orogen. Hc has taught structural geology, mineralogy, general geology, and paleoichnology at several universities. He has studied dinosaur and tertiary vertebrates footprints around Spain and Morocco and is currently researching tracksites there.

This book was designed by Jamison Cockerham and set in type
by Tony Brewer at Indiana University Press and printed by
Sheridan Books, Inc.

The fonts are Electra, designed by William A. Dwiggins
in 1935, Frutiger, designed by Adrian Frutiger in 1975,
and Futura, designed by Paul Renner in 1927. All were
published by Adobe Systems Incorporated.